Optimization in Civil Engineering

Optimization in Civil Engineering

Editor:

Dr. Tanjina Nur, Ph.D.

ArclerPress

www.arclerpress.com

Optimization in Civil Engineering

Editor: Dr. Tanjina Nur, Ph.D.

708 3rd Avenue, 6th Floor
New York
NY 10017
United States of America
www.arclerpress.com

ISBN: 978-1-68094-382-5

Library of Congress Control Number: 2016951068

About the Editor

Dr. Tanjina Nur, Ph.D.

Tanjina finished her PhD in Civil and Environmental Engineering in 2014 from University of Technology Sydney (UTS). Now she is working as Post-Doctoral Researcher in the Centre for Technology in Water and Wastewater (CTWW) and published about eight International journal papers with 80 citations. Her research interest is wastewater treatment technology using adsorption process.

Contents

List of Contributors

Dezhi Li
Department of Construction and Real Estate, Southeast University, Nanjing 210096, China

Yanchao Chen
Department of Construction and Real Estate, Southeast University, Nanjing 210096, China

Hongxia Chen
School of Civil Engineering, Sanjiang University, Nanjing 210012, China

Eddie Chi Man Hui
Department of Building and Real Estate, the Hong Kong Polytechnic University, Kowloon, Hong Kong, China

Kai Guo
Department of Construction and Real Estate, Southeast University, Nanjing 210096, China

Chao Zeng
School of Civil Engineering & Architecture, Chongqing Jiaotong University, Chongqing 400074, China

Boming Tang
School of Civil Engineering & Architecture, Chongqing Jiaotong University, Chongqing 400074, China

Tangzhi Liu
School of Civil Engineering & Architecture, Chongqing Jiaotong University, Chongqing 400074, China

Gwanghee Heo
Department of Civil and Environment Engineering, Konyang University, 121 Daehak-ro, Nonsan, Chungnam 320-711, Republic of Korea

Joonryong Jeon
Department of Civil and Environment Engineering, Konyang University, 121 Daehak-ro, Nonsan, Chungnam 320-711, Republic of Korea

Yanlong Li
State Key Laboratory Base of Eco-Hydraulic Engineering in Arid Area, Xi'an University of Technology, Xi'an 710048, China

Jing Wang
State Key Laboratory Base of Eco-Hydraulic Engineering in Arid Area, Xi'an University of Technology, Xi'an 710048, China

Zengguang Xu
State Key Laboratory Base of Eco-Hydraulic Engineering in Arid Area, Xi'an University of Technology, Xi'an 710048, China

Yi Ji
Key Laboratory of Beijing for Water Quality Science and Water Environment Recovery Engineering, College of Architecture and Civil Engineering, Beijing University of Technology, Beijing 100124, China
State Key Laboratory of Simulation and Regulation of Water Cycle in River Basin, China Institute of Water Resources and Hydropower Research, Beijing 100038, China

Xiaohui Lei
Key Laboratory of Beijing for Water Quality Science and Water Environment Recovery Engineering, College of Architecture and Civil Engineering, Beijing University of Technology, Beijing 100124, China

Siyu Cai
State Key Laboratory of Simulation and Regulation of Water Cycle in River Basin, China Institute of Water Resources and Hydropower Research, Beijing 100038, China

Xu Wang
State Key Laboratory of Simulation and Regulation of Water Cycle in River Basin, China Institute of Water Resources and Hydropower Research, Beijing 100038, China

A. Burak Göktepe
Kolin Const. Co., Akkoy Dam & HEPP Construction, Kurtun, Gumushane, Turkey

A. Hilmi Lav
Istanbul Technical University, Civil Eng. Department, Ayazaga, 34469, Istanbul, Turkey

Selim Altun
Ege University, Department of Civil Engineering, 35100, Bornova, Izmir, Turkey

Gökhan Altıntaş
Celal Bayar University, Department of Civil Engineering, Manisa, Turkey

Murat Kucukvar
Department of Industrial Engineering, Istanbul Sehir University, Uskudar, Istanbul 34662, Turkey

Gokhan Egilmez
Department of Civil, Mechanical and Environmental Engineering, University of New Haven, West Haven, CT 06516, USA

Omer Tatari
Department of Civil, Environmental, and Construction Engineering, University of Central Florida, Orlando, FL 32816, USA

Tae Hyoung Kim
Building and Urban Research Institute, Korea Institute of Civil Engineering and Building Technology, Daehwa-dong 283, Goyandae-Ro, ILsanseo-Gu, Goyang-Si 10223, Korea

Sung Ho Tae
School of Architecture & Architectural Engineering, Hanyang University, 1271 Sa 3-dong, Sangrok-Gu, Ansan-Si 15588, Korea

Sung Joon Suk
Department of Construction Management, Western Carolina University, Cullowhee, NC 28723, USA

George Ford
Department of Construction Management, Western Carolina University, Cullowhee, NC 28723, USA

Keun Hyek Yang
Department of Plant & Architectural Engineering, Kyonggi University, 154-42 Gwanggyosan-ro, Jangan-gu, Suwon-Si 16227, Korea

Barış Şimşek
Department of Chemical Engineering, Faculty of Engineering, Ankara University, 06100, Tandoğan, Ankara, Turkey

Y. Tansel İç
Department of Industrial Engineering, Faculty of Engineering, Baskent University, 06810, Baglica, Etimesgut, Ankara, Turkey

Emir H.Şimşek
Department of Chemical Engineering, Faculty of Engineering, Ankara University, 06100, Tandoğan, Ankara, Turkey

J. Rex
Department of Civil Engineering, SSM Institute of Engineering and Technology, Dindigul, Tamil Nadu 624 002, India

B. Kameshwari
Department of Civil Engineering, RVS College of Engineering and Technology, Dindigul, Tamil Nadu 624 005, India

Ali Reza Ghanizadeh
Department of Civil Engineering, Sirjan University of Technology, Sirjan 78137, Iran

Jiangong Chen
College of Civil Engineering, Chongqing University, Chongqing 400045, China

Key Laboratory of New Technology for Construction of Cities in Mountain Area, Chongqing University, Ministry of Education, Chongqing 400045, China

Zejun Yang
College of Civil Engineering, Chongqing University, Chongqing 400045, China
Key Laboratory of New Technology for Construction of Cities in Mountain Area, Chongqing University, Ministry of Education, Chongqing 400045, China

Richeng Hu
College of Civil Engineering, Chongqing University, Chongqing 400045, China
Key Laboratory of New Technology for Construction of Cities in Mountain Area, Chongqing University, Ministry of Education, Chongqing 400045, China

Haiquan Zhang
College of Civil Engineering, Chongqing University, Chongqing 400045, China
Key Laboratory of New Technology for Construction of Cities in Mountain Area, Chongqing University, Ministry of Education, Chongqing 400045, China

Tsung-Ming Yang
Department of Civil Engineering, National Taiwan University, Taipei 10617, Taiwan

Nien-Sheng Hsu
Department of Civil Engineering, National Taiwan University, Taipei 10617, Taiwan

Chih-Chiang Chiu
Wastewater Sewerage System Engineering Division, Water Resources Department, New Taipei City government 234, Taiwan

Hsin-Ju Wang
MWH Americas Inc., Taiwan Branch 106, Taiwan

Zong Woo Geem
Department of Energy IT, Gachon University, Seongnam 461-701, Korea

Thi Thuy Ngo
Department of Civil, Environmental and Architectural Engineering, Korea University, Seoul 136-713, Korea

Do Guen Yoo
K-Water Research Institute, Korea Water Resources Corporation, 200 Sintangin-Ro, Daedeok-Gu, Daejeon 34350, Korea

Yong Sik Lee
Department of Civil, Environmental and Architectural Engineering, Korea University, Seoul 136-713, Korea

Joong Hoon Kim
Department of Civil, Environmental and Architectural Engineering, Korea University, Seoul 136-713, Korea

Chun-tian Cheng
Institute of Hydropower and Hydroinformatics, Dalian University of Technology, Dalian 116024, China

Wen-jing Niu
Institute of Hydropower and Hydroinformatics, Dalian University of Technology, Dalian 116024, China

Zhong-kai Feng
Institute of Hydropower and Hydroinformatics, Dalian University of Technology, Dalian 116024, China

Jian-jian Shen
Institute of Hydropower and Hydroinformatics, Dalian University of Technology, Dalian 116024, China

Kwok-wing Chau
Department of Civil and Environmental Engineering, Hong Kong Polytechnic University, Hong Kong 999077, China

Mustafa Erkan Turan
Department of Civil Engineering Celal Bayar University, 45140 Muradiye, Manisa, Turkey

Mehmet Ali Yurdusev
Department of Civil Engineering Celal Bayar University, 45140 Muradiye, Manisa, Turkey

Hua Wang
Key Laboratory of Integrated Regulation and Resource Development on Shallow Lake of Ministry of Education, College of Environment, Hohai University, Nanjing 210098, China
College of Environment, Hohai University, No.1 Xikang Road, Nanjing 210098, China

Mengan Wu
Key Laboratory of Integrated Regulation and Resource Development on Shallow Lake of Ministry of Education, College of Environment, Hohai University, Nanjing 210098, China
College of Environment, Hohai University, No.1 Xikang Road, Nanjing 210098, China

Yanqing Deng
Hydrology Bureau of Jiangxi Province, Nanchang 330002, China

Chunyan Tang
Division of Hydrologic Sciences, Desert Research Institute, Las Vegas, NV 89119, USA

Rui Yang
Key Laboratory of Integrated Regulation and Resource Development on Shallow Lake of Ministry of Education, College of Environment, Hohai University, Nanjing 210098, China
College of Environment, Hohai University, No.1 Xikang Road, Nanjing 210098, China

Preface

Civil engineering includes structural engineering, geotechnical engineering, water resources engineering, environmental engineering, transportation engineering, and construction management which can be an industrial sector which derives great benefit from the optimization because these techniques can save a lot of costs in public infrastructure construction and management that require enormous budget. This book tries to cover a wide picture how the optimization techniques can be applied to various civil engineering problems such construction and project management, structural engineering, water and environmental engineering, and transportation engineering. At first, we discuss evaluation and optimization of the financial sustainability of public rental housing projects and service radius model and service scope optimization of city public parking garage. Then we present an experimental study of structural identification of bridges using the kinetic energy optimization technique and the direct matrix updating method and design optimization of a concrete face rock-fill dam by using genetic algorithm. Hedging rules for water supply reservoir based on the model of simulation and optimization and fuzzy decision support system to determine swell/shrink factor affecting earthwork optimization of highways are also discussed.

In the middle part of the book, we focus on life cycle assessment and optimization-based decision analysis of construction waste recycling for a university building and an optimization system for concrete life cycle cost and related CO2 emissions. Then a full factorial design based desirability function approach for optimization of properties of C 40/50 concrete class and studies on pumice lightweight aggregate concrete with quarry dust using mathematical modelling aid of optimization techniques are also discussed. We also try to present an optimization model for design of Asphalt pavements and study on the seismic active earth pressure by variational limit equilibrium method.

The last parts of the book describe applying the Taguchi method to river water pollution remediation strategy optimization and multi objective optimization of water distribution networks using fuzzy theory and harmony search. We also try to find out the optimization of upstream detention reservoir facilities for downstream flood mitigation in urban areas and daily reservoir run-off forecasting method using artificial neural network based on quantum-behaved particle swarm optimization. At the end, we describe optimization of open canal cross sections by differential evolution algorithm and surface water quality monitoring site optimization for the largest freshwater lake in China.

Editor
Dr. Tanjina Nur, Ph.D.

INTRODUCTION

Optimization is the process of obtaining the best result under given circumstances. In design, construction and maintenance of any engineering system, engineers have to take many technological and managerial decisions at several stages. The ultimate goal of all such decisions is either to minimize the effort required or to maximize the desired benefit. A number of optimization methods have been developed for solving different types of optimization problems. In this course, after discussing about the optimization problem formulation, Linear Programming, Non Linear Programming, Dynamic Programming techniques are explained in detail along with number of applications in civil engineering. Advanced optimization techniques such as Evolutionary search algorithms, Multi objective optimization are briefly introduced.

A mechanical engineer designs a new engine, or a car suspension or a robot. A civil engineer designs a bridge or a building. A chemical engineer designs a distillation tower or a chemical process. An electrical engineer designs a computer or an integrated circuit. For many reasons, not the least of which is the competitive marketplace, an engineer might not only be interested in a design which works at some sort of nominal level, but is the best design in some way. The process of determining the best design is called optimization. Thus we may wish to design the smallest heat exchanger that accomplishes the desired heat transfer, or we may wish to design the lowest-cost bridge for the site, or we may wish to maximize the load a robot can lift. Often engineering optimization is done implicitly. Using a combination of judgment, experience, modeling, opinions of others, etc. the engineer makes design decisions which, he or she hopes, lead to an optimal design. Some engineers are very good at this. However, if there are many variables to be adjusted with several conflicting objectives and/or constraints, this type of experience-based optimization can fall short of identifying the optimum design. The interactions are too complex and the variables too numerous to intuitively determine the optimum design. In this text we discuss a computer-based approach to design optimization. With this approach, we use the computer to search for the best design according to criteria that we specify. The computer's enormous processing power allows us to evaluate many more design combinations than we could do manually. Further, we employ sophisticated algorithms that enable the computer to efficiently search for the optimum. Often we start the algorithms from the best design we have based on experience and intuition. We can then see if any improvement can be made. In order to employ this type of optimization, several qualifications must be met. First, we must have a quantitative model available to compute the responses of interest. If we wish to maximize heat transfer, we must be able to calculate heat transfer for different design configurations. If we wish to minimize cost, we must be able to calculate cost. Sometimes obtaining such quantitative models is not easy. Obtaining a valid, accurate model of the design problem is the most important step in optimization. It is not uncommon for 90% of the effort in optimizing a design to be spent on developing and validating the quantitative model. Once a good model is obtained, optimization results can often be realized quickly. Fortunately, in engineering we often do have good, predictive models (or at least partial models) for a design problem. For example, we have models to predict the heat transfer or pressure drop in an exchanger. We have models to predict stresses and deflections in a bridge. Although engineering models are usually physical in nature (based on physical principles), we can also use empirical models (based on the results of experiments). It is also perfectly acceptable for models to be solved numerically (using, for example, the finite element method).

Historical Development and Model Building

The existence of optimization methods can be traced to the days of Newton, Lagrange, and Cauchy. The development of differential calculus methods for optimization was possible because of the contributions of Newton and Leibnitz to calculus. The foundations of calculus of variations, which deals with the minimization of functions, were laid by Bernoulli, Euler, Lagrange, and Weistrass. The method of optimization for constrained problems, which involve the addition of unknown multipliers, became known by the name of its inventor, Lagrange. Cauchy made the first application of the steepest descent method to solve unconstrained optimization problems. By the middle of the twentieth century, the high-speed digital computers made implementation of the complex optimization procedures possible and stimulated further research on newer methods. Spectacular advances followed, producing a massive literature on optimization techniques. This advancement also resulted in the emergence of several well defined new areas in optimization theory.

Engineering applications of optimization

To indicate the widespread scope of the subject, some typical applications in different engineering disciplines are given below.

- Design of civil engineering structures such as frames, foundations, bridges, towers, chimneys and dams for minimum cost.
- Design of minimum weight structures for earth quake, wind and other types of random loading.
- Optimal plastic design of frame structures (e.g., to determine the ultimate moment capacity for minimum weight of the frame).
- Design of water resources systems for obtaining maximum benefit.
- Design of optimum pipeline networks for process industry.
- Design of aircraft and aerospace structure for minimum weight
- Finding the optimal trajectories of space vehicles.
- Optimum design of linkages, cams, gears, machine tools, and other mechanical components.
- Selection of machining conditions in metal-cutting processes for minimizing the product cost.
- Design of material handling equipment such as conveyors, trucks and cranes for minimizing cost.
- Design of pumps, turbines and heat transfer equipment for maximum efficiency.
- Optimum design of electrical machinery such as motors, generators and transformers.
- Optimum design of electrical networks.
- Optimum design of control systems.
- Optimum design of chemical processing equipment's and plants.
- Selection of a site for an industry.
- Planning of maintenance and replacement of equipment to reduce operating costs.
- Inventory control.
- Allocation of resources or services among several activities to maximize the benefit.
- Controlling the waiting and idle times in production lines to reduce the cost of production.
- Planning the best strategy to obtain maximum profit in the presence of a competitor.

Optimization Methods: Introduction and Basic Concepts

- Designing the shortest route to be taken by a salesperson to visit various cities in a single tour.
- Optimal production planning, controlling and scheduling.
- Analysis of statistical data and building empirical models to obtain the most accurate representation of the statistical phenomenon.

Art of Modeling: Model Building

Development of an optimization model can be divided into five major phases.

- Data collection
- Problem definition and formulation
- Model development

- Model validation and evaluation of performance
- Model application and interpretation

Data collection may be time consuming but is the fundamental basis of the model-building process. The availability and accuracy of data can have considerable effect on the accuracy of the model and on the ability to evaluate the model.

The problem definition and formulation includes the steps: identification of the decision variables; formulation of the model objective(s) and the formulation of the model constraints. In performing these steps the following are to be considered.

- Identify the important elements that the problem consists of.
- Determine the number of independent variables, the number of equations required to describe the system, and the number of unknown parameters.
- Evaluate the structure and complexity of the model • Select the degree of accuracy required of the model

Model development includes the mathematical description, parameter estimation, input development, and software development. The model development phase is an iterative process that may require returning to the model definition and formulation phase. The model validation and evaluation phase is checking the performance of the model as a whole. Model validation consists of validation of the assumptions and parameters of the model. The performance of the model is to be evaluated using standard performance measures such as Root mean squared error and R^2 value. A sensitivity analysis should be performed to test the model inputs and parameters. This phase also is an iterative process and may require returning to the model definition and formulation phase. One important aspect of this process is that in most cases data used in the formulation process should be different from that used in validation. Another point to keep in mind is that no single validation process is appropriate for all models.

Model application and implementation include the use of the model in the particular area of the solution and the translation of the results into operating instructions issued in understandable form to the individuals who will administer the recommended system. Different modeling techniques are developed to meet the requirements of different types of optimization problems. Major categories of modeling approaches are: classical optimization techniques, linear programming, nonlinear programming, geometric programming, dynamic programming, integer programming, stochastic programming, evolutionary algorithms, etc.

Optimization Problem and Model Formulation

An objective function expresses the main aim of the model which is either to be minimized or maximized. For example, in a manufacturing process, the aim may be to maximize the profit or minimize the cost. In comparing the data prescribed by a user-defined model with the observed data, the aim is minimizing the total deviation of the predictions based on the model from the observed data. In designing a bridge pier, the goal is to maximize the strength and minimize size. A set of unknowns or variables control the value of the objective function. In the manufacturing problem, the variables may include the amounts of different resources used or the time spent on each activity. In fitting-the-data problem, the unknowns are the parameters of the model. In the pier design problem, the variables are the shape and dimensions of the pier. A set of constraints are those which allow the unknowns to take on certain values but exclude others. In the manufacturing problem, one cannot spend negative amount of time on any activity, so one constraint is that the "time" variables are to be non-negative. In the pier design problem, one would probably want to limit the breadth of the base and to constrain its size. The optimization problem is then to find values of the variables that minimize or maximize the objective function while satisfying the constraints.

Statement of an optimization problem

An optimization or a mathematical programming problem can be stated as follows:

$$\text{To find } \mathbf{X} = \begin{pmatrix} x_1 \\ x_2 \\ . \\ . \\ x_n \end{pmatrix} \text{ which minimizes } f(\mathbf{X}) \qquad \ldots\ldots\ldots\ldots\ldots(1)$$

Subject to the constraints

$$g_i(\mathbf{X}) \leq 0, \quad i = 1, 2, \ldots, m$$

$$l_j(\mathbf{X}) = 0, \quad j = 1, 2, \ldots, p$$

where X is an n-dimensional vector called the design vector, f(X) is called the objective function, and $_{gi}(X)$ and $l_j(X)$ are known as inequality and equality constraints, respectively. The number of variables n and the number of constraints m and/or p need not be related in any way. This type problem is called a constrained optimization problem.

If the locus of all points satisfying f(X) = a constant c, is considered, it can form a family of surfaces in the design space called the objective function surfaces. When drawn with the constraint surfaces as shown in Fig 1 we can identify the optimum point (maxima). This is possible graphically only when the number of design variables is two. When we have three or more design variables because of complexity in the objective function surface, we have to solve the problem as a mathematical problem and this visualization is not possible.

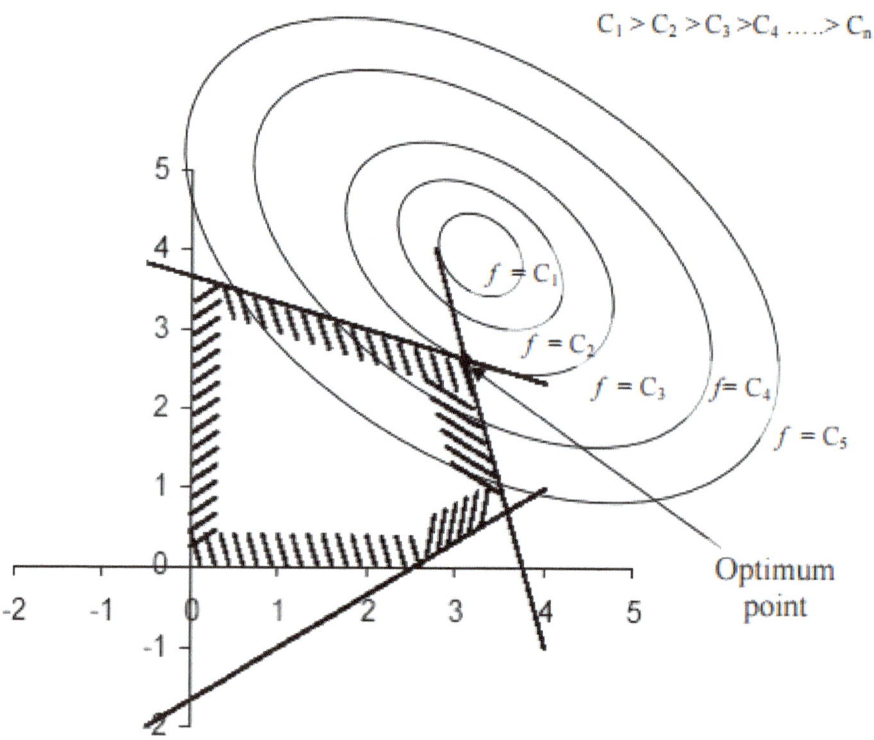

Optimization problems can be defined without any constraints as well.

$$\text{To find } \mathbf{X} = \begin{pmatrix} x_1 \\ x_2 \\ . \\ . \\ x_n \end{pmatrix} \text{ which minimizes } f(\mathbf{X}) \qquad (2)$$

Such problems are called unconstrained optimization problems. The field of unconstrained optimization is quite a large and prominent one, for which a lot of algorithms and software are available

Variables

These are essential. If there are no variables, we cannot define the objective function and the problem constraints. In many practical problems, one cannot choose the design variable arbitrarily. They have to satisfy certain specified functional and other requirements.

Constraints

Constraints are not essential. It's been argued that almost all problems really do have constraints. For example, any variable denoting the "number of objects" in a system can only be useful if it is less than the number of elementary particles in the known universe! In practice though, answers that make good sense in terms of the underlying physical or economic criteria can often be obtained without putting constraints on the variables. Design constraints are restrictions that must be satisfied to produce an acceptable design.

Constraints can be broadly classified as:

1) Behavioral or Functional constraints: These represent limitations on the behavior performance of the system.
2) Geometric or Side constraints: These represent physical limitations on design variables such as availability, fabricability, and transportability.

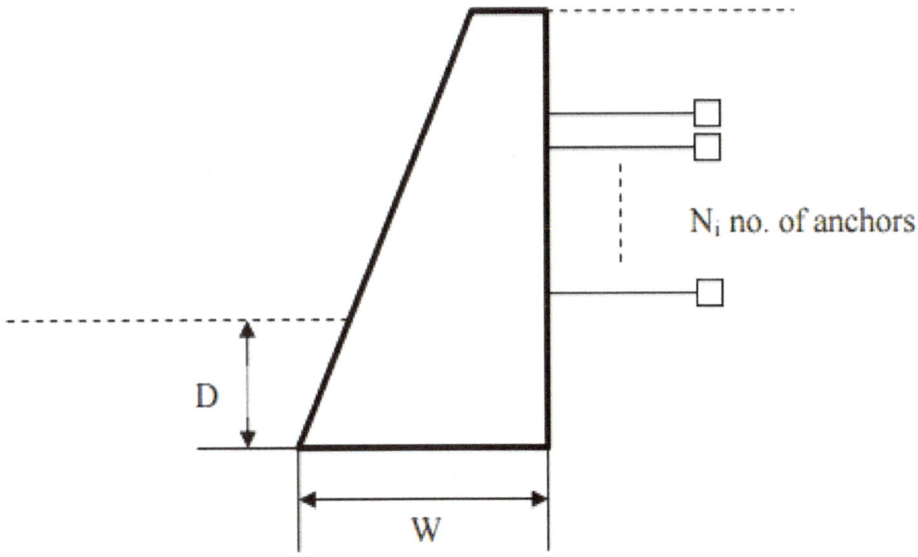

For example, for the retaining wall design shown in the Fig 2, the base width W cannot be taken smaller than a certain value due to stability requirements. The depth D below the ground level depends on the soil pressure coefficients Ka and Kp. Since these constraints depend on the performance of the retaining wall they are called behavioral constraints. The number of anchors provided along a cross section Ni cannot be any real number but has to be a whole number. Similarly thickness of reinforcement used is controlled by supplies from the manufacturer. Hence this is a side constraint.

CHAPTER 1 EVALUATION AND OPTIMIZATION OF THE FINANCIAL SUSTAINABILITY OF PUBLIC RENTAL HOUSING PROJECTS

Dezhi Li , Yanchao Chen , Hongxia Chen , Eddie Chi Man Hui and Kai Guo

Keywords: Public rental housing, affordable housing, private sector, financial sustainability, optimization scenario, China

INTRODUCTION

It is well known that Chinese governments abolished their welfare housing policies and initiated the market-oriented urban housing provision system reform in 1998. Since then, most urban households have been required to satisfy their accommodation needs through the commodity housing market, and are no longer able to rely on their working units (*Danwei* in Chinese). At the same time, China's traditional affordable housing system was created, consisting of the Economical and Comfortable Housing (ECH) program, the Housing Provident Fund (HPF) program, and the Cheap Rental Housing (CRH) program [1]. Among them, ECH was developed for urban lower-middle and middle-income families, and CRH was built for the poorest urban households. HPF is a compulsory housing savings program under which both employers and employees are required to contribute a certain percentage of the employees' wages to a fund for the latter's future house purchase [2]. The significance of this traditional affordable housing program is notable. For example, about 429 million m^2 of ECH had been sold nationwide between 1998 and 2010.

However, there are several shortcomings in China's traditional affordable housing system. First, the nationwide scale and effects of ECH are both limited, because the number of ECH units sold was only about 6% of that of contemporary commodity housing units sold from 1998 to 2010. Second, some wealthier households can purchase and resell ECH illegally, pushing both construction standards and prices of ECH beyond the reach of lower-middle and middle income families [1]. Third, the availability of CRH is very low, and thus few people could benefit from it. Along with China's booming price in the commodity housing market in recent years, housing shortages have become pressing socio-economic issues, particularly for those who can neither afford commodity housing nor meet the eligibility requirements for traditional affordable housing [3].

To address the aforementioned housing problems and improve the Chinese traditional affordable housing system, as well as to stimulate economic growth, a new affordable housing program entitled Public Rental Housing (PRH) was proposed officially by then-Premier Wen Jiabao's "Report on the Work of the Government" at the Second Session of the Eleventh National People's Congress on 5 March 2009. Since Wen's announcement, PRH has been assigned as the very core of China's new affordable housing system in several official documents such as the National 12th Five-Year Plan (2011–2015). In December 2013, the Chinese central government officially announced the merging of CRH into PRH starting 2014. As a result, ambitious plans for the development of PRH have been rapidly devised by the central government and many local governments. For instance, the Henan provincial government promised to merge CRH into PRH and cover the indemnified targets of ECH by PRH since 2014, in order to house 20% of the urban population in the whole province with PRH. Similarly, in February 2015, the Beijing municipal government announced the abolishment of ECH and the development of PRH starting in 2016. In recent years, nationwide PRH projects have been carried out by local governments with subsidies from the central government, resulting in evitable and evident government failure (e.g., huge

Citation: Dezhi Li , Yanchao Chen, Hongxia Chen, Eddie Chi Man Hui and Kai Guo, Evaluation and Optimization of the Financial Sustainability of Public Rental Housing Projects, doi:10.3390/su8040330.

fiscal burden and low supply efficiency) [4]. In fact, Chinese governments are well aware of this situation, and have issued several preferential policies to accelerate the private sector's participation in the provision of PRH since 2009 [5].

Unfortunately, except a few private real estate developers that have participated in PRH projects as agents, most of the private sector in China is unwilling to develop and own PRH projects, with the main reason being the fear of the financial unsustainability of such projects [6,7]. Is this fear rational or not? If yes, what measures should be taken to optimize the financial status of PRH projects from the perspective of the private sector? Furthermore, which measure is the most cost-effective? These urgent topics are to be discussed in this paper through a case study of a PRH project in Nanjing city, the capital of Jiangsu province in eastern China. The rest of this paper is organized as follows. To begin, recent studies concerning the private sector's provision of affordable housing in both developed countries and developing countries (including China) are reviewed, followed by the section where basic information of the studied project is described. Then, its financial sustainability is evaluated with profitability and debt-repayment ability indices, from the perspective of the private sector. In the section that follows, four possible scenarios to optimize the financial sustainability of the studied project are proposed and compared, and the most cost-effective one is selected. The last section concludes this paper with policy suggestions and possible future works.

LITERATURE REVIEW

In recent years, many developed countries have reduced governments' direct provision of PRH while increasing the private sector's provision in this regard, such as the U.S., the U.K., the Netherlands, and Ireland. Many researchers have paid close attention to this trend and analyzed its reasons, effects and/or influences. For instance, one of American HOPE VI's objectives is promoting the mixed-finance partnerships between public, private and non-profit sectors, through a case study in Louisville, Kentucky [8]. Recent Irish experiences in increasing private sector provision of social housing with the aid of housing allowances are assessed [9]. The controversial use of the Private Finance Initiative (PFI) as part of the Labour Government's Decent Homes program in the U.K., namely the transformation of social housing from the "public housing model" of the welfare state era to the "social housing model" of today is addressed [10]. The factors which affect the market orientation of semi-public service organizations in their transition from public to private are explained with a case study in Dutch housing associations [11]. Italian housing policies have shifted from central state provision towards a multi-level local governance organized into partnerships between regions, municipalities, third-sector and private investors since the turn of the new millennium [12]. The changing roles of private not-for-profit housing associations in English and Dutch housing provision are also explored, revealing that aspects of the work and identity of housing associations shift between public and private domains [13].

Several developing countries (such as India, Nigeria, and Malaysia) have also attempted to provide PRH for low-income populations through the private sector, which has been discussed ardently in recent years. For example, the contradictions in enabling private developers to provide affordable housing are examined through a case study in India [14]. The contribution of Public-Private Partnerships (PPP) in improving accessibility of low-income earners to housing in Nigeria is evaluated, suggesting that government provision of land to private developers at zero cost and lowering the high building standards will ensure better results [15]. A case study on the contribution and challenges of the private sector's participation in housing in Nigeria is undertaken, indicating that such participation has the potential for improving housing delivery [16]. The role of the private sector's participation in housing development for low-income households in both Malaysia and Nigeria is compared, revealing that the success of the private sector depends on the existence of a favorable socio-economic environment and of an effective institutional and regulatory framework [17]. Applying the multi-case approach and case studies, the control mechanisms used by public agencies in Malaysia under PPP arrangements to ensure private partner compliance are examined, identifying the governance archetypes which prevailed in previous PPP relationships [18]. Obtaining data through interviews and questionnaire surveys, as well as secondary data sources, a shift from a state-led public housing provision to an enabling approach did stimulate the activities of private house-builders and primary mortgage institutions in Jos, a city in north-central Nigeria [19].

In China, the necessity of accelerating the private sector's provision of PRH has also been widely recognized by a lot of scholars, though their suggested means vary. For example, a dual housing provision structure (namely PRH and commodity housing) is believed to be the future of the Chinese real estate industry, and private real estate developers should possess full property rights of PRH projects and run them [6]. Shortage in funds is declared as the biggest barrier to the development of PRH, and thus a Build-Operate-Transfer (BOT) model to promote the private sector's provision of PRH is designed [20]. A theoretical PPP financing model to attract the private sector's participation in the provision of PRH in China is proposed [21]. The applicability of three financing models, including Real Estate Investment Trusts (REITs), BOT, and Public Intermediary Private Partnerships (PIPPs) are qualitatively analyzed and compared, revealing that the private sector's provision with government subsidies is an important way to build new PRH projects [22]. A strength, weakness, opportunity, and threat (SWOT) analysis of public housing delivery by public-private partnerships in China from the perspective of the public sector is conducted, suggesting that one of the strongest hindrances is "low profits for the private sector to participate in PPP housing" [7]. Taking Beijing as a case study, the provision of PRH with the incremental upgrading of "villages in the city" are linked through the adoption of partnerships between multilevel governments, the collective organizations, villagers, migrants, the informal sector and other stakeholders [23].

From those studies, it is clear that many countries value the private sector's provision of PRH for low-income households. The necessity of promoting privately owned PRH projects in China has also been broadly identified, and a number of quantitative analyses have been carried out from the perspective of the public sector. However, the financial sustainability of PRH projects in China has rarely been evaluated from the perspective of the private sector, despite its importance amongst private developers. This knowledge gap is to be bridged through investigating the financial sustainability and optimization of a PRH project in the city of Nanjing, China.

THE FRAMEWORK

To accelerate the construction process and enhance the supply efficiency of planned affordable housing, an agent-construction system has been adopted by the Nanjing municipal government since 2010. Agent companies are usually traditional real estate developers, with rich experience in developing commodity housing and commercial buildings. They can obtain 3% in profits and 1%~2% in agent fees on the basis of total costs, which are to be completely undertaken by the government [24]. As a result, agent companies have no initiative to reduce the cost of affordable housing, and the financial condition of Nanjing's municipal government worsens with the adoption of agent-construction system, which may end up crushing the local government financially and in turn halt the provision of affordable housing. To avoid such a predicament and achieve a sustainable provision of affordable housing, especially PRH, it is urgent to explore profitable approaches of PRH projects from the perspective of the private sector, and further promote the private sector's participation in these projects.

The studied case is a small PRH project in the Daishan affordable housing zone (hereafter Daishan Zone), which is one of the four largest affordable housing zones in Nanjing. It is located in the urban-rural fringe of Yuhuatai District in southern Nanjing, covering an area of 53,546.92 m^2, which is only about 4.28% of the total residential land area in Daishan Zone. Other residential areas are mainly used for ECH and CRH projects. The building structure of the studied case is frame-shear wall, and other indices are shown in Table 1. In order to conduct the following financial sustainability evaluation successfully, some assumptions have to be made as follows.

Table 1: Indices of the studied PRH project.

Index Name		Index Value	Index Name	Index Value
Aboveground building area (m^2)	PRH buildings	203,478	Floors of PRH building	12
	Commercial buildings	10,710	Floors of commercial building	1
Underground building area (m^2)	Civil air defense works	15,800	PRH suits	3960
	Equipment storage	20,012	Parking spaces	402

(1) Although the studied case is actually built by agent companies and owned by the Nanjing municipal government, it is assumed to be built, owned and operated by a private real estate developer called L Company, with the project enjoying all PRH-related policy incentives and satisfying all PRH-related regulations (e.g., design standard, tenant selection and rent price criteria). L Company is a real private real estate developer in Nanjing interested in participating in the provision of PRH.

(2) Considering its relatively small scale, for the sake of convenience, this PRH project is assumed to be built from January 2016 through December 2017. Thus, the studied project could be available to qualified PRH tenants by January 2018.

(3) In urban China, residential and commercial building owners can only have land use rights for 70 years and 40 years, respectively. However, to reduce long-term uncertainties and be consistent between the two building types, the operation period of PRH buildings and of commercial buildings are both assumed to be 40 years. In fact, the discounted income from PRH buildings in the final 30 years does not noticeably influence the financial sustainability evaluation.

(4) PRH buildings are well decorated and furnished to attract tenants, while commercial buildings only have basic decorations.

(5) The rent of PRH flats is 70% as much as that of surrounding private rental housing flats, as decreed in *Regulations of Public Rental Housing in Nanjing City*.

(6) The rent of commercial buildings in this PRH project is the same as that of surrounding market-oriented commercial buildings, which can be estimated by market surveys and interviews.

(7) To be conservative, 90% of PRH suites, accessory commercial buildings, and parking spaces are leased at average rents. Besides, considering the level of inflation and economic growth in Nanjing, their rents are assumed to increase by 10% after 5 years.

FINANCIAL SUSTAINABILITY EVALUATION

Estimated Costs

Prior to evaluating the financial sustainability of the studied case, its construction and operation costs must be estimated first. Based on relative laws and market conditions, the cost in the construction phase has been estimated and illustrated inTable 2. It is noteworthy that the feasibility report of a government investment project in China is necessary to obtain the official approval by authorities. The feasibility report must be presented by a qualified consulting company, which generally charges 0.2%~3.0% of the total cost or 0.12~2.50 million CNY. Based on a market inquiry, the estimated cost of the feasibility study is 0.20 million CNY.

Table 2: Cost items in the construction phase.

No.	Cost Item	Amount (Million CNY)	Calculation Base
1	Land use rights	250.00	Unit price of similar land nearby
2	Pre-construction works	8.02	
2–1	Submission for approval	1.25	5 CNY per total building area
2–2	Survey and design service	5.50	22 CNY per total building area
2–3	Feasibility study	0.20	Regulations and market survey
2–4	Temporary facilities	1.07	20 CNY per land area
3	Construction and installation	375.00	
3–1	Construction works	250.00	1,000 CNY per total building area
3–2	Equipment and installation	125.00	500 CNY per total building area
4	Infrastructure facilities	48.42	
4–1	Road and sewage pipes	2.68	50 CNY per land area
4–2	Drinking water pipes	5.35	25 CNY per above ground building area
4–3	Electricity implement	25.00	100 CNY per total building area
4–4	Gas pipes	11.59	2,600 CNY per suite
4–5	Cable television system	2.14	10 CNY per above ground building area
4–6	Road lamp	0.32	6 CNY per land area
4–7	Landscape construction	1.34	25 CNY per land area
5	Expenses for other and unpredictable works	17.38	
5–1	Expenses for other works	3.75	15 CNY per building area
5–2	Expenses for unpredictable works	13.63	2% of the sum of cost item No. 1–4
6	Indirect and management expenses	20.96	
6–1	Indirect expenses	13.98	2% of the sum of cost item No. 1–5
6–2	Management expenses	6.99	1%of the sum of cost item No. 1–5
7	Financial expenses	31.53	
7–1	Loan interest	29.68	Annual interest rate (6.55%)
7–3	Other financial expense	4.60	1% of credit ceiling

From Table 2, it is clear that construction costs are estimated to be 751.32 million CNY. The costliest item is construction and installation, followed by land use rights, infrastructure facilities, and financial expenses. These four items amount to almost 94% of the total construction cost. Among them, construction and installation cost and infrastructure facilities cost are almost fixed, since they determine the quality of the studied PRH project. The cost of land use rights and that of financial expenses are about 1/3 and 4% of the total construction cost, which are significant yet changeable items. As a result, the cost of land use rights and that of financial expenses are two key items for future financial optimization.

After obtaining the estimated cost in the construction phase, the cost of this studied PRH project in its operation phase is then estimated. It comprises three components, namely running cost, loan interest, and depreciation. As regards the running cost, it mainly involves managements' wages and welfare, maintenance costs, and advertisements for accessory commercial buildings. In general terms, the annual running cost of a PRH project in China is about 3 CNY/m^2 [25]. Since the total building area of the studied PRH project is 250,000 m^2, the annual running cost of this PRH project is thus estimated at 0.75 million CNY.

To estimate the loan interests of the studied PRH project in its operation phase, the fundraising plan must be formulated beforehand. Similar to listed real estate companies in China, L Company plans to obtain a loan as much as 60% of the total investment, which is 453.28 million CNY. The equity capital, 302.18 million CNY, is mainly used to obtain land use rights and to pay pre-construction works, and the remainder is spent in the second half of the first construction year. Based on the construction schedule and other assumptions in Section 3, the investment and fundraising plan is formulated and illustrated in Table 3.

The last component of estimated cost in the operation phase is depreciation, which directly influences the payback period and profit. However, it is noteworthy that the depreciation would not be included as a negative cash flow in the Net Present Value (NPV) calculation. Since PRH buildings and accessory commercial buildings of the project are both for rental purposes, the depreciable asset equals the whole project and the depreciation base equals the original value of fixed assets, namely the estimated investment in Table 3. The minimum depreciation period of the project is 20 years, according to *Regulation on the Implementation of the Enterprise Income Tax Law of P. R. China*, effective since 1 January 2008. Since the operation periods of PRH buildings and accessory commercial buildings are both assumed to be 40 years, so should the assumed depreciation period of the studied case. The residual value rate is another compulsory parameter, which is determined as 5% in *Notice of the State Administration of Taxation on Strengthening Administration on the Repealed Items of Enterprise Income Tax Administrative Examination and Approval,* effective since 18 June 2003. In line with *Regulation on the Implementation of the Enterprise Income Tax Law of P. R. China*, a straight-line method is adopted to calculate the annual depreciation as 17.84 = (751.32 × (1 − 5%)/40) million CNY.

Table 3: Investment and fundraising plan (Unit: million CNY).

No.	Item	Sum	Year 2016		Year 2017	
			Jan–Jun	Jul–Dec	Jan–Jun	Jul–Dec
1	Total investment	751.32	366.14	112.25	112.25	160.68
1–1	Land use rights	250.00	250.00	0.00	0.00	0.00
1–2	Pre-construction works	8.02	8.02	0.00	0.00	0.00
1–3	Construction and installation	375.00	93.75	93.75	93.75	93.75
1–4	Infrastructure facilities	48.42	0.00	0.00	0.00	48.42
1–5	Miscellaneous and unpredictable works	17.38	4.34	4.34	4.34	4.34
1–6	Indirect and management expenses	20.96	5.24	5.24	5.24	5.24
1–7	Financial expenses	35.67	8.91	8.92	8.92	8.92
2	Raising fund	751.32	365.88	158.55	112.25	114.39
2–1	Equity capital	264.41	258.02	6.39	0.00	0.00
2–2	Loan fund	486.91	112.25	101.73	112.25	160.68

Estimated Income

According to Section 3, the studied case has three sources of income, namely PRH buildings, accessory commercial buildings, and parking spaces. Since the building area and operation period of PRH buildings have been determined (see Table 1), the main task of estimating the income generated through PRH buildings is to forecast the unit rent, which is prohibited to exceed 70% of the market level, as indicated in Section 3. Thus, the rental level of the surrounding private housing flats is vital for estimating the income of the studied PRH buildings.

Few private rental housings are found next to the studied case, since it is built at the urban-rural fringe, as indicated in Section 3. By expanding the radius gradually, about 60 flats in private rental housing are found within a radius of 2 km. Among them, 28 flats are quite similar to PRH flats of the studied project in many aspects, such as building area per flat, decoration standard, and surrounding infrastructures. The average monthly rent of these 28 flats is 15.01 CNY per building area. As a result, the average monthly rent of the studied PRH is about 10.51 (=70% × 15.01) CNY per building area, and the income of the studied PRH buildings in the first year can be estimated as approximately 23.10 million CNY. Considering the assumption in Section 3 that the rent of PRH buildings would increase by 10% after 5 years, the annual rent of the studied PRH buildings from 2021 to 2025 would become 25.41 million CNY. Likewise, the annual rent of the studied PRH building in the following 30 years could be estimated. The total revenue from rental income for the studied PRH buildings in 40 years would be about 1.32 billion CNY.

Similar to the estimation of the income from PRH buildings, the monthly rent per building area is the starting point to estimate the income from accessory commercial buildings. Since comparable commercial buildings next to the studied case cannot be found, the survey radius is expanded to the whole Yuhuatai District, in which the monthly rent per building area varies from 40 CNY to 136 CNY with an average of about 100 CNY. Therefore, the monthly rent per building area of those commercial buildings under study is assumed as 100 CNY. Then, incomes generated by those commercial buildings in the first year and in the whole 40 years are 9.29 million CNY and 531.08 million CNY, respectively.

Likewise, the monthly rent per parking space can be predicated as 400 CNY, based on the assumptions in Section 3 and market survey. Then, incomes generated by those parking spaces in the first year and the whole 40 years are 1.74 million CNY and 99.30 million CNY, respectively. The total income of the studied project in the first year and in the whole 40-year period, can therefore be estimated at 34.12 million CNY and 1.95 billion CNY, respectively.

Estimated Tax Deduction

Besides the estimated cost, three types of taxes are required to be paid in the operation phase of the studied PRH project, namely transaction tax, income tax, and value-added tax. Since all PRH buildings, commercial buildings and parking spaces of the studied project are only for rental purposes, there would be no payable value-added tax. All required tax forms and their tax amounts are presented in Table 4. With regard to tax deduction for PRH, they are calculated in light of *The Note of Preferential Tax Policies on Supporting Public Rental Housing's Construction and Operation*, issued by the Ministry of Finance and State Tax Administration of the P. R. China in September 2010. However, it should be noted that tax deduction policies are only applicable to PRH buildings but not to accessory commercial buildings and parking spaces.

Table 4: Tax amount before and after reduction.

No.	Tax Denomination	Basis	Tax Rate	Tax Deduction for PRH Buildings	Tax amount (Million CNY)	
					Before Deduction	After Deduction
1	Transaction tax				256.63	117.71
1–1	Business tax	Turnover	5%	5%	97.55	31.52
1–2	City maintenance and construction tax	Business tax	7%	7%	6.83	2.21
1–3	Education surtax	Business tax	3%	3%	2.93	0.95
1–4	Land use tax	Land area	7 CNY/m^2	7 CNY/m^2	18.90	6.11
1–5	Building tax (for residence)	Building value	4%	4%	52.83	0.00
1–6	Building tax (for operation)	Rent income	12%	0	75.65	75.65
1–7	Stamp tax	Business income	0.05%	0.05%	0.98	0.32
1–8	Transaction service charges	Business income	0.05%	0	0.98	0.98
2	Income tax		25%	0	168.29	195.13

In fact, many central government institutions have either individually or jointly promulgated other preferential policies in various areas, such as land acquisitions, bank loans, corporate bond issuances, and investment grants [5]. Nonetheless, these preferential policies are often too general and difficult to be implemented. Some provincial/city level governments have also issued their own preferential policies to promote private sector provision of PRH, such as the agent-construction system in Hangzhou, municipal-level tax reduction in Jinan, rental subsidies in Zhengzhou, free land supply in Chongqing *etc.* [5]. Since the studied PRH project is located in Nanjing, where agent-construction has recently been issued, only tax deduction policies of PRH buildings in the studied project are considered (other preferential policies are to be referred to as possible options for future optimization).

From Table 4, it is clear that total transaction tax in the operation phase significantly decreases from about 256.63 million CNY to about 117.71 million CNY, which constitutes a 54.13% reduction. Since the income tax is based on net income and thus is negatively correlated with transaction tax, it increases by about 26.84 million CNY (or 15.95%). As a whole, the amount of tax reduction is about 112.07 (=138.91 − 26.84) million CNY or about 550.79 CNY per PRH building area, which is over 1/3 of the unit cost of construction and installation (Table 2). That is to say, the total and unit tax reduction are both notable. The following question can then be asked: Is the studied case financially feasible with such a tax reduction? This is to be explored in the next section.

EVALUATION INDICES AND RESULTS

After estimating the cost and income of the studied PRH project, the corresponding financial feasibility evaluation indices, which mainly include profitability evaluation indices and debt-repayment ability evaluation indices, can then be calculated. Based on these indices, the financial feasibility of the project can thus be assessed.

The profitability evaluation indices of a construction project can be divided into two categories: (1) static evaluation indices; and (2) dynamic evaluation indices [26]. The latter is calculated on the basis of discounted cash flow, while the former is computed on the basis of undiscounted cash flow statement and income statement. Subsequently, some important parameters are obtained, such as total profit (369.34 million CNY) and after-tax profit (179.63 million CNY). The static profitability evaluation indices can then be calculated, which are shown in Table 5. Because the return on investment is very high, this studied PRH project seems rather attractive to the private sector. However, the annual return on total investment is smaller than the on-going risk-free rate of interest, e.g., the annual interest rate of a five-year period term deposit, which is 4.75% since 5 July 2012. In other words, the private sector will obtain less profit from investing in PRH than from depositing in banks. These results render private developers in a difficult situation as to whether they should be involved in the provision of PRH or not. In this case, when the time value and the discounted cash flow of this project are factored in, what will the profitability evaluation results look like?? Is this studied case attractive for private developers or not?

Table 5: Profitability evaluation indices.

Static Profitability Evaluation		Dynamic Profitability Evaluation	
Indices	Value	Indices	Value
Payback period of total investment (year)	34	Dynamic payback period of total investment (year)	NA
Return on total investment (%)	49.16	NPV of total investment (million CNY)	-371.04
Annual return on total investment (%)	1.23	IRR of total investment (%)	3.97

To obtain the dynamic indices of profitability evaluation of the project, the hurdle cut-off rate is needed. There are four recommended mainstream methods to predict the hurdle cut-off rate, including the Capital Asset Pricing Model (CAPM), the Weighted Average Cost of Capital (WACC), the Typical Projects Simulation Method (TPSM), and the Delphi method [26]. Since only a few PRH projects are developed by private developers in Nanjing, CAPM and WACC, as well as TPSM, they do not seem applicable here. Thus, the Delphi method is considered the most suitable. Based on interviews with five experts and two conductors from the private sector, 10% is determined as the hurdle cut-off rate of the project, taking into account the inflation level, opportunity cost, interest rate of term deposit, and other factors. With this parameter, the dynamic profitability evaluation indices can be computed and also shown in Table 5. It is clear that the dynamic payback period of total investment is "NA (Not Available)". In other words, the total investment cannot be recovered in 40 years. In addition, the NPV of total investment is significantly negative, once again proving that this PRH project is not financially feasible. Finally, the Internal Rate of Return (IRR) of total investment is much lower than the hurdle cut-off rate. The dynamic profitability evaluation suggests that this studied PRH project is not financially feasible, and thus is not attractive to the private sector.

As regards to the debt-repayment ability, there are several evaluation indices, such as Interest Coverage Ratio (ICR), Debt Service Coverage Ratio (DSCR), Loan of Asset Ratio (LOAR), and Loan Repayment Period (LRP) [27]. Since all loans of this project are obtained from local commercial banks, no foreign investment is involved. Hence, only LRP is necessary [27], which can be calculated with the following equation:

$$I_d = \sum_{t=1}^{LPR} R_t \tag{1}$$

where I_d is the sum of principal and interest, and R_t is the repayment capital at year t.

From Table 3, it is clear that 179.93 million CNY and 270.86 million CNY will be borrowed from local commercial banks in year 2016 and in year 2017, respectively, with an annual interest rate at 6.55%, as shown in Table 2. The capital for repayment derives from undistributed profits and depreciation, and the method of maximum capacity in repaying principal and interest is adopted [27], since L Company wants to repay the loan (thus reducing the debt) as quickly as possible. Regarding the undistributed profit, it is the total after-tax profit subtracted by statutory surplus reserve and statutory welfare reserve. According to China's tax laws, both statutory surplus reserve and statutory welfare reserve are not payable unless the undistributed profit is positive, as they are 10% and 5% of the undistributed profit respectively. As for depreciation, its annual value is 17.84 million CNY (see Section 4.1). The statement of the repayment of both principal and interest can then be drawn, which shows that the LRP is 34.39 years. After an interview with two conductors of L Company, this LRP is unacceptable, because the operation phase is only assumed to be 40 years. Moreover, as illustrated in Table 5, the project is not profitable from the perspective of the private sector. It must therefore be optimized if it is to attract private sector participation.

FINANCIAL SUSTAINABILITY OPTIMIZATION OF THE STUDIED CASE

Selecting Optimization Options

There are many factors affecting the evaluation of a project's financial sustainability, mainly involving its costs, incomes, and taxes. Because tax deduction has been considered and taxation is levied on the basis of current laws, rates of relative taxes are essentially fixed. Therefore, taxes are not the focal point in the optimization process. Furthermore, since the overall static evaluation indices of this project are quite attractive, as shown in Table 5, while all the dynamic evaluation indices are unsatisfactory, time value is found to be a key factor in determining the evaluation of its financial sustainability. As a result, cost reduction and income maximization as soon as possible should be the direction towards optimization.

As indicated in Section 4.2, the income of the project is derived from its PRH buildings, accessory commercial buildings, and parking spaces. As regards commercial buildings and parking spaces, their market prices are adopted and their amounts are restricted according to its planning control document. Concerning the PRH buildings, the total building area is also fixed by the planning control document. The only optimization measure is thus through the adjustment of the rent of PRH buildings to market level with local government's housing allowance to subsidize the difference between preferential rent and market rent.

To optimize the financial sustainability of the project through cost reduction, the cost in the construction phase should be emphasized, since they are spent earlier. Owing to the findings in Section 4.1, the cost of land use rights and financial expenses in the construction phase are two key items for future financial optimization. As for the cost in the operation phase, only running cost, loan interest, and depreciation are taken into account. Among them, since the depreciation is essentially fixed under current laws and the running cost is too small, the financial expense (*i.e.*, loan interest) becomes another alternative for possible optimization.

In summary, there are three possible optimization measures, including (1) adjusting the rent of PRH buildings to market level with local government's housing allowance; (2) reducing the cost of land use rights; and (3) reducing the life-cycle financial expenses. In fact, these three optimization measures have been applied to a certain degree by some local governments of China, as mentioned in Section 4.3. Each measure leads to one scenario, while the fourth scenario is the combination of these three measures. The optimization effects of these four scenarios are to be examined and compared in the following section.

Assessing Optimization Results

(1) *Scenario I:* It allows L Company to increase the PRH rent to market level with the local government's subsidies to offset the difference. In this situation, the total amount of subsidies received in 40 years when pre-discounting is estimated to be 565.45 million CNY; post-discounting is estimated at 93.11 million CNY. The optimization results of Scenario I indicate that the NPV is −287.09 million CNY and thus the project is still financially infeasible from the viewpoint of dynamic profitability (Table 6). As a result, in order to make this project financially feasible, the local government should provide at least 380.20 (=93.11 + 287.09) million CNY in subsidies to L Company in the year 2016 (or equivalence in time value).

Table 6: Optimization results of four scenarios.

Indices		Scenario I	Scenario II	Scenario III	Scenario IV
Static profitability indices	Payback period of total investment (year)	20.42	25.68	25.47	20.02
	Return on investment (%)	423.51	253.35	156.13	424.35
	Annual return on total investment (%)	10.59	6.33	3.90	10.61
Dynamic profitability indices	Dynamic payback period of total investment (year)	NA	NA	NA	NA
	NPV of total investment (million CNY)	−287.09	−124.00	−260.01	−24.16
	IRR of total investment (%)	5.45	6.87	4.06	9.40

(2) *Scenario II:* In this situation, the cost of land use rights is reduced to zero and the necessary equity capital is spent equally in 2016. Hence, the received subsidy is equal to the cost of land use rights. The optimization results of Scenario II (Table 6) indicate that the studied project is still financially infeasible from the perspective of dynamic profitability. To make the NPV positive, at least 374.00 (=250.00 + 124.00) million CNY in subsidies should be provided by the local government in the year 2016(or equivalence in time value).

(3) *Scenario III:* The local government is to pay all the life-cycle finance expenses of this project on behalf of L Company. The overall finance expense pre-discounting is 751.32 million CNY and post-discounting is 251.01 million CNY. The optimization results of Scenario III show that the

project, similar to Scenarios I and II, is financially infeasible (Table 6). Under these circumstances, in order to make this project financially feasible, the local government should provide at least 511.02 (=260.01 + 251.01) million CNY in subsidies to L Company in the year 2016 (or equivalence in time value).

(4) *Scenario IV:* It is the combination of the above three scenarios. Therefore, the total amount of pre-discounting subsidy is 1,566.77 million CNY and that of post-discounting subsidy is 594.12 million CNY. Comparing the different optimization results shown in Table 6, it is obvious that Scenario IV is the best among all four scenarios, and static profitability evaluation indices seem rather attractive. However, this studied PRH project is still not financially feasible from the perspective of dynamic profitability. To make the studied project financially feasible, at least 618.28 (=594.12 + 24.16) million CNY in subsidies should be provided in the year 2016 (or equivalence in time value).

Comparing the amount of subsidies involved in each scenario, it is obvious that Scenario II is the most cost-effective. Indeed, it is also the simplest and most operational scenario, since the other three scenarios are comparatively more time-consuming and ineffective. As a result, the measures optimizing the financial status in Scenario II are advised to eliminate the fear of the financial unsustainability for the private sector. In other words, it is recommended that the Nanjing municipal government provide free land use rights and 374.00 million CNY in subsidies to L Company in the year 2016, in exchange for L Company's involvement in the development and operations of the studied PRH project.

CONCLUSIONS

This paper has evaluated the financial sustainability of a Public Rental Housing (PRH) project in Nanjing from the perspective of the private sector, based on the estimation of its costs, incomes, and tax deduction. The results obtained from the dynamic profitability evaluation indices show that the project is financially infeasible, although the results from the static profitability evaluation and the debt-repayment ability evaluation seem rather attractive to the private sector. Through analyzing the significant influencing factors of costs and incomes, three optimization options and corresponding four optimization scenarios have been proposed. From the perspective of cost-effectiveness, the optimization scenario with free land use rights and a subsidy of 374 million CNY in the first year (or equivalence in time value) is strongly recommended.

The evaluation results of this studied PRH project represent actual financial situations and corresponding development predicaments of countless PRH projects in China. First, the widely adopted agent-construction system has proved ineffective in controlling the costs of PRH projects, although this system can attract the private sector's involvement. The privately owned mode is suggested, because the private sector is interested in controlling the total cost of PRH projects now. In addition, this mode will tap into the expertise of the private sector (especially private real estate developers) in building and operating residential communities. Second, common PRH projects are financially infeasible under the present preferential policies, and thus cannot effectively attract the for-profit private sector. To improve this situation and accelerate the development of privately owned PRH projects in China, more preferential policies and greater efforts are required. Since the static profitability indices are attractive and problems exist in dynamic profitability indices, time value proves to be the key influencing factor. The effects of preferential policies will therefore be greater when they are implemented earlier. The most cost-effective optimization measure of the studied case is to provide free land use rights with additional subsidies at the beginning. A comprehensive concession contract is the prerequisite that guarantees the involvement of the private sector in the development of the PRH project (as a privately owned one), enjoying all PRH-related preferential policies and observing whole PRH-related regulations. Third, new valuation methods should be constructed for privately owned PRH projects, since the private sector may obtain option premiums from uncertainties and corresponding management flexibility, which cannot be taken into account in traditional financial feasibility evaluation methods on the basis of discounted cash flows. Therefore, constructing a real option-based valuation model for privately owned PRH projects is a possible direction for future research.

ACKNOWLEDGMENTS

This research is jointly supported by National Natural Science Foundation of China (No. 71301024), Ministry of Education of the People's Republic of China (No. 20120092120019 & No. 15YJC790067), Priority Academic Program Development of Jiangsu Higher Education Institutions (PAPD) and the Fundamental Research Funds for the Central Universities. Authors would like to thank Mr. Cao Yi, a bachelor's student of Southeast University, for his help in data collection and primary analysis.

AUTHOR CONTRIBUTIONS

Author Dezhi Li contributed the largest part to this work, including the methodology and writing; Author Yanchao Chen contributed to case study; Author Hongxia Chen contributed to literature review; Author Eddie Chi Man Hui contributed to linguistic check and modification; Author Kai Guo contributed to data collection.

CONFLICTS OF INTEREST

The authors declare no conflict of interest.

REFERENCES

1. Deng, L.; Shen, Q.Y.; Wang, L. The emerging housing policy framework in China. *J. Plan. Lit.* 2011, *26*, 168–183.
2. Chen, J.; Deng, L. Financing Affordable Housing Through Compulsory Saving: The Two-Decade Experience of Housing Provident Fund in China. *Hous. Stud.* 2014, *29*, 937–958.
3. Yao, S.J.; Luo, D.; Wang, J.L. Housing Development and Urbanisation in China. *World Econ.* 2014, *37*, 481–500.
4. Li, D.Z.; Chen, H.X.; Hui, E.C. M.; Xiao, C.; Cui, Q.B.; Li, Q.M. A real option-based valuation model for privately-owned public rental housing projects in China.

Habitat Int. 2014, *43*, 125–132.

5. Li, D.Z.; Chen, H.X.; Huang, Z.G.; Wu, H.O. Incentive policies and their optimization of the social capital's provision of public rental housing in China. *Mod. Manag. Sci.* 2012, *3*, 43–45. (In Chinese)
6. Cheng, D.T. Planning on the social objectives of China's housing policy and the reconstruction of the supply system.*Economist* 2010, *12*, 50–57. (In Chinese)
7. Yuan, J.F.; Guang, M.; Wang, X.X.; Li, Q.M.; Skibniewski, M.J. Quantitative SWOT analysis of public housing delivery by public-private partnerships in China based on the perspective of the public sector. *ASCE J. Manag. Eng.* 2012, *28*, 407–420.
8. Hanlon, J. Success by design: HOPE VI, new urbanism, and the neoliberal transformation of public housing in the United States. *Environ. Plan. A* 2010, *42*, 80–98.
9. Norris, M.; Coates, D. Private sector provision of social housing: An assessment of recent Irish experiments. *Public Money Manag.* 2010, *30*, 19–26.
10. Hodkinson, S. The private finance initiative in English Council Housing Regeneration: A privatisation too far? *Hou. Stud.* 2011, *26*, 911–932.
11. Kok, R.A.W.; Driessen, P.H. Antecedents of market orientation in semi-public service organizations: A study of Dutch housing associations. *Serv. Ind. J.* 2012, *32*, 1901–1921.
12. Mugnano, S.; Palvarini, P. "Sharing space without hanging together": A case study of social mix policy in Milan. *Cities*2013, *35*, 417–422.
13. Blessing, A. Public, Private, or In-Between? The Legitimacy of Social Enterprises in the Housing Market. *Voluntas* 2015,*26*, 198–221.
14. Mukhija, V. The contradictions in enabling private developers of affordable housing: A cautionary case from Ahmedabad, India. *Urban Stud.* 2004, *41*, 2231–2244.
15. Ibem, E.O. The contribution of Public-Private Partnerships (PPPs) to improving accessibility of low-income earners to housing in southern Nigeria. *J. Hous. Built Environ.* 2011, *26*, 201–217.
16. Adegun, O.B.; Taiwo, A.A. Contribution and challenges of the private sector's participation in housing in Nigeria: Case study of Akure, Ondo state. *J. Hous. Built Environ.* 2011, *26*, 457–467.
17. Abdullahi, B.C.; Aziz, W. The role of private sector participation in achieving anticipated outcomes for low-income group: A comparative analysis of housing sector between Malaysia and Nigeria. *Afr. J. Bus. Manag.* 2011, *5*, 6859–6890.
18. Abdul-Aziz, A.R. Control mechanisms exercised in Malaysian housing public-private partnerships. *Construct. Manag. Econ.* 2012, *30*, 37–55.
19. Daniel, M.M.; Hunt, R.J. Changing housing policies and housing provision in Jos, Nigeria. *Habitat Int.* 2014, *42*, 203–213.
20. Cao, X.L.; Hou, Y.X. Application of BOT in public rental housing. *J. Eng. Manag.* 2011, *25*, 570–574. (In Chinese)
21. Chen, D.Q.; Zheng, S.S. The PPP financing model and pricing mechanism of public rental housing. *Construct. Econ.*2011, *4*, 12–16. (In Chinese)
22. Shen, J.; Xie, S.S. Study on financing models of public rental housing. *Inq. Econ. Issues* 2011, *1*, 87–93. (In Chinese)
23. Lin, Y.L.; De Meulder, B.; Cai, X.X.; Hu, H.D.; Lai, Y.N. Linking social housing provision for rural migrants with the redevelopment of "villages in the city": A case study of Beijing. *Cities* 2014, *40*, 111–119.
24. Li, D.Z.; Wu, H.O.; Cui, M.; Xiao, C.; Yang, H. Investment control of affordable housing and its empirical analysis under agent construction system. *Mod. Manag. Sci.* 2011, *11*, 77–79. (In Chinese)
25. Wang, C.P. The Study on the Multi-Subject Construction Pattern of Public Rental Housing. Master's Thesis, Taiyuan University of Science and Technology, Taiyuan, China, May 2012. (In Chinese).
26. National Development and Reform Commission of China and Ministry of Construction of China. *Economical Evaluation Methods and Parameters of Construction Projects*; China Planning Press: Beijing, China, 2006.
27. Liu, Y.M. *Engineering Economics*; Tsinghua University Press: Beijing, China, 2006. (In Chinese)

CHAPTER 2 SERVICE RADIUS MODEL AND SERVICE SCOPE OPTIMIZATION OF CITY PUBLIC PARKING GARAGE

Chao Zeng, Boming Tang, and Tangzhi Liu

INTRODUCTION

The best layout for public parking garages should be "dispersed, nearby, and convenient," as the main function is to undertake the overflow of vehicles when the parking spots are saturated within the service radius. However, central district resources are scarce, so public parking garages are usually situated in the corner of the central areas with a service radius from 200 m to 500 m or 1000 m, chosen according to Chinese current codes or local standards. A concentric circle service scope simply taken with the service radius fails to consider city characteristics and differences between parking behaviors. Thus, most of them have an unreasonable service radius and limitations on corner locations, which lead to low utilization of public parking garages and a waste of resources or difficulty in alleviating parking congestion.

Current transportation researches still lack practical tools and methodologies to analyze the service radius and service scope of public parking garages. The latest research mainly focuses on the public parking site selection, drivers' parking choice behaviors, and the impact of curbside parking regulation on public parking garage demand. In the field of public parking site selection, a group web-based MC-SDSS (the integration of GIS and Multicriteria Decision Analysis capabilities into the web platform offers an effective Multicriteria Spatial Decision Support System) devoted to solving parking site selection problems is presented ("as discussed by Jelokhani-Niaraki and Malczewski [1]"). Based on GIS, a novel algorithm and model for estimating city parking patterns are presented ("as discussed by Levy and Benenson [2]"). In the field of drivers' parking choice behaviors, radial distance to parking, and availability of P&R sites in the direction of travel, gender, age, and income are found to be important factors in parking usage of city center ("as discussed by Clayton et al. [3]"). Upon arrival, parking search is shown to be dominant amongst drivers' strategies. Uncertain parking availability has been found to be important for parking location choice ("as discussed by Chaniotakis and Pel [4]"). A multinomial regression analysis shows that personal (gender, education, and country of residence) and trip (visit frequency) characteristics are significantly related to the distinguished habitual parking behavior levels ("as discussed by van der Waerden et al. [5]"). The research finds that, firstly, vehicle age is a key variable when choosing parking places in Spain; second, the perception of the parking charge is fairly heterogeneous, depending on both the drivers' income levels and whether or not they are local residents ("as discussed by Ibeas et al. [6]"). The results show that parking prices affect car usage; meanwhile, time limitation and the amount of parking demands affect the choice on the type of parking ("as discussed by Simićević et al. [7]"). In the field of the impact of curbside parking regulation on public parking garage demand, the research shows that the characteristics of curbside parking spaces (allowance and time limits) play a role in garage demand determination, yet pricing is a much more efficient trigger for behavioral change ("as discussed by Gragera and Albalate [8]"). Through establishing competition models between private parking garages and public curbside parking, the research found that drivers differ in their parking durations, searching-time costs, and walking-time costs ("as

Citation: Chao Zeng, Boming Tang, and Tangzhi Liu, "Service Radius Model and Service Scope Optimization of City Public Parking Garage," Mathematical Problems in Engineering, vol. 2016, Article ID 1217985, 11 pages, 2016. doi:10.1155/2016/1217985.

discussed by Inci and Lindsey [9]"). The optimal model for public management can explain how cities maximize the public benefits of their parking assets and provide a reasonable price for off-street parking ("as discussed by Pierce et al. [10]").

Researches focused on the service radius and service scope of public parking garage are far less well developed. And related studies mainly concentrate on two aspects: the statistical analysis of the walking distance after parking ("as discussed elsewhere [11, 12]") and establishing an optimization service radius model according to parking choice probability models ("as discussed by Yu Yang et al. [13]"). The methods for determining service radius of public parking garage are mostly simple statistical analysis of the walking distance after parking ("as discussed by USA City Land Use Institute [14]") or adopting the minimum average of the walking distance after parking as the service radius which has been put forward by Japanese researchers ("as discussed by Hiroyuki [15]").

At present, the definition of public parking service radius is different in China and abroad, lacking scientific quantitative calculation methods. In China, for example, the codes provide limiting values on the service radius of public parking garage, but these values are inconsistent with each other and the service radius defined mostly refers to the early foreign datum, so it lacks scientific theoretical research. Besides, the current service scope put forward by research institutes in China and overseas is a concentric circle with a center point of parking garage, which is not in good accordance with the actual situation as it lacks consideration of city resistance and accessibility.

In view of the abovementioned facts, this study puts forward an optimization model for public parking service radius based on proposing a functional value model and a modified model of parking demand by drivers with satisfactory theory. This model can meet the requirements of both drivers and owners. In consideration of the nonuniformed distribution of city resistance, this paper also puts forward a modified model of service scope based on the theory of moment balance according to the theory of accessibility. Finally, the public parking garage of Kuixing Building in China Chongqing's CBD is taken as an example to verify practical application effects of the model by comparing the effects of the optimization model with the present situation.

INFLUENCING FACTORS OF PUBLIC PARKING SERVICE RADIUS

Influencing factors of the public parking service radius consist of two aspects: macro level and micro level. The former level includes city characteristics, land exploitation and utilization, regional traffic condition, traffic management measures, parking policy, and parking service quality, and the latter level contains parking behavior characteristics, and so forth.

City Scale. City scale has an influence on the public parking service radius in some degrees. The larger the city is, the longer the trip distance becomes, and thus drivers tend to tolerate longer walking distance.

City Resistance. Walking resistance from mountains, waters, buildings, and roads in the city will greatly influence the public parking service radius. Partial blockage caused by the above resistance will lead to a shorter service radius. City Land Exploitation and Utilization. The different characteristics of lands determine the amount of parking demands. Parking demand is relatively concentrated in areas which have been deeply exploited. The concentration degree of demands is negatively related to the service radius.

Accessibility. Accessibility can be divided into two aspects: accessibility of the route to public parking and accessibility of walking to the destination after parking. Accessibility is to some extent positively related to the service radius.

Characteristics of Parking Behaviors. Different parking purposes lead to various walking distances after parking, for example, a commuting driver, who needs a long time for parking, with a longer walking distance after parking, can accept the larger service radius.

Service Quality. With the increasing deterioration of the traffic environment, drivers have higher requirements for parking services. The smart, humanistic, and convenient service will have stronger attraction to drivers.

Influence of Nearby Parking Lots. Drivers always want to park their cars in the location closest to their final walking destination. The utilization and convenience degree of nearby parking garages are in competition with public parking garages.

Policy Condition. Favorable policy conditions are the premise for extending service radius. Generally, congestion fee, P&R preferential transfer, and relative preferential charging standards are adopted in China and abroad to promote attractive rates.

Of the abovementioned factors, the most significant influencing factors to public parking service radius are city resistance, land exploitation, utilization, and parking behavior characteristics.

SERVICE RADIUS MODEL OF PUBLIC PARKING

Due to different regional conditions, each public parking should have a service radius according to its own local condition. However, currently, the fixed value in codes we adopt limits its practical significance.

As for determination of public parking service radius, on the one hand, owners mostly care about service ability and comprehensive operation costs, hoping to seek an optimal relationship between these two aspects and maximize the function value of public parking. On the other hand, from the perspectives of drivers, whether to choose a public parking is largely dependent on their satisfaction degree. Therefore, this study will put forward a service radius model of public parking with consideration of both satisfaction of drivers and requirements of owners.

Basic Hypotheses

The basic hypotheses to establish the optimization model of service radius are as follows.

Basic Hypothesis 1. Location and construction cost of public parking garage have been determined.

Basic Hypothesis 2. The initial service radius (R_0) has been determined according to codes.

Basic Hypothesis 3. The nature of the land utilization and the index of construction in the study area have been known:

the legal curb parking and the matched parking spots within the initial service radius (R_0) have been identified.

Basic Hypothesis 4. Space resistance is assumed to be uniformly distributed.

Function Value Model of Public Parking Meeting the Needs of Owners

Definition of Model Indexes

(i) Service Ability (S) of Public Parking Garage. Referring to related research results of site selection optimization and logistics engineering, service ability of public parking garage is defined as the comprehensive characterization of capital investment, spot scale, and comprehensive service index (measured by intelligence, humanization, and convenience) ("as discussed elsewhere [16, 17]"). The service ability of public parking garage is expressed by ("as discussed elsewhere [18, 19]")

$$S_1 = IB^a P^b. \tag{1}$$

In (1), S_1 represents service ability of public parking; I stands for comprehensive services index; B stands for capital investment for public parking; P represents spot scale of public parking; a and b stand for the input-output elastic coefficients of capital and spot, respectively ("as discussed elsewhere [18, 19]").

In an actual situation, service ability of public parking which has fixed parking space will be improved to some extent along with the expansion of the service scope ("as discussed by Hongzhi et al. [20]"), and owners hope to attract more parking vehicles by enhancing service ability and expanding the service scope. Service radius is positively related to service ability, and S_1 should be further amended as follows:

$$S = S_1 r^\lambda = IB^a P^b r^\lambda. \tag{2}$$

In (2), S represents the actual service ability of public parking; r stands for service radius; a and b stand for the input-output elastic coefficients of capital and spot, respectively; the value range of λ is (0, 3).

(ii) Reasonable Scale (P) Meeting the Needs of Construction Parties. The round service scope determined by the initial service radius of public parking garage (R_0) is small; thus, firstly, we use the parking generation rate model to calculate the total demands (D_i) within the initial service scope of public parking (i). Let k be the average of the parking demand generation rate per unit of land within the initial service scope of public parking (i). Hence,

$$k = \frac{D_i}{\pi R_0^2}. \tag{3}$$

Then, parking demand (P_i) in an area with a radius (r) within the initial service radius of the public parking garage is

$$P_i = k\pi r^2. \tag{4}$$

Public parking garage mainly undertakes the overflow when the matched and curb parking spots become saturated within service scope, so the overflow is taken as the upper limit of the spot demands. Therefore,

$$P_{\max i} = P_i - \sum_{i=1}^{n} p_i = k\pi r^2 - \sum_{i=1}^{n} p_i. \tag{5}$$

In (5), $P_{\max i}$ represents the upper limit of public parking garage's (i) spot demands; n and pi, respectively, stand for the number of legal curb and matched parking garages and the spot amount of each public parking garage within the initial service scope of public parking garage (i).

The competitive relationship and service scope overlap between the matched and curb public parking garage within the service scope should be considered. A reasonable scale is determined on the basis of inspecting the utilization rate and turnover rate of other public parking garages within the service scope as follows:

$$P = \frac{P_{\max i}}{\theta\varepsilon} = \frac{\left(k\pi r^2 - \sum_{i=1}^{n} p_i\right)}{\theta\varepsilon}. \tag{6}$$

In (6), P stands for the reasonable scale after optimization of public parking; θ and ε represent the average utilization rate and turnover rate within service scope of public parking garage.

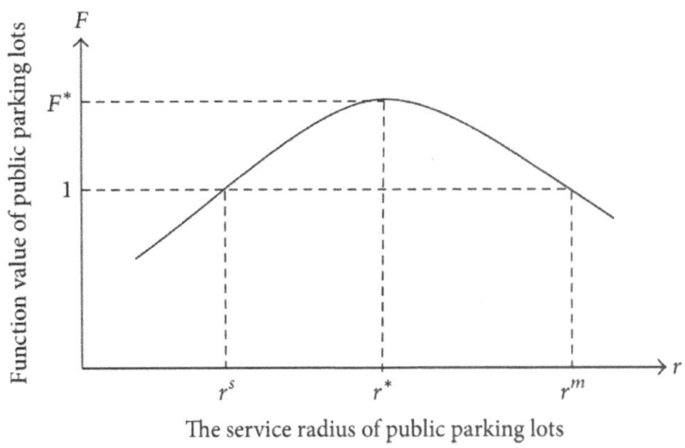

Figure 1: Public parking function value changes with the service radius.

(iii) Comprehensive Operation Costs (C) of Public Parking Garage. Comprehensive operation costs of public parking garage consist of the fixed costs C_1 determined by designed scale and the variable costs changing with demand. C_1 includes the operation costs of parking facilities and rent or depreciation of facilities. Variable costs include energy consumption, equipment maintenance, and labor wages. Variable costs can be calculated via direct costs (c) generated by each spot and the total amounts of spots, and then the comprehensive operation costs are

$$C = C_1 + cP = C_1 + c\left[\frac{\left(k\pi r^2 - \sum_{i=1}^{n} p_i\right)}{\theta\varepsilon}\right]. \tag{7}$$

Model Establishment

(i) Function Value (F) of Public Parking Garage. Function value of public parking garage is defined as the dynamic characteristic reflecting service ability and comprehensive operational costs of public parking garage ("as discussed by Hamacher and Nickel [21]"). Hence,

$$F = \frac{S}{C} = \frac{IB^a P^b r^\lambda}{C_1 + c\left[\left(k\pi r^2 - \sum_{i=1}^{n} P_i\right)/\theta\varepsilon\right]}. \tag{8}$$

From (8), if other factors remain constant, the service radius will directly affect the function value of parking garage, and the related curve is shown in Figure 1. r^* stands for the optimal service radius only with consideration of the demands of owners, namely, to achieve efficient utilization of resources. $[r^s, r^m]$ represents the effective operational region of the public parking garage. In this region, the function value of public parking $F>1$; that is to say, the value created is higher than the operation costs, which do ensure a sustainable development.

(ii) Make the Derivation of (8) Equal to Zero. This is done as follows:

$$\frac{dF}{dr} = \lambda\left[C_1 - \frac{c}{\theta\varepsilon}\sum_{i=1}^{n} P_i\right] - \frac{k\pi r^2 c}{\theta\varepsilon}(2-\lambda) = 0. \tag{9}$$

The optimal service radius (r^*) of public parking garage which meets the demands of owners after simplification is

$$r^* = \sqrt{\frac{\lambda\left(C_1\theta\varepsilon - c\sum_{i=1}^{n} P_i\right)}{ck\pi(2-\lambda)}}. \tag{10}$$

We can deduce C_1 from (7) and then put it in (10). Through combination of these two equations, the result shows that r^* is positively correlated with comprehensive operational costs. From owners' prospective, the higher the costs, the larger the number of parking cars that they expect, for only in this way can profits be increased, and then the service radius will be enlarged accordingly. And r^* is negatively correlated with the parking generation rate in the region. Therefore, if the parking generation rate around the public parking garage is larger, a small service radius can meet the operation demands.

Modified Model with Consideration of Satisfaction and Demands of Drivers.

Whatever the expectations of owners are, to a great extent, satisfaction degree is a vital factor for drivers to choose a public parking garage. Determining the service radius from the functional view of owners is just an ideal description of public parking operation state. Therefore, we need to modify the model according to the characteristics of drivers' demands.

There are many indexes influencing parking demand satisfaction of drivers, but the walking distance after parking is the most significant acknowledged factor which influences parking choice and parking satisfaction. Therefore, satisfaction degree of walking distance after parking is used as the evaluation index for analysis and quantization.

Satisfaction Model of Walking Distance after Parking.

Let t_i be the shortest walking distance which is from a public parking lot to a parking-demanded site (i); L_i and S_i are the longest walking distance in the case where drivers feel very satisfied and the tolerable walking distance in the case where drivers feel very unsatisfied, respectively. (t_i) is the satisfaction degree of drivers for a walking distance after parking in the public parking lot. Accordingly, $L_i \leq S_i$, so $(t_i) \in [0, 1]$ ("as discussed by Siskos et al. [22]").

At present, distance or time satisfaction theory is usually applied to the logistics center and the study of emergency facilities location ("as discussed elsewhere [23, 24]"). Common customer satisfaction models are linear function, concave convex function, ridge type function, decreasing semi-Cauchy type function, and exponential decay Sigmoid function ("as discussed by Yanfei [25]"). Through trail and fitting of prophase research, a ridge type or decreasing semiCauchy type time satisfaction model has been proved to be fitter for the satisfaction degree model of walking distance after parking

(i) Ridge Walking Distance Satisfaction Model. This function has little change near thresholds L_i and S_i, as the slope is larger

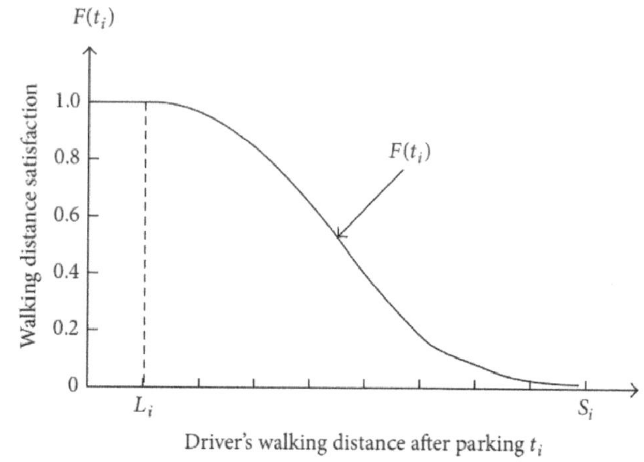

Figure 2: The curve of ridge walking distance satisfaction.

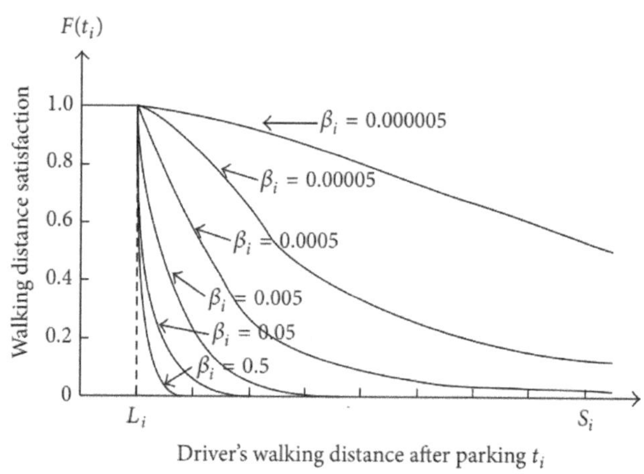

Figure 3: The curve of walking distance satisfaction.

in the middle of the curve. The functions are as follows (see Figure 2):

$$F(t_i) = \begin{cases} 1, & t_i \leq L_i \\ \frac{1}{2} + \frac{1}{2}\cos\left(\frac{\pi}{S_i - L_i}\left(t_i - \frac{L_i + S_i}{2}\right) + \frac{\pi}{2}\right), & L_i < t_i \leq S_i \\ 0, & t_i > S_i. \end{cases} \tag{11}$$

(ii) Decreasing Semi-Cauchy Time Satisfaction Model. This is represented as follows:

$$F(t_i) = \begin{cases} 1, & t_i \le L_i \\ \dfrac{1}{1+\beta_i (t_i - L_i)^2}, & t_i > L_i, \end{cases} \tag{12}$$

where β_i is a positive time sensitive coefficient. The functions are presented in Figure 3.

Calculation of Thresholds L_i, S_i

(i) The Longest Acceptable Walking Distance L_i in the Case of Being Very Satisfied. American research shows that a parking lot should guarantee that 85~95% of users are within their tolerable service radius ("as discussed by Yali [26]"). After investigating parking behaviors, we count and summarize the data of walking distance after parking and then decide to use the 85th percentile walking distance after parking as the longest acceptable walking distance L_{i1}

Besides, the difference of parking purposes has great influence on the satisfactory walking distance of drivers. Thus, calculating synthetic satisfactory walking distance under different parking purposes is necessary.

Let $X = \{x_1, x_2, ..., _n\}$ be N independent parking purposes; the occupied percentage and tolerable walking distances of each parking purpose are $W = \{w_1, w_2, ..., w_n\}$ and $T = \{t_1, t_2, ..., t_n\}$, respectively, and L_{i2} is calculated by

$$L_{i2} = \frac{w_1 t_1 + w_2 t_2 + \cdots + w_n t_n}{w_1 + w_2 + \cdots + w_n} = \frac{\sum_{i=1}^{n} w_i t_i}{\sum_{i=1}^{n} w_i}. \tag{13}$$

Take $L_i = \min(L_{i1}, L_{i2})$ as the longest acceptable walking distance in the case of being very satisfied.

(ii) Tolerable Walking Distance (S_i) in the Case of Being Very Unsatisfied. Chinese codes generally stipulate that the public parking service radius should not exceed 500 m, and the longest service radius is within 1000 m, so we take S_{i1} = 500 m here. The public parking optimal service radius (r^*) with consideration of owners is usually much larger than the satisfactory walking distance of drivers; thus, $S_{i2} = r^*$. Hence,

$$S_i = \max(S_{i1}, S_{i2}). \tag{14}$$

Difference Regulatory Factors with Consideration of Drivers' Satisfaction.

We can use (11) and (12) to establish a satisfaction model (t_i) where the driver demand-satisfied set $V = \{f(t_1), f(t_2), \ldots, f(t_n)\}$. The central limit theorem shows that the above large-sample investigation ($n \ge 30$) obeys normal distribution; mean value of sample $\overline{f(t)}$ and variance S^2 are as follows:

$$\overline{f(t)} = \frac{1}{n}\sum_{i=1}^{n} f(t_i),$$

$$S^2 = \frac{1}{n-1}\sum_{i=1}^{n}\left[f(t_i) - \overline{f(t)}\right]^2. \tag{15}$$

Considering the different characteristics of parking behavior, a public parking service radius should not be a fixed value, and a dynamic interval is more practical to the actual situation. Let $f \in [0, 1]$ be a difference regulatory factor which considers characteristics of drivers' satisfaction, and confidence interval which reflects differences of characteristics of drivers' satisfaction is used as the regulatory factor f when the confidence range is $1-\alpha$. Hence,

$$f \in \left[\overline{f(t)} \pm \frac{S}{\sqrt{n}} t_{\alpha/2}(n-1)\right], \tag{16}$$

where $t_{\alpha/2}$ is the $\alpha/2$ quantile of standard normal distribution.

Table 1: City resistance relative value.

Land use type	Relative resistance	Influencing factors on walking	Relative resistance
Road, square	1	Grade crossing	1.1
Commercial land	1.1	Grade separation	1.2
Residential land	1.3	Mountain/slope land with escalator	1.1
Public green space	1.5	Mountain/slope land with staircase	1.2
Land for municipal facilities	1.8	Mountain/slope land with elevator	1
Land for storage	2	Mountain, river with no crossing facilities	+∞
Industrial land	5	Height difference with no crossing facilities	+∞

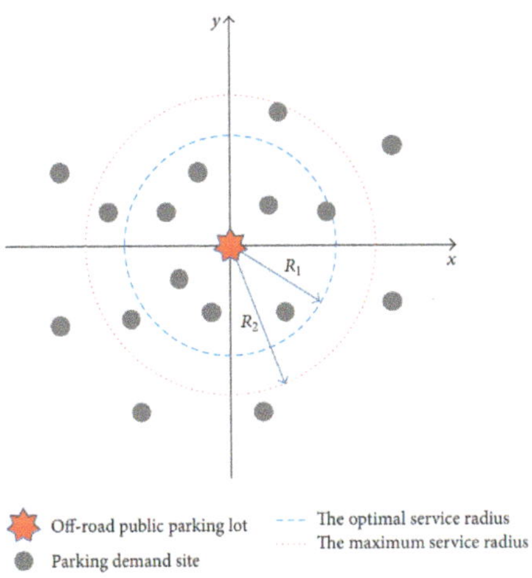

Figure 4: Dynamic service radius of public parking.

Public Parking Service Radius Optimization Model with Consideration of Demands of Drivers and Owners.

By the above analysis, the public parking optimal service radius model with consideration of owners and characteristics of drivers' demands is as follows:

$$R = r^* \times f = r^* \times \left[\overline{f(t)} \pm \frac{S}{\sqrt{n}} t_{\alpha/2}(n-1) \right] \Longrightarrow$$

$$\begin{cases} R_1 = r^* \times f = r^* \times \left[\overline{f(t)} - \frac{S}{\sqrt{n}} t_{\alpha/2}(n-1) \right] \\ R_2 = r^* \times f = r^* \times \left[\overline{f(t)} + \frac{S}{\sqrt{n}} t_{\alpha/2}(n-1) \right]. \end{cases} \tag{17}$$

The public parking service radius calculated from (17) is a dynamic interval with consideration of demand satisfaction characteristics of owners and drivers in the case of assuming that space resistance is uniformly distributed and is used in concentric circles of which circle center is public parking (i) (Figure 4). The lower limit of the dynamic interval R_1 is the optimal service radius, and the upper limit R_2 is the maximum service radius

OPTIMIZATION OF PUBLIC PARKING SERVICE SCOPE

The dynamic concentric circles service radius proposed by the above study is obtained by assuming the uniform distribution of space resistance. However, in reality, the accessibility of public parking is influenced by resistance caused by mountains, waters, buildings, and so forth, which results in a great change of the service scope within the public parking service radius. Especially in a mountainous city, it is more important to consider the influence of topographic height differences on accessibility. So we need to modify the service scope, making it accord with the actual situation better.

Walking Accessibility of Public Parking Garage

Without consideration of differences of space resistance distribution, based on the idea that the shortest distance between two points is a straight line, accessibility indexes of public parking garage show a concentric circle-like distribution. And if we take influence of comprehensive impedance factors into consideration such as city resistance and walking resistance, accessibility indexes will change greatly. Then, the difficulty degree to which drivers overcome resistance to go to destinations after parking is measured by the walking distance or time.

City Resistance and Walking Impedance

Walking accessibility can be extremely reduced by mountains without pedestrian passageway, water without a bridge, and topographic height differences that are hard to cross, and drivers will suffer different level resistance because they need to walk through areas with all kinds of features by different crossing facilities. Let uniform distributed city resistance be γ and nonuniform distributed walking impedance $\gamma' = \phi\gamma$, where ϕ is the nonuniform distributed city resistance modification coefficient. Combined with related research datum ("as discussed by Xiaolai [27]") and parking survey datum in Chongqing, reference value of ϕ is as in Table 1.

Resistance Distribution Matrix and Calculation of Route Resistance

In combination with the site situation and on the basis of city land utilization, walking influence, and natural barrier, the initial service scope is divided into a square grid with a certain interval. This division should center around public parking garage and then use dij and γ'_{ij} to express walking distance and relative space impedance value of each square grid area, respectively.

Let $L = \{l_1, l_2, ..., l_t\}$ be t available routes leading to a certain point of the service boundary for drivers after stopping theirs car in the public parking, and use $R_e(l_t)$ to express the accumulative resistance value of t routes.

Service Radius Modified Model Based on the Principle of Moment Balance.

Taking public parking as a centroid, divide the circular boundary with r radius into x ($x \to +\infty$) globules with the same volume V. The density of these globules is space resistance ρ. The globules and centroids are connected by connecting rods whose mass is negligible and the moment from globule to centroid is $M_i = F_i L_i = \rho_i V g r$ ($i = 1, 2, \dots, x$), where g is gravitational acceleration. As space resistances are uniformly distributed, all moment values M are the same, so the total moment of system is $\sum M = 0$, which keeps the system in a stationary state. To explain this easily, we take eight intersection points between eight direction lines in geography and service radius as an example (Figure 7).

When space resistance is distributed no uniformly, the value ρ of globule will change. With insurance of no migration of centroid and a stationary balance state of system, we need to change the length of each connecting rod (arm of force) r_i' to make $M_i = M_i'$ and keep moment in balance: $\sum M$=0.

Besides, there is a special circumstance that the barrier of natural resistance cannot be crossed within service radius ($\rho i = +\infty$). To reflect an actual situation, r_i' represents the distance from centroid to the points of natural barrier boundary (Figure 5(a)). Then, the balance condition is $M_i = M_i' + E_i$, where E stands for the moment produced by natural resistance.

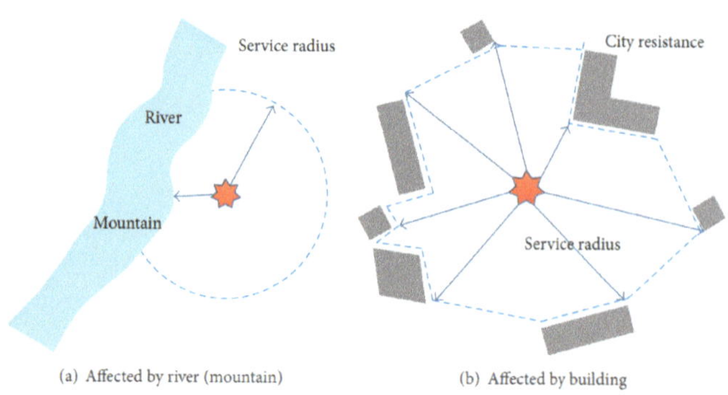

Figure 5: Effect of resistance to the service radius of city parking

According to the above analysis, when $\rho_i \neq +\infty$, we establish a modified service radius model r_i':

$$\begin{aligned}
&\rho_i = r\gamma \\
&R_{ei}(l_t) = \sum d_{ij}\gamma_{ij} = \gamma\sum d_{ij}\phi_{ij}, \quad l_t \in (l_1, l_2, \ldots, l_t) \\
&\rho_i' = \min R_{ei}(l_t) = \min\left(\gamma\sum d_{ij}\phi_{ij}\right) \\
&M_i = F_i L_i = \rho_i v g r = \gamma v g r^2 \\
&M_i' = F_i' L_i' = \rho_i' v g r_i'
\end{aligned} \tag{18}$$

$$\text{st.} \begin{cases} M_i = M_i' \\ \sum M = 0 \end{cases}$$

$$\Longrightarrow r_1' = \frac{\rho_i r}{\rho_i'} = \frac{r^2}{\min\left(\sum d_{ij}\phi_{ij}\right)}. \tag{19}$$

In (18) and (19), ρi represents the density of the globule i; r is the service radius; γ is the uniform distributed city resistance; $R_{ei}(l_t)$ represents the accumulative resistance value of the route t; d_{ij} and γ_{ij}' represent walking distance and relative space impedance value of each square grid area, respectively. ϕ is the nonuniform distributed city resistance modification coefficient; m_i is the moment from globule to centroid when space resistance is uniformly distributed; F_i and L_i are the space resistance acting on the globule i and the relevant level-arm when space resistance is uniformly distributed; and M_i' is the moment from globule to centroid when space resistance is uniformly distributed; F_i' and L_i' are the space resistance acting on the globule i and the relevant level-arm when space resistance is uniformly distributed.

According to (19), we modify the circular distribution scope under uniform distribution of space resistance into an irregular polygon service scope with public parking garage as base point, which is more coincident with actual situation.

Table 2: Satisfaction and the shortest walking distance of each point in demand.

Demand site	1	2	3	4	5	6	7	8
t_i (m)	184	204	106	148	431	490	442	394
$f(t_i)$	1.000	0.998	1.000	1.000	0.357	0.158	0.316	0.500
Demand site	9	10	11	12	13	14	15	16
t_i (m)	249	311	324	345	487	496	381	367
$f(t_i)$	0.954	0.804	0.761	0.688	0.167	0.141	0.551	0.606
Demand site	17	18	19	20	21	22	23	24
t_i (m)	208	248	233	312	245	338	427	499
$f(t_i)$	0.997	0.956	0.977	0.800	0.960	0.713	0.372	0.133

CASE OF APPLICATION

Case Background

We carry out our experiment in Kuixing Parking Building, an 8-storey self-propelled tridimensional public parking building (925 spots) in Jiefangbei CBD, Chongqing, China, whose use characteristics are complex, and the parking demands here are large. However, in practical operation, although demands for parking spots exceed supplies in CBD, the spot utilization of Kuixing Building fails to reach expectation, and high vacancy rate leads to waste of resource. So, we use the models and methods proposed by this paper to redetermine the service radius of this public parking and optimize parking spot scale and then compare them to the present situation.

Application of Service Radius Model

According to the codes, service radius of this public parking garage is $R_0 = 500$ m; the total demand within initial service scope is $D_i = 18055$ spots; then, we get $k = 2.3$ spots/100 m^2. Fixed cost is $C_1 = 50$ million yuan and direct cost of unit spot is $c = 0.0026$ million yuan/spot; there are $n = 24$ various public parking spots within the initial service scope, so the total amount of spots is 13991.The regulatory factor is $\lambda = 0.7$; spot utilization is $\theta = 84\%$; turnover rate is $\varepsilon = 3.8$. According to (10), we can obtain the optimal service radius $r^* = 594$ m which meets the demands of owners. By conducting an investigation about parking behavior, customers' satisfaction degree on walking distance within initial service scope can be fitted by a mountain type model, where $L_i = 194$ m and $S_i = 594$ m. The shortest walking distance t_i from 24 parking-demanded places within the initial service scope to the public parking and values of satisfaction function obtained are as in Table 2.

According to (15), $\overline{f(t)} = 0.663$ and $S^2 = 0.103$ can be obtained. Then, let the confidence level be 95%, and regulatory factor $f \in (0.619, 0.707)$ is carried out. Finally, bring parameter $\overline{f(t)}$ into (17) and make out the service radius of this public parking garage: $R_1 = 368$ m and $R_2 = 420$ m. The dynamic interval is $R \in (368$ m, 420 m$)$.

Application of Modified Service Scope Model

In combination with site situation of this public parking, 8 direction lines in geography are taken as the datum line to construct 16 points with modified radius (Figure 9). Suppose the uniform distributed city resistance is 1, and then calculate the relative value of nonuniform distributed city resistance by Table 1. Next, establish the distribution matrix of resistance (Figure 6) and calculate the accumulative resistance value of each route to get ρ_i'. The results modified according to Section 4.4 are shown in Table 3 and Figure 8.

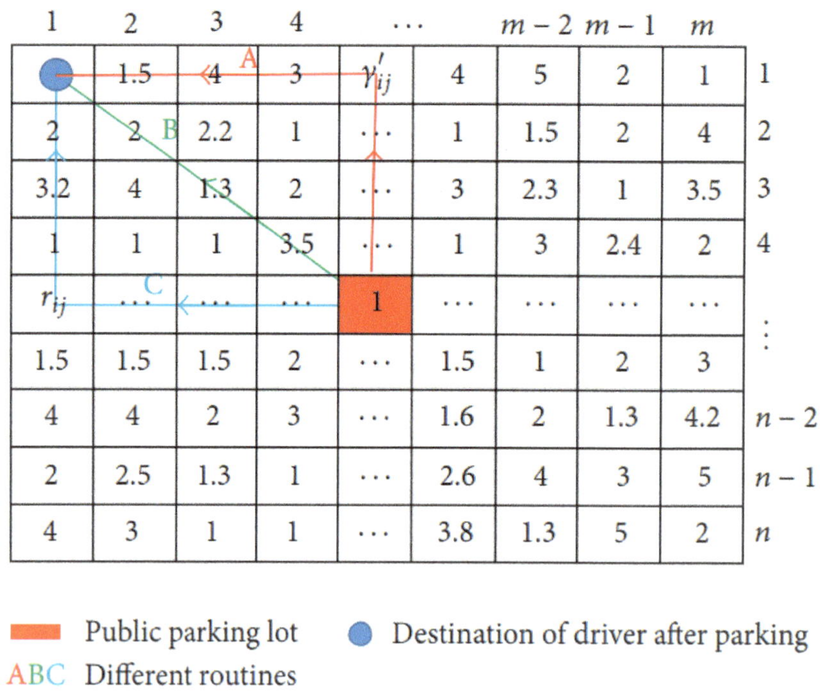

Figure 6: Resistance spatial distribution matrix.

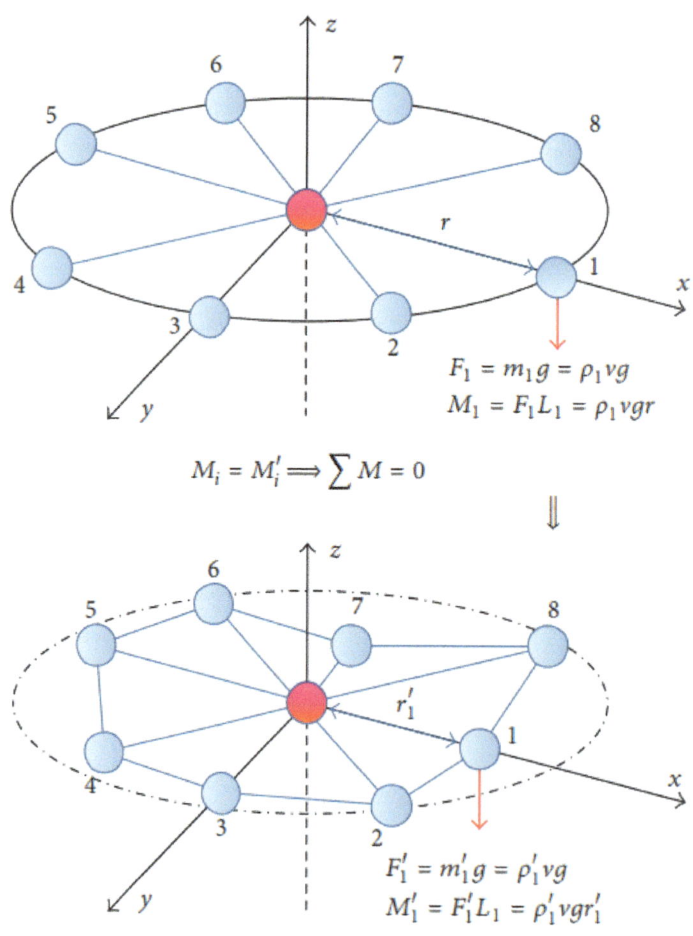

Figure 7: Radius optimization based on the principle of moment balance.

Result Analysis.

Before service scope optimization, the utilization of 925 parking spots in this public parking during workdays and holidays is shown in Figure 9.

Figure 8: Diagram of modified service scope.

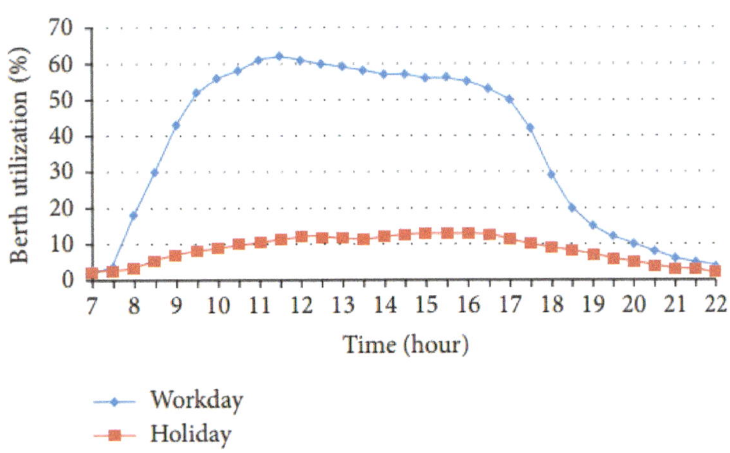

Figure 9: Utilization rate of Kuixing Public Parking Building.

Our analysis from the view of owners, with consideration of construction and operation, is as follows. The amount of parking spots is redetermined on the basis of the modified service scope. According to the study in Section 3.2, there are 722 spots after optimization. The maximum utilization rate of workday parking is improved from 62% to 74.9% and of holidays from 12.8% to 16.4%. It shows that the utilization of parking lot is improved significantly. So the main reason for the low utilization of this public parking is the large quantity of parking spots supply caused by the overdesigned service radius. By optimizing the service scope, the reoptimized spot supply is much more in accord with the present utilization. Therefore, we cannot blindly pursue a large service radius and more parking spots when constructing public parking.

Table 3: Modified calculation table for service scope.

Datum point	1	2	3	4	5	6	7	8
ρ'_{i1}	—	—	—	433	478	499	386	481
R'_1 (m)	229	266	358	313	283	271	351	282
ρ'_{i2}	—	—	—	456	521	590	432	525
R'_2 (m)	229	266	358	387	339	299	408	336
Datum point	9	10	11	12	13	14	15	16
ρ'_{i1}	441	485	461	500	519	387	—	—
R'_1 (m)	307	279	294	271	261	350	281	253
ρ'_{i2}	562	521	575	552	624	459	—	—
R'_2 (m)	314	339	307	320	283	384	281	253

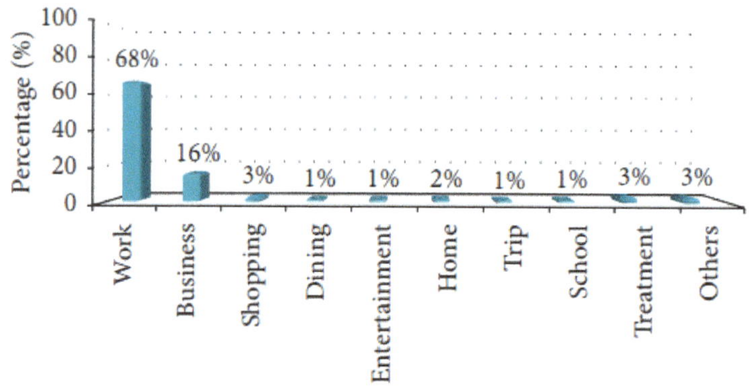

Figure 10: Parking purposes of drivers.

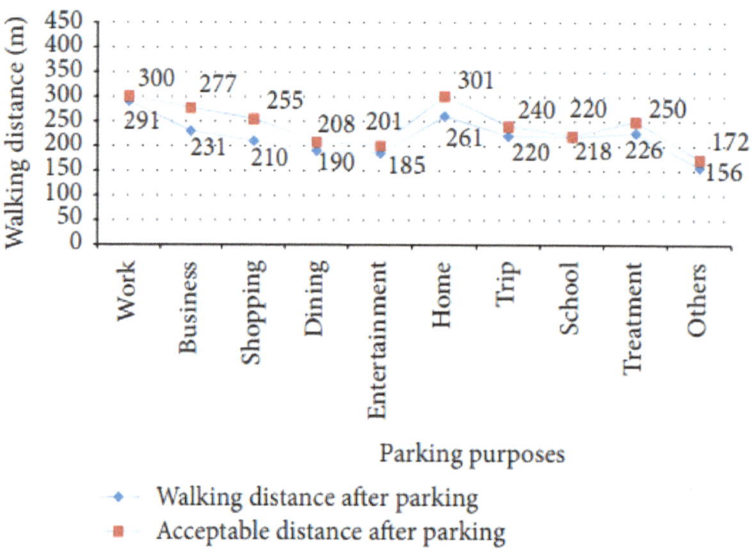

Figure 11: Walking distance and acceptable distance of drivers with different parking purposes.

A moderate scale and dispersed layout will effectively relieve hard-parking.

Our analysis from the view of drivers, with consideration of satisfaction in walking distance after parking, is as follows. According to Table 3 and Figure 8, the optimal service radius is R_1 = 229 m and the maximum service radius is R_2 = 408 m. We have made RP (Revealed Preference Survey) and SP (Stated Preference Survey) investigation of drivers in Kuixing Parking Building. And Figure 10 shows the results of parking purposes; Figure 11 shows the results of walking distance and acceptable distance after parking from drivers with different parking purposes

Figures 10 and 11 show that, as for this parking building, those who park for working, business, and shopping account for 87% of the total amount of drivers and they are the main service object. For 95% of drivers except for parking users for the purpose of dining and entertainment and a few drivers who have other purposes, the walking distances and acceptable distances after parking are within both the optimal service radius and the maximum service radius. The results meet the requirement mentioned above that parking lot should guarantee that 85~95% of users are within their tolerable service radius. Therefore, we can conclude that the models and methods proposed by this paper are in good accordance with the results obtained from the investigation of the driver's parking behavior. And the present service radius of this parking building is overlarge, exceeding the maximum acceptable service radius of drivers and failing to meet the satisfaction in walking distance of drivers, which is one of the reasons why the utilization of the parking building cannot reach the expectation.

CONCLUDING REMARKS AND FUTURE RESEARCH

A scientifically rational service radius of public parking has a great significance for reasonably determining parking spot scale and improving utilization of public parking. The innovative highlights of this research are as follows:(1)Based on the theory of value engineering and satisfaction, the optimization model for public parking garage service radius is established, which satisfies the requirements of both drivers and owners.(2) According to the accessibility theory and principle of moment balance, we modify the circular service scope to an irregular polygon service scope with consideration of city resistance and walking impedance; thus, it is more coincident with the actual situation.(3)This study proposes new thoughts and methods to determine service radius and scope of off-street public parking garages, remedying the defects of determining service radius according to experience.(4)The application of research results improves the whole operational efficiency of public parking on the basis of considering the demand satisfaction of both owners and drivers, which promotes the sustainable development of city and alleviates the public parking pressure.

Furthermore, due to the related replacement between parking fee and walking distance, there is the possibility of adjusting the public parking service scope by means of policies. Therefore, how to introduce policy factors into this model and method will be the future direction of this study.

COMPETING INTERESTS

All the authors of this paper declare that they have no financial or personal relationships with other people or organizations that can inappropriately influence their work, and there is no professional or any other personal interest of any nature or kind in any product, service, and/or company that could be construed as influencing the position presented in the paper.

ACKNOWLEDGMENTS

This research was financially supported by the National Natural Science Foundation of China (E080701) and the Municipal Technology Plan of Chongqing (2013-11).

REFERENCES

1. M. Jelokhani-Niaraki and J. Malczewski, "A group multicriteria spatial decision support system for parking site selection problem: a case study," Land Use Policy, vol. 42, pp. 492–508, 2015.
2. N. Levy and I. Benenson, "GIS-based method for assessing city parking patterns," Journal of Transport Geography, vol. 46, pp. 220–231, 2015.
3. W. Clayton, E. Ben-Elia, G. Parkhurst, and M. Ricci, "Where to park? A behavioural comparison of bus Park and Ride and city centre car park usage in Bath, UK," Journal of Transport Geography, vol. 36, pp. 124–133, 2014.
4. E. Chaniotakis and A. J. Pel, "Drivers' parking location choice under uncertain parking availability and search times: a stated preference experiment," Transportation Research Part A: Policy and Practice, vol. 82, pp. 228–239, 2015
5. P. van der Waerden, H. Timmermans, and A. N. R. da Silva, "The influence of personal and trip characteristics on habitual parking behavior," Case Studies on Transport Policy, vol. 3, no. 1, pp. 33–36, 2015.
6. Ibeas, L. Dell'Olio, M. Bordagaray, and J. D. D. Ortúzar, "Modelling parking choices considering user heterogeneity," Transportation Research Part A: Policy and Practice, vol. 70, pp. 41–49, 2014.
7. J. Simićević, S. Vukanović, and N. Milosavljević, "The effect of parking charges and time limit to car usage and parking behaviour," Transport Policy, vol. 30, pp. 125–131, 2013.
8. Gragera and D. Albalate, "The impact of curbside parking regulation on garage demand," Transport Policy, vol. 47, pp. 160–168, 2016.
9. E. Inci and R. Lindsey, "Garage and curbside parking competition with search congestion," Regional Science and Urban Economics, vol. 54, pp. 49–59, 2015.
10. G. Pierce, H. Willson, and D. Shoup, "Optimizing the use of public garages: pricing parking by demand," Transport Policy, vol. 44, pp. 89–95, 2015.
11. M. C. Childs, Parking Space, The McGraw-Hall Compaines, New York, NY, USA, 1999.
12. M. B. W. Kobus, E. Gutiérrez-i-Puigarnau, P. Rietveld, and J. N. Van Ommeren, "The on-street parking premium and car drivers' choice between street and garage

parking," Regional Science and Urban Economics, vol. 43, no. 2, pp. 395–403, 2013.

13. Y. Yang, H. Hao, and L. Yun-Xia, "Parking choice behavior-based optimization model of service radius of road outside public parking," in Proceedings of the 19th Cross Straits Symposium on Urban Traffic, pp. 428–434, Guiyang, China, 2011.
14. USA City Land Use Institute, Z. Sheng, S. Ping, and W. Li-Jun, Foreign Design of Parking, Intellectual Property Publishing Home, Beijing, China, 4th edition, 2003.
15. Y. G. Hiroyuki, Business and Investment of Multistory Parking, Urban Co., Tokyo, Japan, 1991.
16. R. Mahmud and V. Indriasari, "Facility location models development to maximize total service area," Theoretical and Empirical Researches in Urban Management, vol. 4, no. 1, pp. 87–100, 2009.
17. Tamir, "The k-centrum multi-facility location problem," Discrete Applied Mathematics, vol. 109, no. 3, pp. 293–307, 2001.
18. W. Fanian, Research on Service Radius of Logistics Node Based on the Satisfactory Degree Theory, Xidian University, 2010.
19. Z. Drezner and C. Scott, "Locating a service facility with some unserviced demand," IMA Journal of Management Mathematics, vol. 17, no. 4, pp. 359–371, 2006.
20. G. Hongzhi, Z. Yuhong, C. Hongbo, and L. Xiaoming, "A methord on forecasting the scale of a logistics center," China Civil Engineering Journal, vol. 38, no. 5, pp. 96–99, 2005.
21. H. W. Hamacher and S. Nickel, "Classification of location models," Location Science, vol. 6, no. 1–4, pp. 229–242, 1998.
22. Y. Siskos, E. Grigoroudis, C. Zopounidis, and O. Saurais, "Measuring customer satisfaction using a collective preference disaggregation model," Journal of Global Optimization, vol. 12, no. 2, pp. 175–195, 1998.
23. M. A. Yun-Feng, M. Zhang, and J. Yang, "Definition and application of time satisfaction function in logistics facility location," Logistics Technology, vol. 9, pp. 26–29, 2005.
24. F. Lei and H. Jianmin, "Optimal location model and algorithm of urban emergency systems," Journal of Management Sciences in China, vol. 2, pp. 12–15, 2005.
25. S. Yanfei, Research on Time-Satisfaction Logistic Facility Location and Distribution System, Zhejiang University of Technology, Hangzhou, China, 2007.
26. L. Yali, Research on Supply-Demand of Parking Planning in Urban Central District, Southwest Jiaotong University, Chengdu, China, 2008.
27. Z. Xiaolai, GIS-Based Researches on Service Radius of Urban Public Park, Huazhong Agricultural University, Wu Han, China, 2007.

CHAPTER 3 AN EXPERIMENTAL STUDY OF STRUCTURAL IDENTIFICATION OF BRIDGES USING THE KINETIC ENERGY OPTIMIZATION TECHNIQUE AND THE DIRECT MATRIX UPDATING METHOD

Gwanghee Heo and Joonryong Jeon

INTRODUCTION

During their service period, structures are exposed not only to gradual aging but also to unspecified harmful environmental effects, such as earthquakes, strong winds, impacts, and structural instability due to external forces that can give rise to various structural defects. These defects can in turn lead to reduced life expectancy and unexpected serious structural damage. Thus, from a long-term perspective, continuous structural health monitoring (SHM), systematic evaluation, and maintenance efforts are pivotal, especially for structures that are vulnerable to vibration occurring during service, such as cable-stayed bridges. For these efforts, a technique for accurate structural identification (SI) is of utmost importance.

SI is defined as the process of defining a mathematical model of a given structure using measured physical information acquired from the actual target structure. Since Lie and Yao (1978) [1] introduced the concept of SI into structural engineering, many experiments have been conducted to identify a mathematical model that can represent the behavioral characteristics (equations of motion) of a target structure [2–5]. Despite such efforts, the SI technique has been applied in everyday practice in only a few cases and with limited effectiveness. It has not been adapted to convenient everyday application in general structural design and maintenance [6–8]. Appropriate utilization of the SI technique not only allows for thorough SHM of the target structure but also reduces the amount of effort required for evaluation and maintenance of the structure and can contribute to ensuring structural safety.

As explained above, to conduct SI, physical structural information must be obtained from the target structure through experimentation. In such experimentation, if the number of sensor locations can be selectively reduced such that all measurement locations are optimized to the target mode, both efficiency and economic viability can be improved from an administrative viewpoint. Soundness monitoring, which requires long-term, real-time measurements, involves figure-intensive data processing and requires repetitive operations. Thus, maximizing the quality of the monitored information while minimizing the number of sensors is of particular importance [9–16]. To achieve this, Kammer [13] used estimated error covariance matrix (EECM) to develop a new approach called the effective independence method (EIM). This method incorporates the concept of spatial independence and is typically applied to truss structures. EIM guarantees linear independence, and its calculations are very effective with regard to sensor location selection.

On the other hand, Gwanghee [12] proposed the kinetic energy optimization technique (KEOT) based on principles of structural kinetic energy under the premise that optimizing the

Citation: Gwanghee Heo and Joonryong Jeon, "An Experimental Study of Structural Identification of Bridges Using the Kinetic Energy Optimization Technique and the Direct Matrix Updating Method," Shock and Vibration, vol. 2016, Article ID 3287976, 13 pages, 2016. doi:10.1155/2016/3287976.

composition and location of sensors by maximizing the kinetic energy (KE) of a structure would be the most effective way of determining damage in structures that feature low and close interval frequencies and high attenuation [9–11]. This method acquires optimal modal information by maximizing the kinetic energy measured from the structural system, thus ensuring the independence of the measuring mode used in KEOT. These types of advantages became handy when applied to large structures, thus proving the usefulness of KEOT [12].

Moreover, for SI, the mathematical model of the target structure must be defined using physical information measured from the target structure [17, 18]. In this process, an FE model that represents the structural state of the target structure is developed based on the measured data of the target structure. In addition, if the numerical modal parameters can be derived, they can be used to define the state of the baseline structure. At this point, it is typical for a certain margin of error to come into play with respect to the modeling. This is because the method of FE model interpretation can vary depending on factors such as lifespan, economics, expected value of the structural model, and the fact that it is constructed based on initial architectural drawings. FE models developed to particularly resolve dynamic structural problems must effectively predict changes in modal parameters caused by alterations in design parameters. In this regard, FE model improvement not only provides dynamic characteristics that conform to the target structure but also can be utilized in the future as an objective and reasonable standard based on which aging and localized damage to a structure can be measured [19, 20]. In accordance with these needs, a range of FE model improvement methods [21] have been introduced. Namely, Friswell and Mottershead [22] have systematically organized classical theories such as the direct matrix updating method (DMUM) [23, 24] and the error matrix updating method (EMUM) [25, 26] and mathematically verified them to prove their validity. Thus, in this study, experiments were conducted to evaluate SI techniques aimed at accurately determining the modal parameters of a target structure based on optimal measurement locations (i.e., minimum number of sensors), which are the required parameters for efficient and practical SHM. To this end, the optimal sensor locations were selected based on a model of a cable-stayed bridge and then the responses from the selected measurement locations were used to improve the FE model. Taking into account the structural characteristic of cable-stayed bridges, which feature relatively natural frequency of narrow interval and short cycle, KEOT was applied to select the optimal sensor locations. Moreover, to improve the FE model, DMUM was utilized because it requires fewer calculations, has high accuracy, and is practical for everyday use.

In conclusion, for structures such as cable-stayed bridges where the characteristic flexing behavior creates a natural frequency of narrow interval and short cycle, situations where multiple sensors are required because of structural complexity, and situations where there exist uncertainties in the FE model used for interpreting the data, KEOT and DMUM can facilitate rational and effective SI by enabling identification of the optimal sensor locations and improvement of the FE model. Ultimately, this study proves through experimentation that a combination of KEOT and DMUM is the ideal hybrid method to provide both practical and effective SHM.

OPTIMAL SENSOR LOCATION AND FINITE ELEMENT (FE) MODEL UPDATE

Kinetic Energy Optimization Technique (KEOT)

In this study, the KEOT method was applied and evaluated for its effectiveness in selecting optimal sensor locations. KEOT recommends selecting optimal measurement locations by utilizing the strain-kinetic energy of the structure [9–12]. To do this, Kammer's EIM approach [13] involves optimizing and selecting a set of target modes in order to identify the structure based on FE analysis. An initial candidate set of transducer locations is also selected. These locations are then ranked based on their contribution to the linear independence of the corresponding FE model target mode partitions, and locations that do not contribute are removed from the candidate set. The described energy optimization technique algorithm is a modification of EIM, and it is designed to improve the modal information and maximize the measured kinetic energy of the structural system. The spatial independence of the identified mode shapes is satisfied by the sensing configuration obtained using the KEOT algorithm [11, 12]. The kinetic energy in the system is

$$\mathrm{KE} = \Phi^T M \Phi, \tag{1}$$

where Φ is the measured mode shape vector. After decomposing the mass matrix M into upper and lower triangular Cholesky factors, the kinetic energy matrix can be derived as

$$\mathrm{KE} = \Psi^T \Psi, \tag{2}$$

where $\Psi = U\Phi$ and $M = LU$. The matrices L and U denote the lower and upper triangular Cholesky factors

The projection of the mode shapes onto the reduced configuration is denoted by

$$\overline{\Phi} = \text{projection}\,(\Phi)\,,$$
$$\overline{\Psi} = \text{projection}\,(\Psi)\,. \tag{3}$$

Similarly, the energy measured by the reduced set of transducers is obtained from the initial energy by removing the contribution of all transducers that have been eliminated:

$$\overline{\mathrm{KE}} = \overline{\Psi}^T \overline{\Psi}. \tag{4}$$

The objective of the transducer placement is to find a reduced configuration that maximizes the measurement of the kinetic energy of the structure. It is desirable to stop eliminating the transducers when doing so results in a rank deficiency of the energy matrix. Assuming that the mass matrix is nonsingular, the column rank of the quantity $\overline{\mathrm{KE}}$ is equal to the number of linearly independent projected vectors in matrix $\overline{\Phi}$. The problem is solved iteratively by the following procedure. First, the eigenvalues Λ and eigenvectors ψ of the energy matrix are extracted:

$$\overline{\mathrm{KE}}\psi = \Lambda\psi. \tag{5}$$

Computing the eigenpairs at each iteration of the energy optimizing technique (EOT) procedure does not significantly increase the computational cost because the matrix $\overline{KE}$ is a square, symmetric, and positive-definite matrix of size N. Then, using an approach similar to Kammer's EIM [13], the fractional contributions of each remaining transducer are assembled into the KEOT vector:

$$\text{KEOT} = \sum_{i=1,\ldots,m} \left[\overline{\Psi}\psi\Lambda^{-1/2}\right]^2. \tag{6}$$

The transducer location with the minimum contribution in the KEOT vector is then selected for removal. Subsequently, the contribution of the removed transducer to the kinetic energy matrix is deleted, and the new matrix is checked for rank deficiency. If the removal of the transducer produces a rank deficiency, it implies that the transducer location in question cannot be removed. If removing the transducer does not produce a rank deficiency, the transducer location is removed from the candidate set and the process is repeated until the required number of transducers is ascertained. Because it can be verified that the quantity between the brackets in (6) represents a linear combination of the measured mode shapes that is designed to produce orthogonal vectors,

$$\left[\overline{\Psi}\psi\Lambda^{-1/2}\right]^T \left[\overline{\Psi}\psi\Lambda^{-1/2}\right] = I. \tag{7}$$

Furthermore, each KEOT of the vector is a heuristic measure of the contribution of each transducer to the measured energy. The normalization factor $\Lambda^{-1/2}$ prevents the contributions of high frequency modes from dominating the low frequency modes. In theory, the number of remaining transducers is equal to the size of the target modal set. However, the apparent rank is often increased due to noise in the experimental data, and more than M transducers are required to identify N independent modes.

Direct Matrix Updating Method (DMUM)

In general, when a modified structure is added to or removed from the initial structure, changes occur in the dynamic characteristics of the structure. Dynamic characteristics before and after a structural change can be derived as eigenvalue formulas shown below:

$$([K] - \Lambda [M]) \{\Phi\} = 0,$$
$$\left(([K] + [\Delta K]) - \overline{\Lambda}([M] + [\Delta M])\right)\left\{\overline{\Phi}\right\} = 0, \tag{8}$$

where $[K]$ is the stiffness matrix, $[M]$ is the mass matrix, and $[\Delta K]$ and $[\Delta M]$ are changes in the stiffness and mass of the structure, respectively, due to structural changes. The elements Λ and $\{\Phi\}$ are eigenvalue and eigenvector, respectively, and $\overline{\Lambda}$ and $\{\overline{\Phi}\}$ are changes in eigenvalue and eigenvector due to structural changes, respectively. Regarding the derivation methods available for calculating changes in stiffness $[\Delta K]$ and mass $[\Delta M]$ due to structural changes, this study selected and applied DMUM considering its effectiveness in FE model improvement as well as its convenience and practicality for everyday use. Here, using Lagrange multipliers, objective functions, composed to limit the range of change in the stiffness and mass matrices while simultaneously satisfying the eigenvalue, are shown below as (9). DMUM is advantageous from the perspective of usability because it can calculate the changes in stiffness and mass from just a single matrix operation [23, 24]:

$$\epsilon_K = \left\| [K_A]^{-1/2} ([K_U] - [K_A]) [K_A]^{-1/2} \right\|,$$
$$\epsilon_M = \left\| [M_A]^{-1/2} ([M_U] - [M_A]) [M_A]^{-1/2} \right\|, \tag{9}$$

where $[K_A]$ and $[M_A]$ are the stiffness and mass matrices before structural change and $[K_U]$ and $[M_U]$ are the stiffness and mass matrices after structural change. The correlation between $[K_A]$ and $[K_U]$ can be expressed as shown in (10), and the correlation between $[M_A]$ and $[M_U]$ can be expressed as shown in (11):

$$[K_U] = [K_A] + [\Delta K] \tag{10}$$

$$[M_U] = [M_A] + [\Delta M] \tag{11}$$

The element $[\Delta K]$ in (10) can be defined as (14) using the interaction formulas of (12) and (13) [27]:

$$[K_U] - [K_U]^T = 0, \tag{12}$$

$$[\Phi_X]^T [K_U] [\Phi_X] - [\Lambda_X] = 0, \tag{13}$$

$$\begin{aligned}[\Delta K] = &- [K_A] [\Phi_X] [\Phi_X]^T [M_A] \\ &- [M_A] [\Phi_X] [\Phi_X]^T [K_A] \\ &+ [M_A] [\Phi_X] [\Phi_X]^T [K_A] [K_A] [\Phi_X] [\Phi_X]^T [M_A] \\ &+ [M_A] [\Phi_X] [\Lambda_X] [\Phi_X]^T [M_A].\end{aligned} \tag{14}$$

The element $[\Delta M]$ in (11) can be defined as (17) using the interaction formulas of (15) and (16) [28]:

$$[M_U] - [M_U]^T = 0, \tag{15}$$

$$[\Phi_X]^T [M_U] [\Phi_X] - [I] = 0, \tag{16}$$

$$\begin{aligned}[\Delta M] = &- [M_A] [\Phi_X] \left([\Phi_X]^T [M_A] [\Phi_X]\right) \\ &+ [\Phi_X]^T [K_A] \\ &+ \left([I] - [\Phi_X]^T [M_A] [\Phi_X]\right)\left([\Phi_X]^T [M_A] [\Phi_X]\right).\end{aligned} \tag{17}$$

In this study, the objective of FE model improvement is FE interpretation and modal experiment results, so the subscript

A in each equation is the result value based on analysis and subscript X is the result value based on experimentation.

FE ANALYSIS AND MODAL TEST OF MODEL STRUCTURE

Setup of the Model Structure (Cable-Stayed Bridge)

The second moment of inertia of the cable-stayed bridge model used in this study was reduced as much as possible so that the bridge slab would be sensitive to vibrations caused by external forces. Moreover, to make FE model configuration more convenient, the cross-sectional shape and the materials used for the members were simplified and unified. Further, to reduce the possibility of a production error, cross section dimensions were equalized and initial changes were minimized. The model used in this study is a three-span continuous beam structure with a total length of 4220 mm, a width of 170 mm, and a pylon height of 1000 mm. The cables are 0.8 mm steel wires, with ten wires connected to corresponding points on the crossbeam from each side of the pylon. The anchors located at each end of the model were composed of rollers, and the anchors of the pylons were composed of rollers and hinges. Finally, 1 kg of additional mass was added to the top of the bridge slab at each of the 39 points where the wires join the cross beam to maximize the flexibility of the bridge without creating any initial deflections. The model built using the components explained above is shown in Figure 1, and detailed design data is shown in Table 1.

Table 1: Design data of the cable-stayed bridge model.

Category	Data
Total length	4.22 m
Length of the center span	2.22 m
Length of each side span	1.0 m
Width of the superstructure	0.17 m
Pylon height	1.00 m
Pylon anchor type	Roller and hinge (Pin)
Bridge end anchor type	Roller
Diameter of cable	0.8 mm

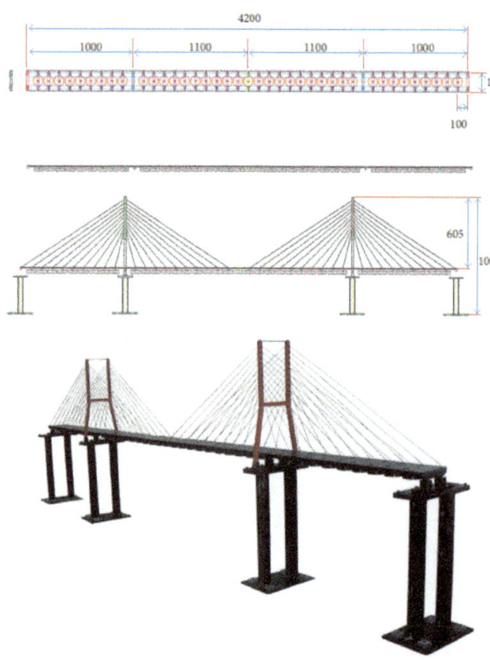

Figure 1: View of the cable-stayed bridge model.

During the assembly of the model, the cables were tensioned taking into consideration the effect of gravity on the superstructure across all sections. Thus, the self-weight of the superstructure was assumed to be a uniformly distributed load distributed at equal intervals across the cross beams and was utilized as the tensile force of the cables. To quantify the tensile force, a force gauge was utilized when tensioning the cables. After the completion of cable tensioning, a level was used to inspect the horizontality of the superstructure. In addition, the cables were tensioned progressively outward from the pylons and symmetrically left and right in accordance with the cable tensioning order of an actual cable-stayed bridge structure, as shown in Figure 2.

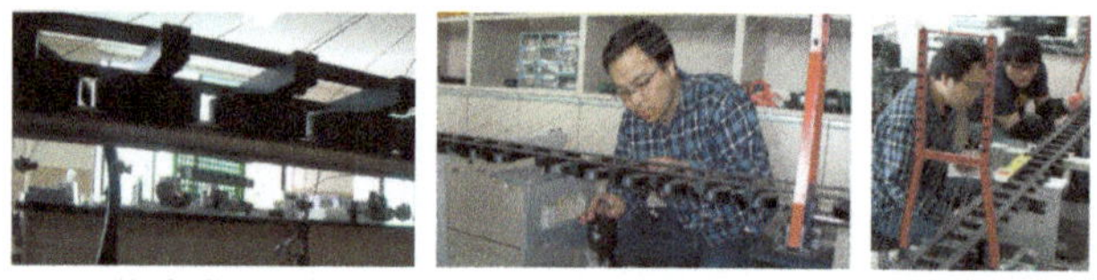

(a) Jig for cable tensioning (b) Tensioning the cables (c) Inspecting the horizontality

Figure 2: Steps for tensioning the cables of the cable-stayed bridge model.

FE Analysis of Model Structure (Cable-Stayed Bridge)

In order to analyze the structural dynamic properties of the produced model bridge, in this paper, FE modeling-based numerical analysis was conducted. For this purpose, I-DEAS from UGS, a commercial structural analysis program, was used, and a three-dimensional detailed FE model was organized as in Figure3 by applying the physical properties of Table 2. Here, 1D beam element was considered for the bridge deck and the tower of the structure, rigid element was considered for the bridge deck boundary conditions, and 1D rod element was considered for cable; as for the floor beam and lumped mass, 1 kg size of lumped mass was considered for each of the 39 nodes, which are the positions of the floor beams on the area of the central bridge deck excluding the pylon and both sides. Next, as for the boundary condition, clamp was considered for the substructure of the pylon, roller was considered for the both sides of superstructure and the right-side pylon connection, and Pin was considered for the left-side pylon connection. Lastly, as for the freedom degree condition of the structure, DOF was given in the direction of -axis with the 39 nodes on the central bridge deck as the base to complete the three-dimensional detailed FE model. In this context, the FE analysis of the model bridge considered a total of two lowest flection modes for the purpose of vertical vibration control, and the natural frequency and mode shape of the model bridge that has been analyzed with Guyan reduction method applied are as in Figure 4.

Table 2: Material properties of the cable-stayed bridge model.

Material property	Modulus of elasticity (kgf/mm^2)	Shear modulus of elasticity (kgf/mm^2)	Poisson's ratio	Unit weight (kgf/mm^3)	Yield strength (kgf/mm^2)
Steel	2.10×10^4	8.10×10^3	0.30	7.85×10^{-6}	40

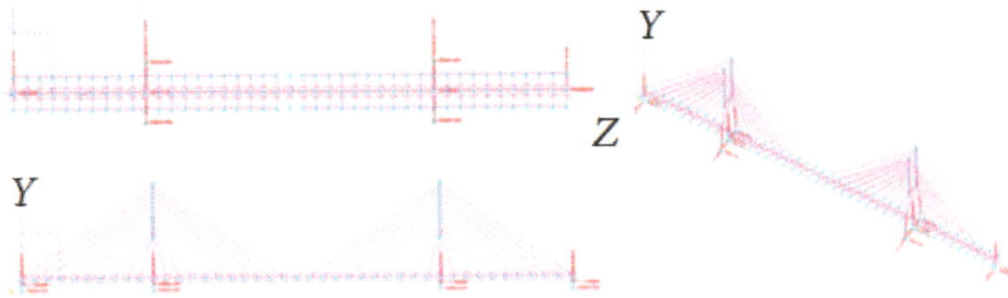

Figure 3: Initial FE modeling of the model bridge using I-DEAS.

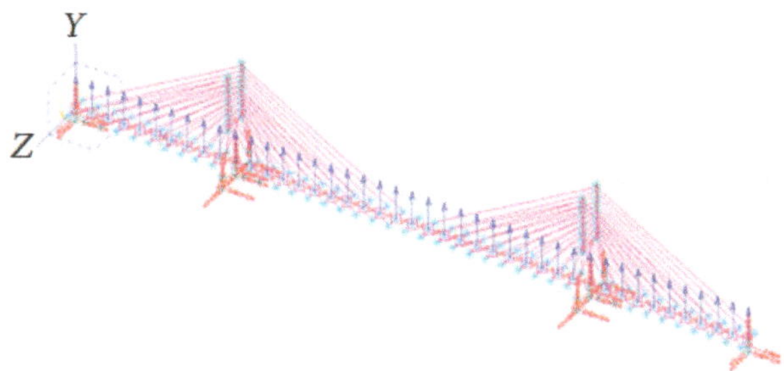

Figure 4: Initial FE modeling of the model bridge using I-DEAS (39 DOF).

In this paper, the DOFs of FE model were reduced into thirty-nine, with application of Guyan reduction method [29] by which a structural analysis of eigenvalues is made by reducing removed DOFs into unremoved ones and in turn composing a matrix of reduced strength and mass as in (18). Therefore, it has become possible to reduce the DOFs by selecting a desired point of measurement and to effectively perform structural FE analysis. Here, each r refers to a retained set and o to an omitted set. Equation (18) is applied for an analysis of eigenvalues, based on an optimal location of measurement determined in Section 4.1:

$$K_r = K_{rr} - K_{ro}K_{oo}^{-1}K_{ro}^T,$$
$$M_r = M_{rr} - M_{ro}K_{oo}^{-1}K_{ro}^T - K_{ro}K_{oo}^{-1}M_{ro}^T + K_{ro}K_{oo}^{-1}M_{oo}K_{oo}^{-1}K_{ro}^T. \quad (18)$$

As shown in Figure 4, based on the total of 39 DOF, Guyan reduction method was applied to interpret the frequency of vibration and the mode shape, with the results shown in Table 3.

Table 3: FE analysis results of the cable-stayed bridge model.

	1st bending	2nd bending	3rd bending
Analysis results	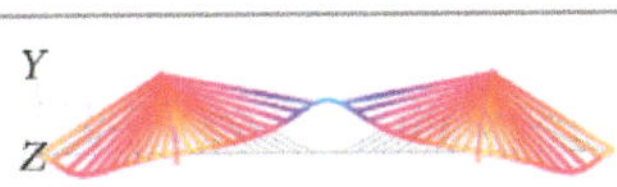 9.1719 Hz	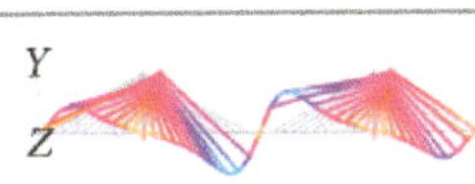 11.2544 Hz	 16.8643 Hz

Modal Test of the Model Structure (Cable-Stayed Bridge)

To confirm the validity of the composed FE model and to analyze the fundamental kinetic characteristics of the model bridge, this study employed a modal test conducted using an impact hammer. An HP-VXI 1432 was used to measure the response signals from the structure, and data was obtained and analyzed by utilizing T-DAS by MTS. In order to obtain acceleration responses of the structure, Dytran model 3134D was used for a total of 39 points on the bridge deck that had been selected in equal intervals. To connect the accelerometer to the experimental bridge model and the measurement system, 10 ft cables were used for each channel. The impact hammer used to create vibratory forces in the structure was a 5850A model produced by Dytran, and the center of the slab was selected as the location of the impact. The size of each frame of the measured frequency response function (FRF) was set to 2048, and the maximum frequency range was set to 35 Hz. Data was collected by averaging the results of 30 tests. Finally, the acquired time and frequency responses are as shown in Figure 5, and the results of the modal test analysis are as shown in Table 4.

Table 4: Results of the modal test of the cable-stayed bridge model.

	1st bending	2nd bending	3rd bending
Modal test results	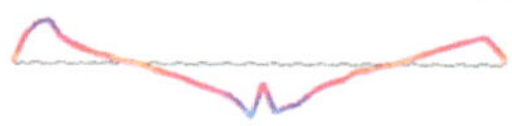 9.0890 Hz	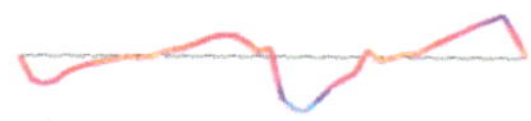 11.2557 Hz	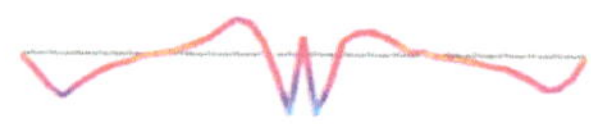 17.1442 Hz

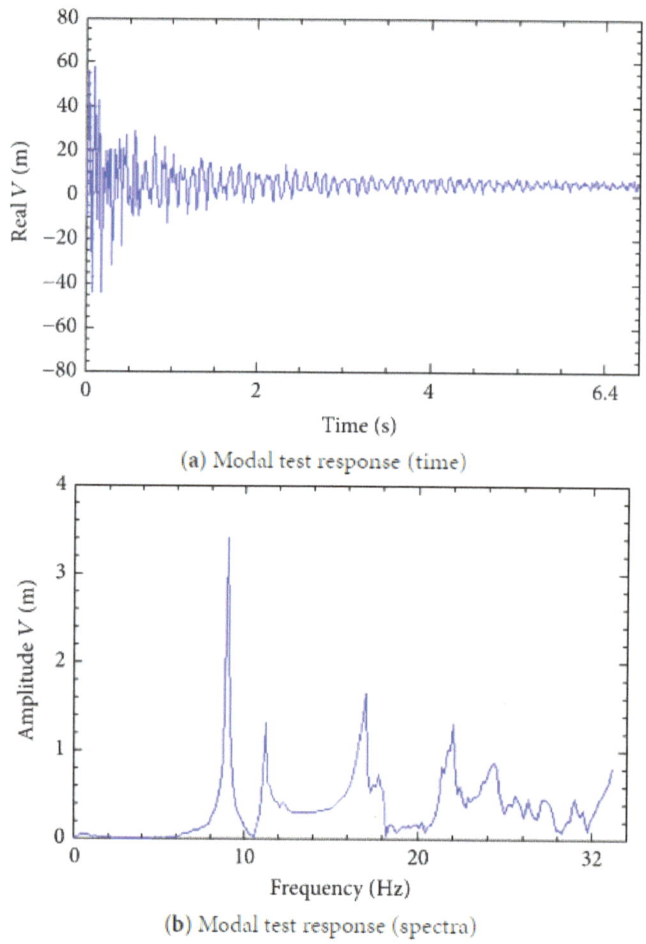

Figure 5: Modal test results.

To confirm the validity of using the FE model to understand the dynamic behavior characteristics of the model bridge and for use in deriving the modal parameters for future SHM and maintenance, this study compares and contrasts the frequency results of the FE model analysis and the modal experiment as shown in Table 5. As observed from the data in the table, reliable results were achieved, with an error ratio of around 1% between the FE analysis and the modal experiment. These results prove that the detailed FE model constructed in this study reliably mimicked the behavioral characteristics of the actual structural model, thereby confirming its validity.

Table 5: Comparison of analysis and experimental frequency.

	Analysis and experimental results		
	FE analysis results (Hz)	Modal experiment results (Hz)	Error ratio (%)
1st bending mode	9.1719	9.0890	0.9120
2nd bending mode	11.2544	11.2557	0.0115
3rd bending mode	16.8643	17.1442	1.0326

The analysis and experimental results above were acquired using a total of 39 DOF and sensors. From the standpoint of long-term continuous SHM and from the practical standpoint of everyday evaluation and maintenance, using the whole structure to acquire signals as done in this study may be inefficient and impractical. Doing so would require a great number of sensors as well as a great deal of labor to install them. Moreover, interpreting and sorting through the resulting massive amount of data would be a laborious task. To resolve this problem, an alternative approach is required. This study therefore considers an approach whereby optimal sensor locations are selectively chosen in order to acquire the maximum information about the behavioral characteristics of the structure from a minimum number of sensors. By selecting the optimal locations, the sensors can thus be positioned at the most rational and efficient points to acquire the target information. Moreover, the sensor locations selected through this approach can later be utilized as real-time sensor locations for actual structures.

OPTIMAL SENSOR LOCATION AND FE MODEL UPDATING OF THE MODEL BRIDGE

Optimal Sensor Location Using the KEOT

The aim of this study is to conduct SI with a limited number of sensors by selecting the optimal sensor locations. To select the optimal sensor locations, we utilized KEOT, which maximizes the structure's strain-kinetic energy to enable consideration of the optimal measurement conditions. The KEOT method has proven to be effective for ascertaining the kinetic parameters of structures that feature low and short-interval frequencies, such as the cable-stayed bridge used in this study [9–12]. In this study, we set out to use KEOT to select the optimal sensor locations to represent the three lowest bending modes among the 39 DOF nodes. The three lowest bending modes are important and interesting modes that can occur during vertical vibration and thus should be the target modes for SI. Also, the optimal sensor locations were selected with a low error rate based on these target modes. Figure 6 and Table 5 show the optimal sensor locations selected and the results of interpretation using the KEOT method compared to the rate of error in frequency evaluated per DOF node. Rank deficiency in the energy matrix did not occur until the 39 sensor locations had been reduced to 2. This study therefore limited the final reduced DOF count to 5 and selected the optimal sensor locations based on analysis of changes in eigenvalue.

DOF \ S/L	Number of sensor locations (total 39 nodes)																																						
	2	3	4	5	6	7	8	9	10	12	13	14	15	16	17	18	19	20	21	22	23	24	25	26	27	28	29	30	31	32	33	34	35	36	37	38	39	40	41	42
30	•	•	•	•	•	•	•					•	•	•	•	•	•	•	•	•	•	•	•	•	•	•	•	•						•	•	•	•	•	•	
25		•	•	•	•	•							•	•	•	•	•	•	•	•	•	•	•	•	•	•	•								•	•	•	•	•	
20		•	•	•	•										•	•	•	•	•	•	•	•	•	•	•	•										•	•	•	•	
15			•	•											•	•	•	•	•	•	•	•	•	•	•												•	•		
10			•													•	•	•	•	•	•	•	•	•																
9			•													•	•		•	•	•	•	•	•																
8			•													•	•		•	•	•		•	•																
7																•	•		•	•	•		•	•																
6																•	•			•	•		•	•																
5																•	•			•			•	•																
4																•	•			•			•																	
3																	•			•			•																	
2																				•			•																	
1																							•																	

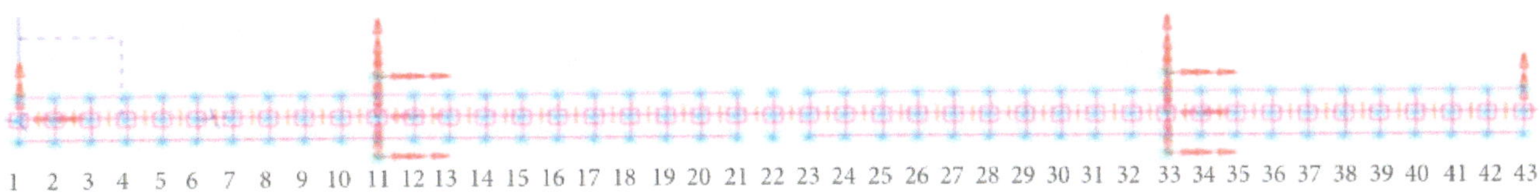

Figure 6: Study of optimal sensor locations selected using the EOT method.

By observing the trends in optimal sensor locations selected through gradual removal of DOF nodes (as shown in Table 6), it can be observed that when the total of 39 locations had been reduced to 10 (at the 75% reduction point), all of the left- and right-side span sensor locations had been removed. Thus, it was concluded that the strain-kinetic energy due to the 1st, 2nd, and 3rd bending mode behaviors was smaller in the side spans than in the center span. On the other hand, within the center span, the remaining sensor locations grouped around nodes 19, 22, and 25 accounted for all the 1st, 2nd, and 3rd bending modes. These three nodes remained until only three sensor locations were left (at the 92.5% reduction point). Thus, it was concluded that nodes 19, 22, and 25 were the essential sensor locations that can most optimally represent the 1st, 2nd, and 3rd bending modes simultaneously. In conclusion, for comprehensive representation of the 1st, 2nd, and 3rd resonance modes, which have the greatest impact on bending, it is essential that these three sensor locations are included.

Table 6: FE analysis results per DOF of cable-stayed bridge model using EOT.

DOF	Target mode Reduction (%)	1st bending Freq. (Hz)	1st bending Error (%)	2nd bending Freq. (Hz)	2nd bending Error (%)	3rd bending Freq. (Hz)	3rd bending Error (%)
39 (full)	—	9.1719	—	11.2544	—	16.8643	—
30	23.07	9.1728	0.0009	11.2567	0.0204	16.8691	0.0284
25	35.89	9.1757	0.0414	11.2651	0.0950	16.8856	0.1263
20	48.71	9.1890	0.1864	11.3049	0.4487	16.9834	0.7062
15	61.53	9.2219	0.5451	11.7537	4.4364	17.2843	2.4904
10	74.35	9.3253	1.6724	11.7537	4.4364	18.1914	7.8692
5	87.17	9.5531	4.1561	12.3951	10.1355	18.8781	11.9412

On the other hand, when sensor locations or DOF nodes are removed using the KEOT method explained above, changes in eigenvalue can be impacted due to the reduction in DOF. This study evaluated changes in eigenvalue when DOF nodes are removed using KEOT. Table 6 shows the results of eigenvalue analysis when the 39 DOF nodes are reduced to 30, 25, 20, 15, 10, and finally 5.

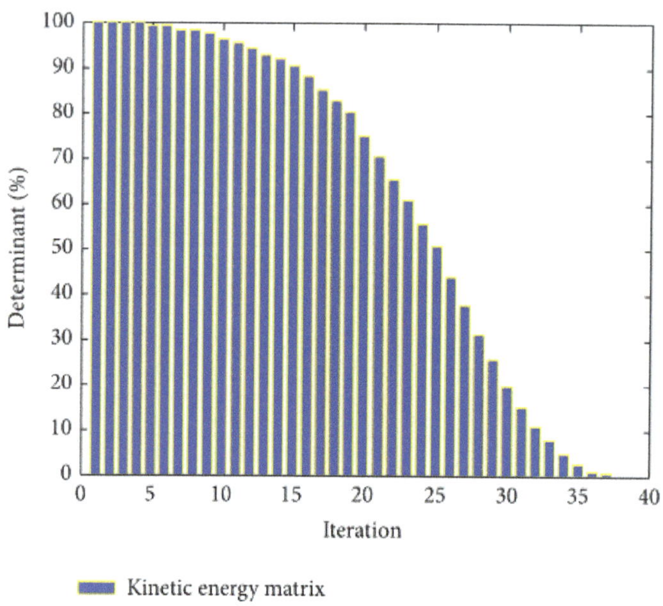

Figure 7: Kinetic energy matrix values to DOF modification.

As shown in Table 6, changes in eigenvalue were clearly observed as the total of 39 DOF nodes was reduced to 25, 25, 20, 15, 10, and 5. Looking closely at the changes in frequency, it was also observed that, with the reduction in the number of DOF nodes, the frequency steadily increased. In the case of the 1st bending mode, the frequency error ratio significantly increased when DOF nodes were reduced to 10 (reduction rate of about 75%) and the error ratio exceeded 4% when they were reduced to 5 (reduction rate of about 87%). In the case of the 2nd bending mode, the frequency error ratio increased significantly when the DOF nodes were reduced to 15 (reduction rate of about 60%). It was close to 4.5% when the nodes were reduced to 10 (reduction rate of about 75%) and exceeded 10% when the nodes were reduced to 5 (reduction rate of about 87%). Moreover, in the case of the 3rd bending mode, the frequency error ratio significantly increased when the DOF nodes were reduced to 15 (reduction rate of about 60%). It was close to 8% when the nodes were reduced to 10 (reduction rate of about 75%) and reached about 12% when the nodes were reduced to 5 (reduction rate of about 87%).

These tendencies are due to the fact that stiffness is the most considered factor in the process of condensing the behavioral characteristics of an entire structure to a limited number of DOF nodes. In conclusion, to derive FE analysis results using Guyan reduction [29], it is important to select the range of allowable error ratio and the corresponding reduction of DOF nodes. Otherwise, excessive reduction in the number of DOF nodes will increase the frequency error ratio and lead to discrepancies in the initial FE analysis and modal test results, ultimately producing distorted frequency results.

Using the above results, a total of seven sensor locations were selected to include the three optimal sensor locations 19, 22, and 25, to limit the frequency error ratio to less than 10% and to clearly observe the shape of the modes. Using the KEOT method, the optimal sensor locations selected were 18, 19, 21, 22, 23, 25, and 26.

Figure 7 illustrates the distribution of kinetic energy for each DOF condition as calculated and derived in this experiment using the KEOT. It can be observed from Figure 7 that as the number of DOF nodes (sensor locations) was gradually reduced, the total kinetic energy contribution of the remaining nodes was also reduced. In the case of the model bridge used in this study, when the number of DOF nodes was reduced to 15, the kinetic energy contribution was 90%. The contribution of kinetic energy decreased significantly with subsequent removals of DOF nodes. Moreover, when the number of DOF nodes was reduced to the seven optimal sensor locations selected in this study, the kinetic energy contribution was about 10%.In this study, we conducted SI on a model of a cable-stayed bridge using seven sensors. As described above, the locations of the seven sensors, selected using the KEOT method, were nodes 18, 19, 21, 22, 23, 25, and 26. The total contribution of kinetic energy from these nodes was about 10%. The selection of the seven sensor locations used in this study was verified by the inclusion of the three DOF nodes (19, 22, and 25) that were considered to be the essential sensor locations for most effectively and simultaneously representing the 1st, 2nd, and 3rd bending modes even at a 92.5% reduction in the number of DOF nodes. The results derived from the FE model and frequency analysis based on the seven selected sensor locations are shown in Table 7.

Table 7: Frequency and mode shape of 7 DOF using KEOT.

DOF	FE model	1st mode shape	2nd mode shape	3rd mode shape
7		 9.4244 Hz	 12.0453 Hz	 18.8065 Hz

Table 7 shows that reducing the number of DOF nodes results in changes in frequency. For quantitative analysis of the changes in frequency, the frequency results and error ratio derived from the initial study with all 39 DOF nodes along with those derived from the modal test are shown in Table 8.

Table 8: Comparison of 7 DOF analysis results and full DOF cases.

	Analysis (7 DOF) versus analysis (39 DOF)			Analysis (7 DOF) versus experimental (39 DOF)		
	Analysis (7 DOF)	Analysis (39 DOF)	Error (%)	Analysis (7 DOF)	Experimental (39 DOF)	Error (%)
1st mode	9.4244 Hz	9.1719 Hz	2.7529	9.4244 Hz	9.0890 Hz	3.6900
2nd mode	12.0453 Hz	11.2544 Hz	7.9160	12.0453 Hz	11.2557 Hz	7.0151
3rd mode	18.8065 Hz	16.8643 Hz	11.5166	18.8065 Hz	17.1442 Hz	9.6959

As observed in Table 8, using seven DOF nodes (approximately 82% reduction of DOF nodes) resulted in an error ratio of approximately 2.7–11.5% compared to that in the results of FE analysis using all 39 DOF nodes. In addition, using seven DOF nodes resulted in an error ratio of about 3.6–9.6% compared to the modal test results. The correlation between the mode shape acquired from the FE analysis using seven DOF nodes and the mode shape acquired from the modal test was further evaluated using the modal assurance criterion (MAC) shown in (19) [30]:

$$\mathrm{MAC}(A, X) = \frac{\left|\sum_{j=1}^{n} (\Phi_X)_j (\Phi_A)_j^*\right|^2}{\left(\sum_{j=1}^{n} (\Phi_X)_j (\Phi_X)_j^*\right)\left(\sum_{j=1}^{n} (\Phi_A)_j (\Phi_A)_j^*\right)}. \quad (19)$$

Here, Φ_A and Φ_X are the mode shapes as calculated and acquired from the analysis and test, respectively. When the two modes are the same, the MAC value equals 1, and when the two modes have no correlation, the MAC value equals 0 [31]. Similarly, Ewins [30] points out in his study that the most highly correlated modes will have a MAC value of around 0.9, with MAC values as low as 0.7 considered fair in some cases. He also asserts that modes with no correlation will have MAC values of around 0.005. Figure 8 and Table 9illustrate that the mode shapes studied in this research showed satisfactory correlations, but the 3rd bending mode had a relatively lower correlation than those of the 1st and 2nd modes.

Table 9: MAC values (7 DOF anal. and exp.).

	FE analysis modes		
	1st	2nd	3rd
Experimental modes			
1st	**0.9886**	0.0000	0.4304
2nd	0.0055	**0.9953**	0.0017
3rd	0.0259	0.0000	**0.7663**

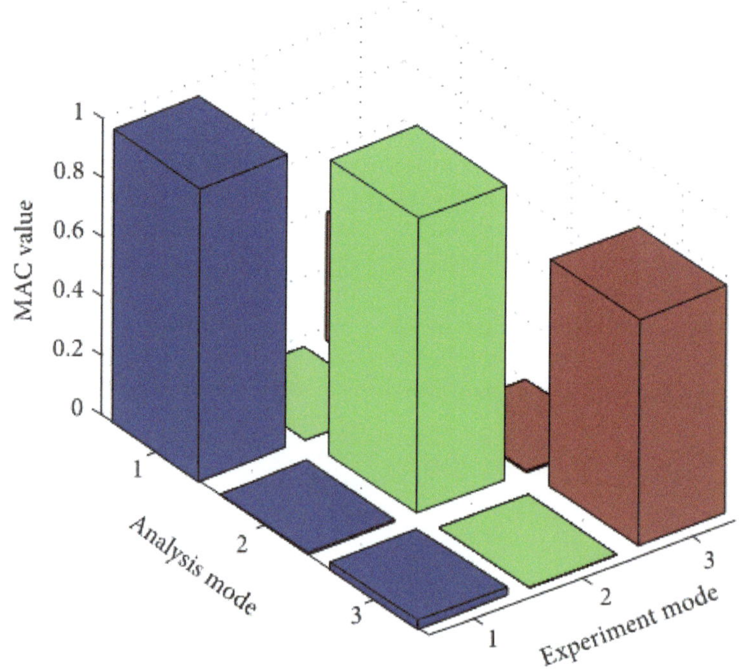

Figure 8: MAC plot (7 DOF anal. and exp.).

The changes in frequency shown in Table 8 and the changes in mode shape shown in Figure 8 and Table 9 are interpretation errors (FE modeling errors) created as a result of reduction of the DOF nodes. To accomplish a thorough SI of a structure as targeted in this study, more accurate modal information needs to be provided regarding the model bridge, and the discrepancies between the analysis results using the seven DOF nodes and the modal test results need to be overcome. Therefore, DMUM was utilized in this study in order to overcome FE modeling errors based on modal test values and to derive a baseline structure that can provide valid modal information.

FE Model Updating Using DMUM

FE model updating is a technique designed to overcome modeling errors in FE analysis and to secure an accurate model for numerical analysis of the current state of the structure taking into account potential construction errors. In FE model updating, the modal test results derived from the actual structure are applied to the FE analysis results to ultimately create an ideal FE model that contains the modal information of the current structure. This updated FE model can reliably reflect the dynamic behavioral characteristics of the actual structure. The updated model also plays a very important role in acquiring numerical analysis results, which is necessary for accurate assessment of structural states. To achieve FE model updating using DMUM, this study utilized frequency, mode shape, and mass and stiffness matrix results derived from the FE analysis along with frequency and mode shape results acquired from the modal test. Table 10 reflects the error ratio between the frequency results derived from the FE model updated using DMUM and the frequency results acquired from the modal test. As shown in Table 10, it was confirmed that the results derived from FE model updating had a significantly lower error rate (close to 0).

Table 10: Comparison of experimental results and updating results.

	Analysis and experimental results using EOT		Error ratio
	FE model update	Experimental	%
1st mode	9.0880 Hz	9.0890 Hz	0.0110
2nd mode	11.2543 Hz	11.2557 Hz	0.0124
3rd mode	17.1418 Hz	17.1442 Hz	0.0139

Next, to verify the effectiveness of DMUM in FE model updating, MAC tests were conducted on the mode shapes. The results of the correlation analysis between the mode shapes derived from the updated FE model and those derived from the modal test are as shown in Figure 9 and Table 11. The results show that using DMUM to update the FE model results in a MAC value of "1," thus demonstrating an outstanding correlation with the modal test results.

Table 11: MAC values (updating and exp.).

	FE analysis modes (update)		
	1st	2nd	3rd
Experiment modes			
1st	**1.0000**	0.0003	0.0225
2nd	0.0002	**1.0000**	0.0005
3rd	0.0226	0.0005	**1.0000**

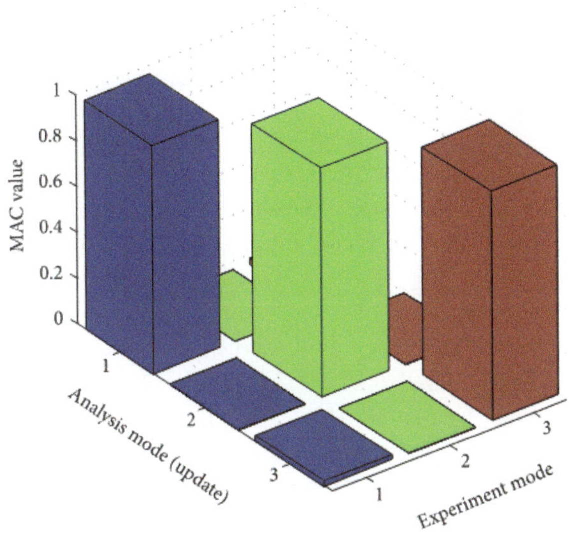

Figure 9: MAC plot (update and exp.).

In addition, in order to visualize the mode correlation evaluated using MAC and to evaluate the correlation between mode shapes before and after FE model updating, the mode shapes derived from analysis using seven DOF nodes or sensor locations and the mode shapes derived from the updated FE model were graphed as shown in Figure 10. As the graphs illustrate, the mode shape from the updated FE model follows the mode shape of the modal test, and not the mode shape of the FE analysis. This verifies the fact that by incorporating the modal information of the structure, FE model updating not only provides more accurate frequency values but also provides more accurate mode shape patterns. Ultimately, this proves that FE model updating using DMUM can overcome modeling errors and it is an effective way to create an accurate model for numerical analysis that can take into account the actual state of the structure.

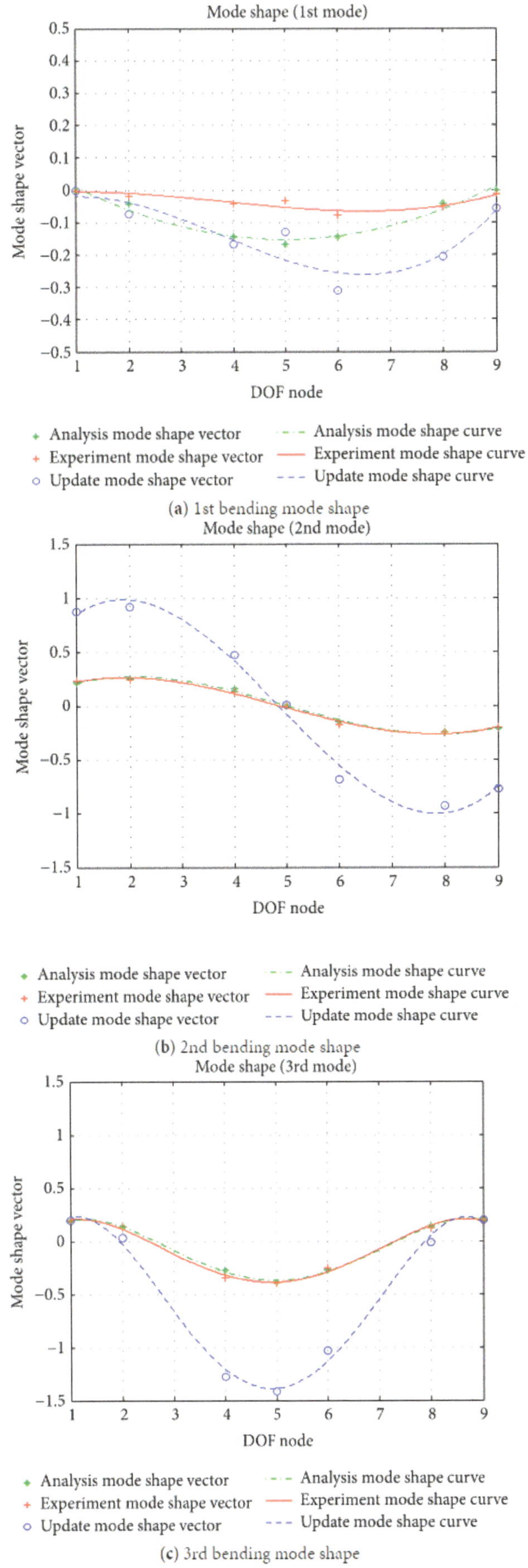

Figure 10: Comparison of 7 DOF analysis, updating, and experimental mode shapes.

CONCLUSION

The aim of this study was to research the formulation of a baseline structure through optimal sensor location selection and FE model updating for conducting SI on cable-stayed bridges. The conclusions are as follows:(1)By comparing the contribution of each DOF node to the total kinetic energy and progressively removing those that do not contribute based on the target modes, the KEOT method was proven to be an effective approach for selecting the optimal number and location of sensors among the multiple sensor locations (DOF) in order to acquire the desired responses. In this study, using seven sensors out of a total of 39 sensor locations to consider the three lowest bending modes (about 20% of the total) was sufficient to effectively acquire relevant structural information on the cable-stayed bridge.(2)In addition, by applying structural information measured from the actual structure and constructing a corresponding FE model to overcome initial errors in FE modeling, the DMUM method was proven to be an effective method for defining a baseline structure that incorporates the behavioral characteristics of the actual structure. DMUM has particularly strong potential for practical use from temporal and economical standpoints because it calculates stiffness and mass values from just a single matrix operation as opposed to other methods that require repetitive operations such as EMUM.(3)In conclusion, for structures such as cable-stayed bridges where the characteristic flexing behavior creates low and short-interval frequencies, situations where multiple sensor locations (DOF) are required because of structural complexity, and situations where SHM and evaluation and maintenance needs to be performed on structures with low and short-interval frequencies, the use of both KEOT and DMUM can be the most effective approach for conducting SI from the standpoints of both economy and usability because they provide accurate results with minimum number of measurement locations.

CONFLICT OF INTERESTS

The authors declare that there is no conflict of interests regarding the publication of this paper.

ACKNOWLEDGMENTS

This research was supported by the Basic Science Research Program through the National Research Foundation of Korea (NRF) funded by the Ministry of Education, Science and Technology (Grant no. NRF-2013R1A2A1A01016192) and funded by the Ministry of Science, ICT and Future Planning (Grant no. NRF-2013R1A1A1063540).

REFERENCES

1. S. C. Lie and J. T. P. Yao, "Structural identification concept," Journal of the Structural Division, vol. 104, no. 12, pp. 1845–1858, 1978.
2. G. F. Sirca Jr. and H. Adeli, "System identification in structural engineering," Scientia Iranica A, vol. 19, no. 6, pp. 1355–1364, 2012.
3. V. Arora, "Structural damping identification method using normal FRFs," International Journal of Solids and Structures, vol. 51, no. 1, pp. 133–143, 2014.
4. Z. D. Zheng, Z. R. Lu, W. H. Chen, and J. K. Liu, "Structural damage identification based on power spectral density sensitivity analysis of dynamic responses," Computers & Structures, vol. 146, pp. 176–184, 2015.
5. D. Dessi and G. Camerlengo, "Damage identification techniques via modal curvature analysis: overview and comparison," Mechanical Systems and Signal Processing, vol. 52-53, pp. 181–205, 2015.
6. C. Farrar and P. J. Cornwell, "Structural health monitoring studies of the alamosa canyon and I-40 bridges," Tech. Rep. LA-13635-MS, Los Alamos National Laboratory, Los Alamos, NM, USA, 1998.
7. F. Benedettini, M. Dilena, and A. Morassi, "Vibration analysis and structural identification of a curved multi-span viaduct," Mechanical Systems and Signal Processing, vol. 54-55, pp. 84–107, 2015.
8. H. Chen, M. Kurt, Y. S. Lee, D. M. McFarland, L. A. Bergman, and A. F. Vakakis, "Experimental system identification of the dynamics of a vibro-impact beam with a view towards structural health monitoring and damage detection," Mechanical Systems and Signal Processing, vol. 46, no. 1, pp. 91–113, 2014.
9. M. L. Wang, G. Heo, and D. Satpathi, "Dynamic characterization of a long span bridge: a finite element based approach," Soil Dynamics and Earthquake Engineering, vol. 16, no. 7-8, pp. 503–512, 1997.
10. M. L. Wang, G. Heo, and D. Satpathi, "A health monitoring system for large structural systems," Smart Materials and Structures, vol. 7, no. 5, pp. 606–616, 1998.
11. G. Heo, M. L. Wang, and D. Satpathi, "Optimal transducer placement for health monitoring of long span bridge," Soil Dynamics and Earthquake Engineering, vol. 16, no. 7-8, pp. 495–502, 1997.
12. H. Gwanghee, An automated health monitoring system for large civil structural system [Ph.D. thesis], University of New Mexico, Albuquerque, NM, USA, 1997.
13. D. C. Kammer, "Effect of model error on sensor placement for on-orbit modal identification of large space structures," Journal of Guidance, Control, and Dynamics, vol. 15, no. 2, pp. 334–341, 1992.
14. F. E. Udwadia, "Methodology for optimum sensor locations for parameter identification in dynamic systems," Journal of Engineering Mechanics, vol. 120, no. 2, pp. 368–387, 1994.
15. R. M. Rao, K. Lakshmi, and S. Krishnakumar, "A generalized optimal sensor placement technique for structural health monitoring and system identification," Procedia Engineering, vol. 86, pp. 529–538, 2014.
16. Z. Y. Shi, S. S. Law, and L. M. Zhang, "Optimum sensor placement of structural damage detection,"Journal of Engineering Mechanics, vol. 126, no. 11, pp. 1173–1179, 2000.
17. T. Pothisiri and K. D. Hjelmstad, "Structural damage detection and assessment from modal response,"Journal of Engineering Mechanics, vol. 129, no. 2, pp. 135–145, 2003.
18. M. Ge and E. M. Lui, "Structural damage identification

using system dynamic properties," Computers and Structures, vol. 83, no. 27, pp. 2185–2196, 2005.

19. T. P. Waters, Finite element model updating using measured frequency response functions [Ph.D. thesis], Department of Aerospace Engineering, University of Bristol, Bristol, UK, 1995.
20. Q. W. Zhang, C. C. Chang, and T. Y. P. Chang, "Finite element model updating for structures with parametric constraints," Earthquake Engineering and Structural Dynamics, vol. 29, no. 7, pp. 927–944, 2000.
21. J. E. Mottershead and M. I. Friswell, "Model updating in structural dynamics: a survey," Journal of Sound and Vibration, vol. 167, no. 2, pp. 347–375, 1993.
22. M. I. Friswell and J. E. Mottershead, Finite Element Model Updating in Structural Dynamics, vol. 38 ofSolid Mechanics and its Applications, Kluwer Academic Publishers, Dordrecht, The Netherlands, 1995.
23. M. Baruch, "Optimization procedure to correct stiffness and flexibility matrices using vibration tests,"AIAA Journal, vol. 16, no. 11, pp. 1208–1210, 1978.
24. Berman and E. J. Nagy, "Improvement of a large analytical model using test data," AIAA Journal, vol. 21, no. 8, pp. 1168–1173, 1983.
25. J. Sidhu and D. J. Ewins, "Correlation of finite element and modal test studies of a practical structure," in Proceedings of the 2nd International Modal Analysis Conference, pp. 156–162, Orlando, Fla, USA, February 1984.
26. D. J. Ewins, J. He, and N. A. J. Lieven, "A review of the error matrix method (EMM) for structural dynamic model comparison," in Proceedings of the ESA Symposium on the Applications of Advanced Structural Materials, Noordwijk, The Netherlands, 1988.
27. J.-S. Fuh and A. Berman, "Comment on 'Stiffness matrix adjustment using mode data'," AIAA Journal, vol. 24, no. 8, pp. 1405–1406, 1986.
28. Berman, "Mass matrix correction using an incomplete set of measured modes," AIAA Journal, vol. 17, no. 10, pp. 1147–1148, 1979.
29. R. J. Guyan, "Reduction of stiffness and mass matrices," AIAA Journal, vol. 3, no. 2, p. 380, 1965.
30. D. J. Ewins, Modal Testing: Theory, Practice and Application, Research Studies Pre, 2000.
31. G. Heo, J. Jeon, C. O. Lee, G. Lee, and W. Lee, "FE model updating for health monitoring of structures and its experimental verification by damage detection," Key Engineering Materials, vol. 321–323, pp. 268–272, 2006.

CHAPTER 4 DESIGN OPTIMIZATION OF A CONCRETE FACE ROCK-FILL DAM BY USING GENETIC ALGORITHM

Yanlong Li, Jing Wang, and Zengguang Xu

INTRODUCTION

The objective function and constraint conditions for the design optimization of a concrete face rock-fill dam (CFRD) are nonlinear functions of design variables. Constraint conditions are implicit functions of design variables; however, deriving these constraint conditions is difficult [1]. The current design optimization methods for CFRD mainly focus on mathematical programming and criterion methods, which include the full stress criterion, complex method, and penalty function method. These methods have several limitations, such as slow convergence and low efficiency [2].

Holland [3] first proposed a genetic algorithm (GA) that has main features operating directly on object structures. The continuity of the GA derivatives and functions has no limitations. GA has inherent implicit parallelism and good global optimization capability. GA also uses the probability optimization method, which can automatically access and guide the optimized search space, can adjust the search direction adaptively, and does not require rule determination.

Ackley [4] proposed a strategy called "stochastic iterated genetic hill-climbing" (SIGH) that uses a complex probability election mechanism composed of "voters" to decide the value of the new individual ("" refers to the size of groups). Experimental results show that four of the six test functions of the SIGH with single-point crossover, namely, the uniform crossover neural GA, exhibited better performance than that of other algorithms. Overall, the SIGH is more competitive in the speed of solving compared with many existing algorithms. In another study, Whitley et al. [5] proposed a crossover operator on the basis of the cross field. This crossover operator is specifically for crossing individuals by using numbers to represent genes. The application of the crossover operator to the travelling salesman problem (TSP) has been verified by experiments [6].

Bersini and Seront [7] combined GA with the simplex method and formed a single operation called "multi-parent crossover operator." This operator produces new individuals through two maternal individuals and an additional individual. The cross results are consistent with those of three individuals with elections. Three crossover operators have been compared with the subpoint crossover and uniform crossover operators. Results show that the three crossover operators have better performance than that of the other two operators.

Many experts and scholars in China have improved the crossover operator of GA. For instance, Dai et al. [8] presented an algorithm of pattern extraction approach to the mutation mechanism for Genetic Algorithm. The direction of the mutation of individual is guided by two consecutive evolution directions of the individual, such as the mutation probability, different mutation operators to search for a variable space, and migration of operators among populations to exchange genetic information and solve the problems of the classical GA converging to a local optimum value. Zhao et al. [9] proposed the building-block-coded parallel GA, which is encoded by gene block. This GA is utilized when the simple GA is used on large-scale combinatorial optimization problems with low searching efficiency. This method is based on the coarse-grained parallel GA and aims to identify possible gene block among the chromosome population. The identified gene block is used as a new gene block to recode the chromosome and to generate chromosomes with shorter lengths. The recoded

Citation: Yanlong Li, Jing Wang, and Zengguang Xu, "Design Optimization of a Concrete Face Rock-Fill Dam by Using Genetic Algorithm," Mathematical Problems in Engineering, vol. 2016, Article ID 4971048, 11 pages, 2016. doi:10.1155/2016/4971048.

chromosome groups are then used as the initial population that progresses to the next round of evolution. Jiang [10] utilized a flexible strategy to maintain the diversity of groups for parallel GA to solve the TSP, which enables the algorithm to cross the local convergence obstacles and reach the evolution of the global optimum direction. Zhang et al. [11] constructed the analysis and inverse analysis system of temperature stress simulation. The system realizes the automation and intellectualization of complex data analysis and preparation work in simulation process and complex data adjustment in the inverse analysis process, which can facilitate the real-time tracking simulation and feedback analysis of concrete temperature stress in construction process. Gu et al. [12] proposed a new method to inverse the actual initial zoning deformation modulus and to determine the inversion objective function for the actual zoning deformation modulus, based on the dam displacement measured data and finite element calculation results. Furthermore, based on the chaos genetic optimization algorithm, the inversion method for zoning deformation modulus of dam, dam foundation, and reservoir basin is proposed. Combined with the project case, the feasibility and validity of the proposed method are verified. The internal relation between concrete dam crack behavior abnormality and statistical change point theory is deeply analyzed by Li et al. [13] from the model structure instability of parametric statistical model and change of sequence distribution law of nonparametric statistical model. And the nonparametric change point diagnosis method of concrete dam crack behavior abnormality is used in the actual project, demonstrating the effectiveness and scientific reasonableness of the method established. Rezaiee-Pajand and Tavakoli [14] introduced an efficient procedure for crack detection in concrete gravity dams. A genetic algorithm and finite element modeling are employed to perform the optimization tasks. Moreover, a genetic algorithm approach for crack identification is proposed, which can identify the location and magnitude of cracks in concrete gravity dams. By minimizing the difference between the analytical responses given by Rezaiee-Pajand and Tavakoli's element and the measured ones, the genetic algorithm identifies the crack.

GA has achieved great success in practical applications. However, on a biological basis, the GA has mathematical flaws [15, 16]. A number of defects exist when GA is used in calculations, which include slow speed and immature convergence. Moreover, GA encounters problems on search efficiency and time [17–19]. These deficiencies seriously impede the promotion and application of GA. In this paper, GA was improved based on adaptive principle [20] and applied for the design optimization of CFRD.

This paper combined with the adaptive principle, based on the genetic algorithm, and applied it to optimal design of the shape of the concrete face rock-fill dam.

MATHEMATICAL MODELS FOR THE DESIGN OPTIMIZATION OF CFRD

Design Variables

In the cross-sectional design of CFRD, the dam height, the width of the dam crest, and other data are unchanged parameters that are determined in accordance with the project planning requirements. However, these parameters are generally not considered as design variables. The angle of the upstream dam slope x_1, the angle between the dam material boundaries and the horizontal direction x2, and the angle of the downstream dam slope x_3 are considered as design variables of the CFRD cross section.

Objective Function

The CFRD sections consist of different material partitions. After considering different comprehensive factors, such as exploitation, transportation, and construction, the unit price of volume formed by different materials was determined. For the unit length section, the unit price of each feed zone was compared. A material was selected, and the unit price of the material was assumed as 1.0. The ratio of the unit price of the selected material to that of the price of other materials is price proportional, which is denoted as C_i. The objective function can be expressed as follows:

$$f = \sum (C_i \times S_i), \quad (1)$$

where f is the project cost of the dam and is the area of the material in the dam cross section.

Constraint Conditions

The water pressure on the upstream panel can be directly determined by considering the structure and state of workability of CFRD. Thus, the sliding force produced by the dead weight of the dam and the water pressure on the panel significantly outweighs the horizontal thrust. However, this phenomenon does not produce an integral sliding problem. During design optimization, the constraint conditions take the following forms:

(1) Geometric constraint conditions $x_l \le x \le x_u$.

(2) Maximum stress constraint conditions $\sigma \le [\sigma]$.

(3) Preventive plastic shear failure conditions $S_L \le [S_L]$.

(4) Dam slope stability constraint conditions $F_s \ge [F_s]$.

x_l is the lower limit of variables, x_u is the upper limit of variables, σ is the stress, S_L is the stress level, and Fs is the safety coefficient of the dam slope stability.

IMPROVED GA WITH THE ADAPTIVE PRINCIPLE

As a kind of stochastic optimization method, the basic GA also have some shortcomings, like the weak local searching ability, the nonhigh optimization accuracy, premature convergence, and many other shortcomings. In order to make up for these deficiencies, this paper combined with the adaptive principle to improve the GA and applied it to optimal design of the shape of the concrete face rock-fill dam; the specific improvement process is as follows.

Parameter Coding

Coding is the most important issue concerning GA application. Several types of coding methods have been proposed, which can be divided into three categories: binary coding method, floating-point coding method, and symbol coding method. The optimization of CFRD has many design variables. If the binary coding method is used, the coding strings will be long, and the coding and decoding processes will be complex, which may affect the computational efficiency. Therefore, the simple floating-point coding with high accuracy and simple process was chosen for this study.

Generation of the Initial Group

To solve the constrained initial group, an initial feasible individual x_1 can be randomly generated if the individual is a boundary point of a feasible region. The regeneration is conducted until the initial feasible individual x_1 is produced, which is the interior point of the feasible region. Thus, another initial individual x2can be randomly generated, as expressed in the following equation:

$$x_2 = a + r_2 \times (b - a), \tag{2}$$

Where
$a = (a_1, a_2, \ldots, a_n)^T$, $r_2 = (r_{21}, r_{22}, \ldots, r_{2n})^T$, and $b = (b_1, b_2, \ldots, b_n)^{\hat{T}}$

We need to determine whether the individual x_2, which is generated by (2), can satisfy the constraint conditions. If the condition is met, then a new initial individual x_3 will be generated. Otherwise, the value of x_2 continually approximates that of x_1; that is, an iteration will be conducted by using the following equation:

$$x_2 = x_1 + \alpha (x_2 - x_1), \tag{3}$$

where α is a coefficient greater than 0 but less than 1 and is generally taken as 0.5

x_2 can become a feasible individual after the constant iterations. The individual x_3 can be generated by using the same method as that of x_2. x_3 can be converted into an initial feasible individual; the iteration continues until m individuals are generated.

Treatment of Fitness Function and Constraint Conditions

To meet the requirements of GA, the fitness function should be nonnegative and the problems should be maximum optimization problems. In the optimization problem of CFRD, the upper and lower limit constraints of design variables are considered during coding. The other constraint conditions are used to deal with the penalty function. Therefore, the objective function in the optimization problem of CFRD is expressed as follows:

$$F(x) = \begin{cases} C_{\max} - f(x) & f(x) < C_{\max} \\ 0 & f(x) \geq C_{\max}, \end{cases} \tag{4}$$

where $C_{\max}$ is a large number greater than $F(x)$ and is taken as 50 000.

Genetic Operation

Selection Operation

The basic idea of selection proportion is that the probability of each individual being selected is proportional to the degree of its fitness. Individuals with a higher degree of fitness have greater probability of passing their characteristics to the next generation.

Crossover Operation

The arithmetic crossover operator was used; that is, two individuals were randomly selected and then combined linearly to generate two new individuals:

$$\begin{aligned} A &= \gamma_0 X_{j1} + (1 - \gamma_0) X_{j2}, \\ B &= \gamma_0 X_{j2} + (1 - \gamma_0) X_{j2}, \end{aligned} \tag{5}$$

where γ_0 is a random number uniformly distributed between 0 and 1. Consider

$$P_c = \begin{cases} k_{13} & |f' - f_{\max}| \leq v_{\text{fix}} \\ \dfrac{k_1 (f_{\max} - f')}{f_{\max} - f_{\text{ave}}} & f' \geq f_{\text{ave}} \\ k_3 & f' < f_{\text{ave}}. \end{cases} \tag{6}$$

Here, $f_{\max}$ is the maximum fitness value among groups, f is the fitness value of the variation individual, f_{ave} is the average fitness value of the groups of each generation, and f' is the larger fitness value between two individuals conducting the interlace operation.

To improve the quality of crossover, the crossover probability was generated by using (6) and P_c was controlled by the degree of evolution. Here, $k_1 = 1$, $k_3 = 0.9$, $k_{13} = 0.5$, and $V_{\text{fix}} = 0.001$.

Mutation Operation

A nonuniform mutation operation was used; that is, every variable can be mutated independently according to the following equation:

$$X_i = \begin{cases} (X_{i\max} - X_i)\left(1 - r^{(1-t/T)^B}\right) & \text{rad} = 0 \\ (X_i - X_{i\min})\left(1 - r^{(1-t/T)^B}\right) & \text{rad} = 1, \end{cases} \tag{7}$$

where $t$is algebra of evolution; r is a random number between 0 and 1; T is the maximum evolution algebra; B is the system parameter, which is generally taken as 2

To improve the quality of mutation, the mutation probability was generated by using (8); that is, P_m was controlled by the degree of evolution. Here, $k_2 = 0.5$, $k_4 = 0.065$, $k_{24} = 0.07$, and $V_{\text{fix}} = 0.001$. Consider

$$P_m = \begin{cases} k_{24} & \left|f' - f_{\max}\right| \le v_{\text{fix}} \\ \dfrac{k_2\left(f_{\max} - f'\right)}{f_{\max} - f_{\text{ave}}} & f' \ge f_{\text{ave}} \\ k_4 & f' < f_{\text{ave}}. \end{cases} \quad (8)$$

OPTIMAL DESIGN PROGRAM OF CFRD BASED ON THE IMPROVED GA

Overall Design of the Program Structure

The program includes three modules, namely, the module of inputting the original data, the module of calculating the optimal design, and the module of outputting the results. The entire program is written by using FORTRAN. The initial data are read in the form of documents and stored in the form of program variables. The final optimization results can be obtained through cycle calculation and the output in the form of files.

Function Design of Each Program Module

Module of Inputting the Original Data

The original data can be divided into three parts. The first part is used to calculate the geometry of CFRD, which includes two aspects, (1) the operation parameters of genetic operation, which include the size of group M and the maximum time of iterations T, and (2) the structural dimensions of CFRD, which include the thickness of panels and cushions, the level of thickness of transition layers, the dam height, and the water level. The second part is used to calculate the stability of the dam slope, which includes the physical and mechanical indexes of dam materials, the upstream and downstream water levels during the construction period, and the earthquake coefficient. The third part is used to calculate the stress and deformation of the dam, which included the load series, the number of structure partitions, and the calculation parameters of Duncan-Chang stress-dependent bulk modulus (E-B) model of each dam material.

Module of the Optimal Design Calculation

The module of the optimal design calculation is the core of the entire program. Given that CFRD has many material partitions, the calculation of the structural reanalysis requires a significant amount of work. Therefore, when writing the program that calculates the optimal design of CFRD, the calculations of slope stability and dam stress deformation must be treated as subprograms, whereas the improved GA must be treated as the main program.

(1) Calculation of Slope Stability. The calculation program of the slope stability utilizes the Bishop method for the analysis of the circular-slip surface. This program can be used not only in analyzing the stability of any designated slip-out point and arbitrary designated sliding arc but also in determining the stability of all possible slide-out points as artificially designated density interpolates between any two slide-out points of the dam slope.

(2) Calculation of Stress Deformation. The nonlinear finite element method is used to analyze the characteristics of the dam stress deformation. The dam materials utilize the nonlinear Duncan-Chang E-B constitutive model during simulation. The simulation construction process calculates and applies loads layer by layer. The filling layer and the filling soil are calculated; however, the resulting displacement and strain are changed to zero, thus leaving the stresses in the equation. The load increment should have the smallest possible value to reflect the nonlinear characteristics of the dam material.

(3) Calculation of Genetic Manipulation. The module of genetic manipulation is the main program of the improved GA optimization program. The genetic manipulation includes selection operation, crossover operation, and mutation operation. Figure 1 shows the calculation flow chart.

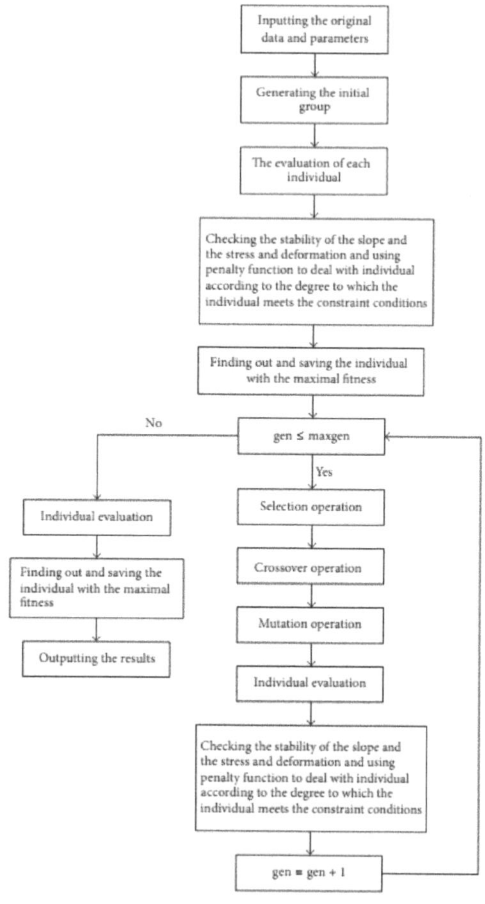

Figure 1: Module flow chart of GA.

Module of Outputting Results

The output data of this optimization program are as follows:(1) The optimal solution and the value of its objective function, in which the solution has to meet the constraint condition, include stability and stress deformation.(2)The safety factor of slope stability, which meets the constraint condition and is under the specified calculation condition, corresponds to the optimal solution.(3)The calculation results of stress and deformation on the dam and panel, in which the results are under the specified calculation condition, must correspond to the optimal solution and must meet the constraint condition. The horizontal displacement is positive when it moves toward the dam downstream, whereas the vertical displacement is positive when it moves toward the dam upstream. Stress is positive when it is compressive and negative when it is tensile.

EXAMPLE ANALYSES

Project Profile

Jishixia CFRD is a large (II) type of hydraulic complex project, in which the dam is a single-level building with crest elevation of 1861 m, crest width of 10 m, and maximum dam height of 100 m. The minimum excavation height of the toe slab is 1761 m, the slope ratio of the upstream dam slope is 1 : 1.4, the comprehensive slope ratio of the downstream dam slope is 1 : 1.71, the normal high water level is 1856 m, and the corresponding downstream water level is 1783.26 m. The grade zones of material in the dam from upstream to downstream consist of the following: concrete face slab, cushion zone, transition zone, main rock-fill area, and downstream rock-fill zone. Figure 2 shows the diagram of the dam design optimization variables.

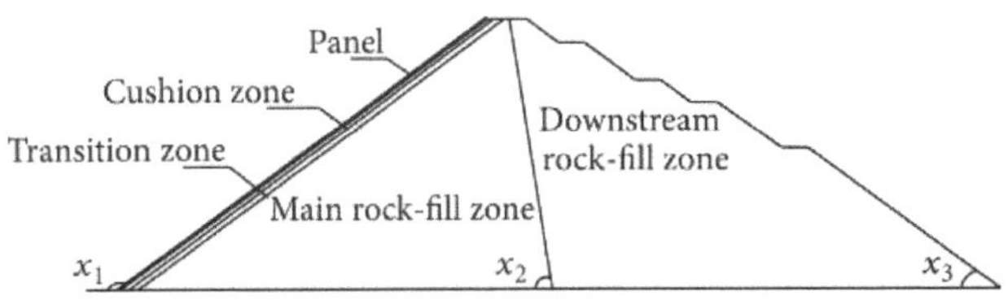

Figure 2: Diagram of the optimal design variables.

Basic Information

The comprehensive parities of the dam materials include the downstream rock-fills, main rock-fills, transition materials, cushion material, and panel concrete, with ratio of 1 : 1.42 : 1.81 : 2.13 : 18.68.

Geometric Constraints. Consider $2.486 \le x_1 \le 2.61$, $x_1 \le x2 \le x_3$, and $0.532 \le x_3 \le 0.656$.

Condition Constraints. The allowable stability safety factor is $[F_s] = 1.35$, the allowable tensile stress is $[\sigma_L] = 1.0$ MPa, the allowable pressure stress is $[\sigma_c] = 10.0$ MPa, and the maximum allowable stress level is $[S_L] = 1.0$.

The geometric constraints were selected based on the experience of completed domestic and international constructions. The safety factor of stability was selected based on the relevant provisions of China in Design Code for Concrete Face Rock-Fill Dams (SL228-98).

Structure analysis must be performed to establish the condition constraints. Dam slope stability analysis, which is conducted by using the Swedish slice method and the finite element analysis of CFRD deformation and stress, should also be conducted. The condition for water storage calculation for the upstream water level is 1856 m, whereas the condition for the downstream water level is 1783.26 m.

The linear elastic constitutive model was used to simulate the concrete face. ε (elastic modulus) = 2×10^{10} pa, μ (Poisson's ratio) = 0.167, and ρ (density) = 2450 kg/m^3. The Duncan-Chang E-B constitutive model was used to simulate the dam materials, as shown in Table 1.

Table 1: Material parameters simulated by using the Duncan-Chang E-B model.

Material	ρ/(kg·m^{-3})	φ/(°)	$\Delta\varphi$/(°)	C	K	K_{ur}	K_b	n	R_f	m
Cushion	2200	50.6	7	0	400	800	520	0.35	0.78	0.22
Transition layer	2180	52.5	10	0	850	1700	550	0.3	0.77	0.20
Main rock-fill zone	2135	56.5	13	0	1070	1300	260	0.36	0.82	0.40
Secondary rock-fill zone	2121	56.5	11	0	310	700	110	0.39	0.59	0.30

Optimizing Results

Analysis of the Optimal Program

The paper uses the genetic algorithm and the complex algorithm to conduct the optimizing design of Jishixia concrete face rock-fill dam, respectively, and the calculation results of these two optimizing design programs are compared with the initial design by listing. The design variables and the results of objective function of the design with genetic algorithm and complex algorithm and the initial design are compared in Table 2. The comparative figure about optimal program with genetic algorithm and the initial design are shown in Figure 3.

Table 2: Comparative results of the initial design and the optimal program.

Program	Iterative step	Design variables/rad			Objective function
		x_1	x_2	x_3	$F(x)$
Initial design	—	2.521	1.373	0.620	23946.91
Complex algorithm	535	2.500	1.560	0.656	22819.05
GA	157	2.497	1.560	0.653	22803.83

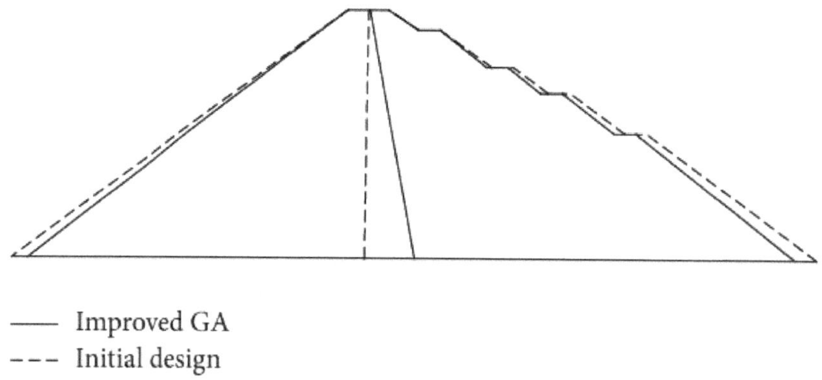

Figure 3: Comparison of the dam section between the initial design and the GA.

(1) Table 2 and Figure 3 show that, after optimization, the upstream and downstream slopes of the dam became steeper, the angle of the upstream slope of the dam decreased from 2.521 rad to 2.497 rad, and the angle of the downstream slope of the dam increased from 0.620 rad to 0.653 rad. Consequently, the area of the dam section decreased, thus decreasing the dam cost. The angle between the dam material boundaries and the horizontal direction increased from 1.373 rad to 1.560 rad. This increase in angle increased the content of the downstream rock-fill, thereby reducing the dam cost. Compared with the initial design scheme, the dam cost was reduced by 4.8% after optimization.

(2) From the comparison optimized results of the complex algorithm and the genetic algorithm, it can be seen that the two optimal results of optimization method are basically the same; it is also shown that the two methods are suitable for the optimization design of the face rock-fill dam. The generation number of the complex algorithm converging to the optimal solution is 535. However, the generation number of genetic algorithm converging to the optimal solution is 157; the consumed time of genetic algorithm is below one-third of complex algorithm. Thereby, the genetic algorithm is more efficient and speedy.

(3) The convergence process for objective function according to generation number is reflected in Figure 4; we can see that the objective function decreases rapidly and then the speed of decrease is reduced and gradually comes to stability in the top 15 generation numbers. Figures 5–7 show the changing process for three parameters of x_1, x_2, and x_3 according to generation number. The trend of parameter x_1 and objective function are basically the same in Figure 5. We can see that parameter x_2 and parameter x_3 rise rapidly and then the speed of rise is reduced and gradually comes to stability with increase in the top 15 generation numbers in Figures 6 and 7.

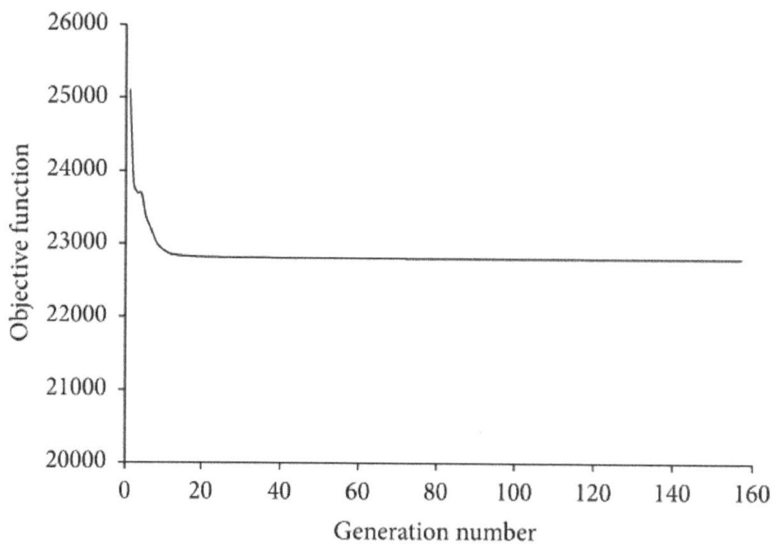

Figure 4: Evolution of objective function along generation number.

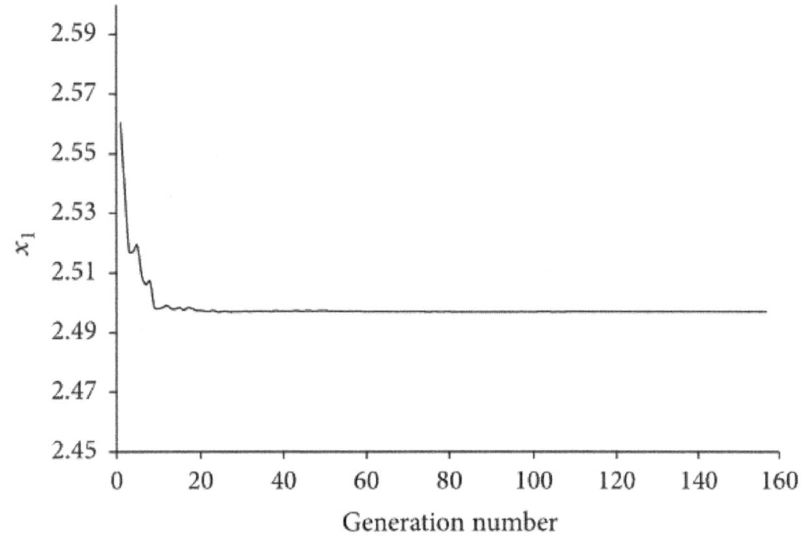

Figure 5: Evolution of x_1 along generation number.

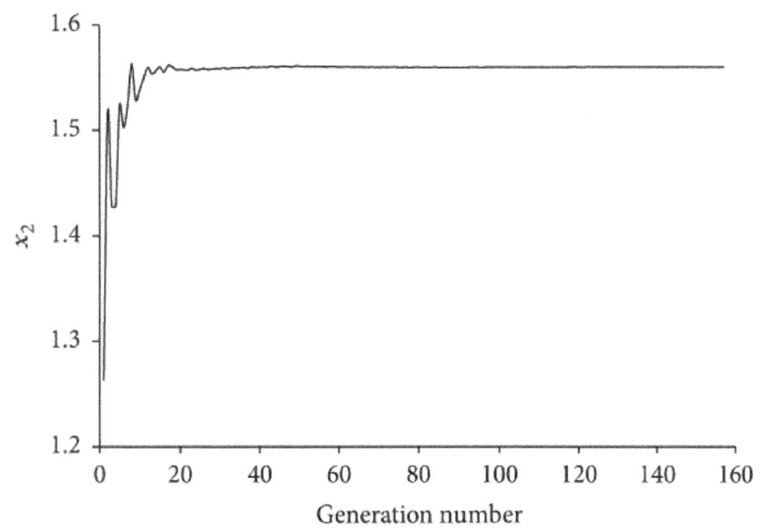

Figure 6: Evolution of x_2 along generation number.

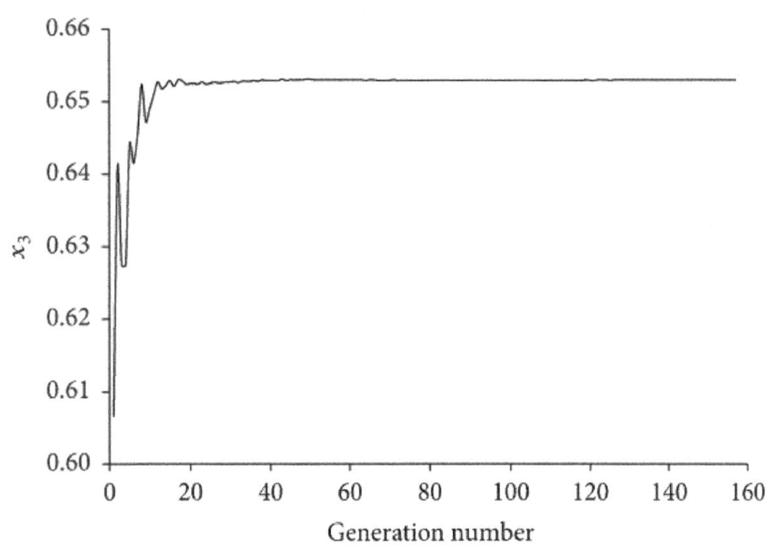

Figure 7: Evolution of x_3 along generation number.

Stress Deformation Analysis of the Dam

Table 3 shows the results, optimum solution, optimal value, and main condition index of the genetic optimization algorithm.

Table 3: Comparative program results before and after optimization.

Program	Conditions	Behavior indicators				
		σ_L/MPa	σ_c/MPa	S_L	δ/m	F_s
Before optimization	Completion period	0.43	1.69	0.68	0.11	1.79
	Impounding period	0.50	1.84	0.58	0.22	
After optimization	Completion period	0.43	1.72	0.83	0.16	1.35
	Impounding period	0.51	1.85	0.72	0.27	

Note: F_s is the dam slope safety factor of stability under normal operating conditions.

(1) Results and Analysis of the Displacement Calculation. The vertical displacements of the dam during completion and impoundment periods are reflected in Figures 8 and 9, respectively.

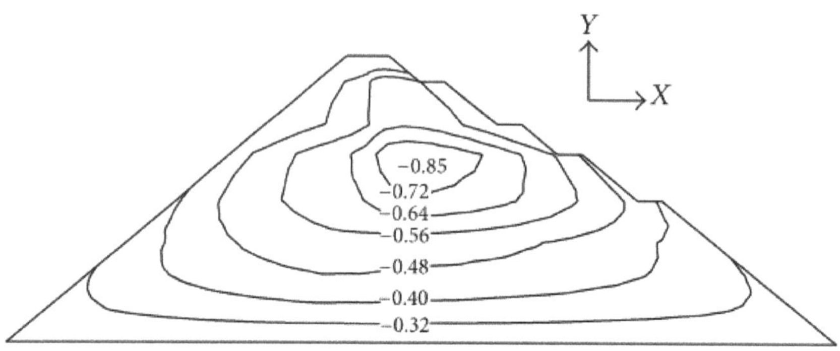

Figure 8: Contour map of the vertical displacement during the completion period (unit: m).

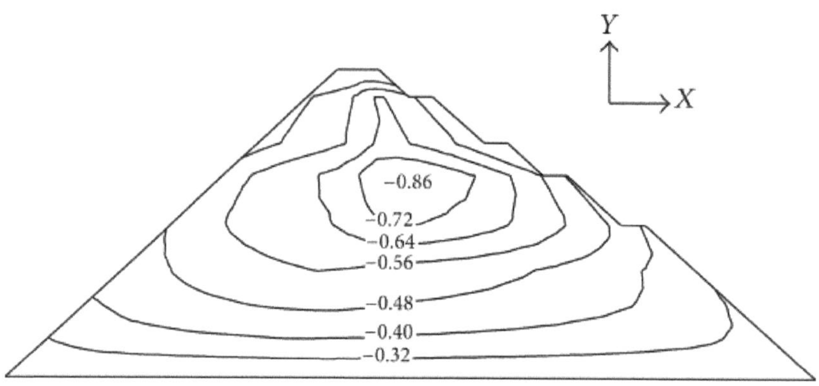

Figure 9: Contour map of the vertical displacement during the impoundment period (unit: m).

After optimization, the largest settlement of the dam during the completion period was 0.85 m, which accounts for about 0.85% of the dam height and occurs in about two-thirds of the dam height. Given that the downstream rock-fill of the dam is relatively weak, the settlement of the dam displacement is biased toward the dam downstream, and the location of the maximum settlement of the dam was in the secondary rock-fill zone. Consequently, the displacement of the dam crest was slightly biased toward the dam downstream. The dam mainly bears dead weight during the completion period. The area of the dam section decreased after optimization, and the dead weight was reduced. Thus, the dam settlement decreased slightly after the completion period compared with that before optimization.

Compared with the completion period, the impoundment period bore water load and the dam settlement increased slightly. In addition, the maximum settlement during the impoundment period was 0.86 m, which accounts for about 0.86% of the dam height. The maximum settlement increased about 0.01% compared with that of the completion period. The maximum settlement occurred in about two-thirds of the dam height, which is biased toward the dam downstream. Compared with the design before optimization, the dam settlement decreased slightly during the impoundment period.

The horizontal displacements of the dam during the completion and impoundment periods are shown in Figures 10 and 11, respectively.

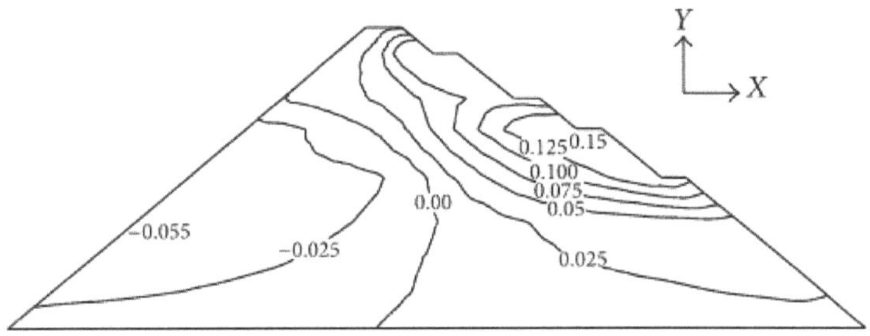

Figure 10: Contour map of the horizontal displacement during the completion period (unit: m).

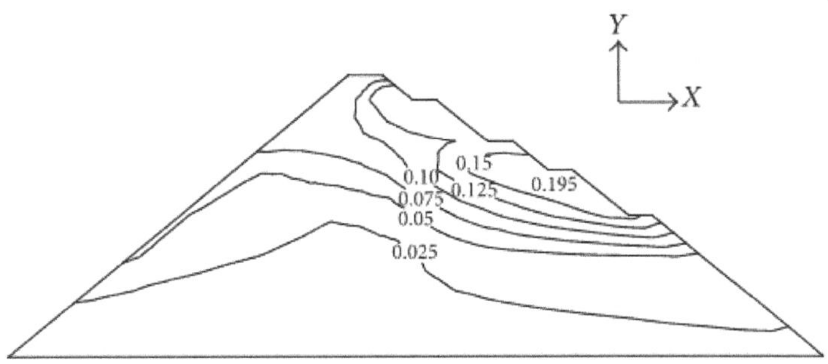

Figure 11: Contour map of the horizontal displacement during the impoundment period (unit: m).

During the completion period, the dam is mainly under the action of dead weight load, the horizontal displacement is not symmetrically distributed, and the upstream part of the dam tends to be displaced upstream. In addition, the maximum value of the displacement is 0.055 m, which occurred in one-third of the dam height near the upstream face. The downstream part of the dam tends to be displaced upstream. The maximum value of displacement is 0.15 m, which occurred in one-half of the dam height near the downstream surface. The tendency of displacement during the impoundment period is the same as that during the completion period. However, under the combined actions of water load and dead weight load, the distribution of dam horizontal displacement significantly changed; the zone of the horizontal displacement directing to the upstream side of the dam was significantly reduced, and the maximum horizontal displacement was reduced to 0.00 m. Moreover, the horizontal displacement directing to the downstream increased, and the maximum horizontal displacement increased to 0.195 m, which is located in one-half of the dam height near the downstream surface.

Compared with the design before optimization, the displacement trends of the dam during the completion and storage periods are the same, but the horizontal upstream and downstream displacement significantly changed. During the completion period, the horizontal upstream displacement of the dam increased to 1.68 cm, whereas the horizontal downstream displacement of the dam increased to 7.8 cm. During the impounding period, the horizontal upstream displacement of the dam increased to 0.78 cm, whereas the horizontal downstream displacement of the dam increased to 7.98 cm. This result is mainly caused by the increase in the proportion of downstream rock-fill in the section after optimization. The relatively weak downstream rock-fill of the dam also caused this increase in horizontal displacement.

(2) Results and Analysis of the Stress Calculation. The major principal stress isoline, minor principal stress isoline, and stress level isoline during the completion and impoundment periods are shown in Figures 12–17.

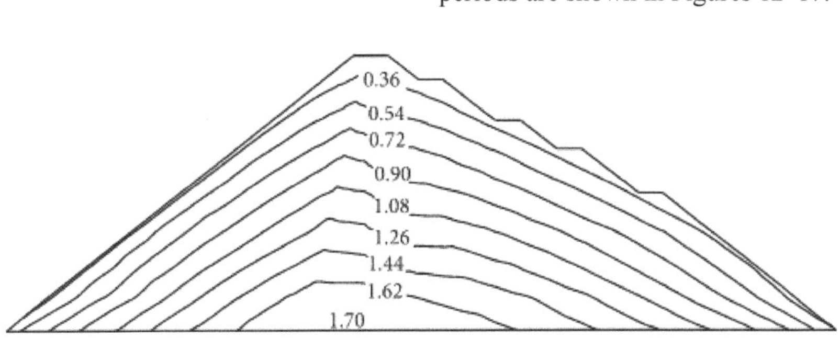

Figure 12: Major principal stress isoline during the completion period (unit: MPa).

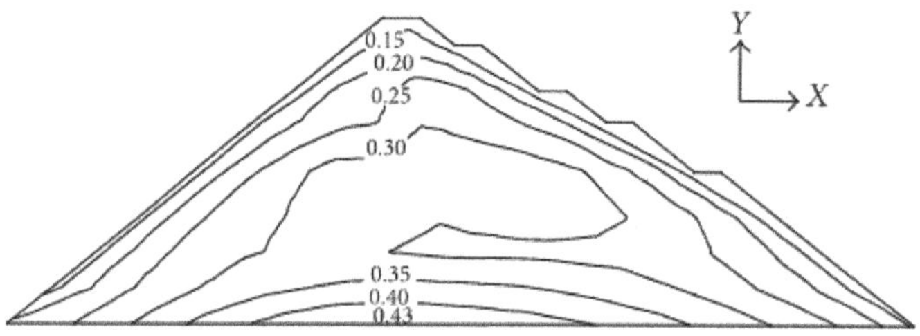

Figure 13: Minor principal stress isoline during the completion period (unit: MPa).

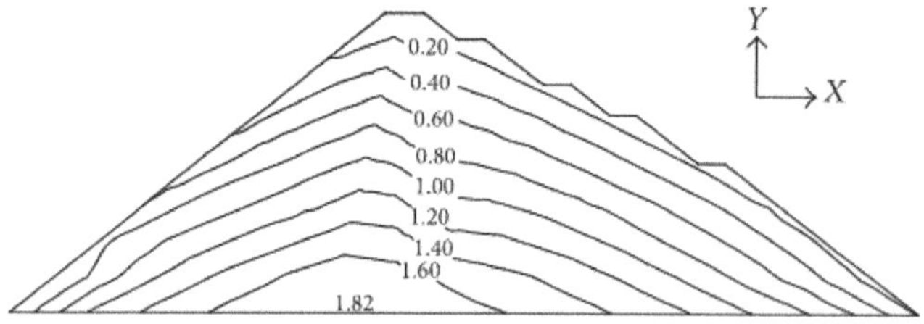

Figure 14: Major principal stress isoline during the impoundment period (unit: MPa).

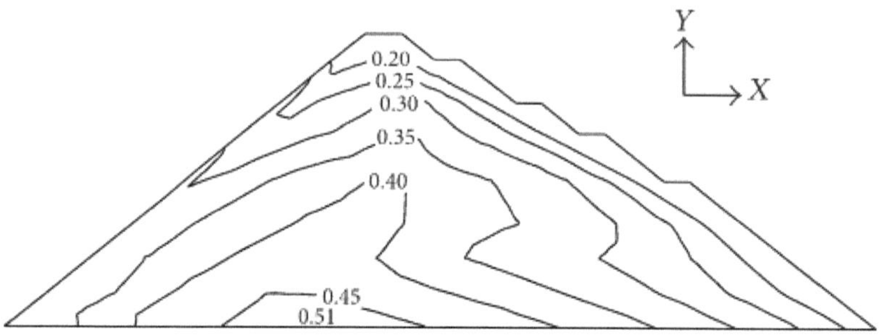

Figure 15: Minor principal stress isoline during the impoundment period (unit: MPa).

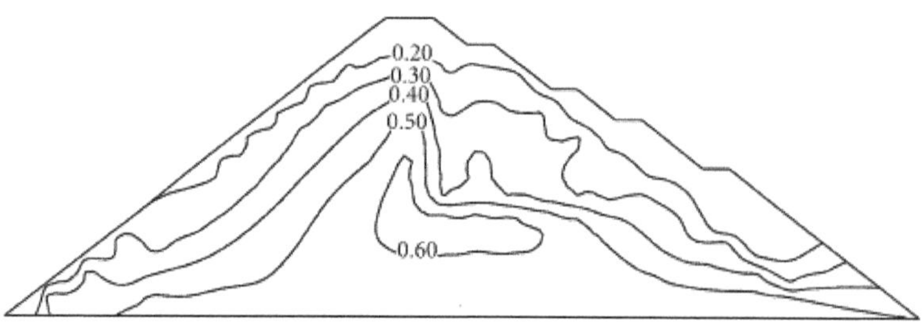

Figure 16: Stress level isoline during the completion period.

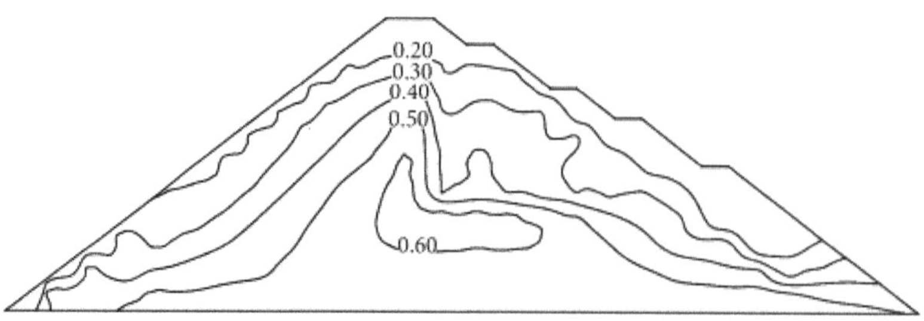

Figure 17: Stress level isoline during the impoundment period.

During the completion period, the maximum major principal stress is 1.70 MPa and the maximum minor principal stress is 0.43 MPa, in which both stresses are located in the dam base and are near the dam axis. Both stresses are compressive. The directions of the major and minor principal stresses are close to that of gravity. The vertical stress component is close to the dead weight stress of the overlying rock-fill. The stress level of the entire dam is less than 1.0, and the maximum value is 0.65.

During the impoundment period, the dam stress is redistributed, the dam is under the combined action of the dead weight load and water load, the dam stress in the upstream side of the dam axis is more than that in the downstream side, and the dam stress level is less than that of the completion period, in which the maximum value was 0.63. This result shows that impounding is favorable for the stability of the concrete face rock-fill dam. The water load near the cushion material is perpendicular to the panel and close to the direction of the minor principal stress in the dead weight stress field. The increment of the minor principal stress is greater than that of the major principal stress. This phenomenon results in directional changes of the principal stress. During the impoundment period, the maximum major principal stress is 1.82 MPa and the maximum minor principal stress is 0.51 MPa. Both stresses are located in the dam base and are near the dam axis. The dam stress during the impoundment period did not significantly change. The dam stress is mainly caused by the dead weight of the rock-fills, and the dam stress generated by the reservoir water pressure is relatively small. The stress level of the entire dam is less than 1.0.

Compared with the initial design, the major and minor principal stresses of the dam after the completion period did not significantly change. The major principal stress increased to 0.02 MPa, whereas the minor principal stress increased to 0.06 MPa. After the impoundment period, the major principal stress decreased to 0.03 MPa, whereas the minor principal stress increased to 0.09 MPa. The stress level increased to 0.02 in the completion period and increased to 0.05 in the impoundment period. The stress level in most parts of the dam was low. Thus, the dam optimal design with the GA is in good working condition.

(3) Calculation Results of the Panel Deflection and Analysis. The deflection curves of the panel during the completion and impoundment periods are shown in Figures 18 and 19, respectively.

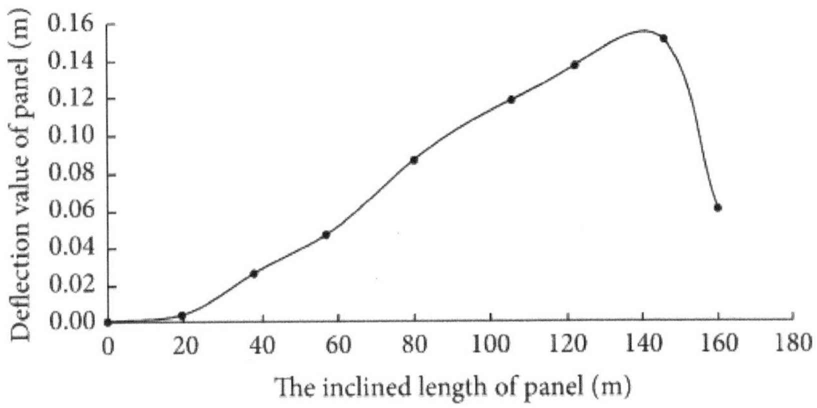

Figure 18: Deflection curves of a panel during the completion period.

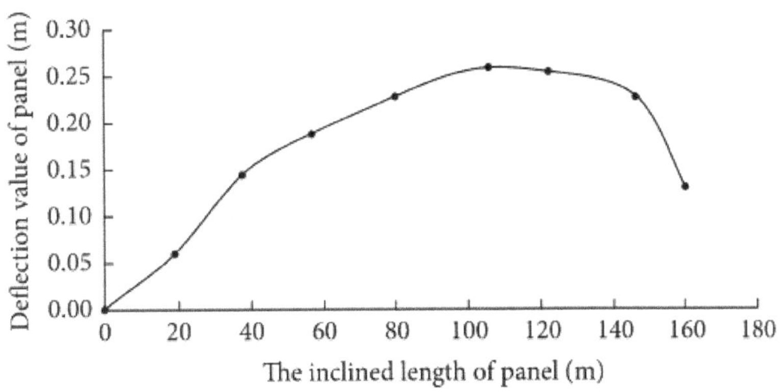

Figure 19: Deflection curves of a panel during the impoundment period.

The following can be seen from Table 3 and Figures 18 and 19. The maximum deflection of the panel during the completion period is 15.59 cm, which is located in the upper part of the panel. In the impoundment period, the maximum deflection of the panel increased to 26.71 cm under the effect of the upstream water pressure. The entire panel deformed downstream, and the maximum deflection was located in the middle of the panel, which gradually decreases in the upper and lower parts of the panel. Analysis results show that the panel deflection of the Jishixia project is relatively small, which could not trigger the peripheral joints to produce large displacement and can meet the requirements of the design.

Compared with the initial design, the deflection of the panel increased after optimization. The deflection of the panel increased to 4.34 cm in the completion period and increased to 4.48 cm in the impoundment period.

CONCLUSION

We conclude the following:(1)We improved GA with the adaptive principle; by using the improved GA optimization program to conduct the optimal design of Jishixia CFRD, we have determined that the GA could not only solve the global optimal solution problem of complex optimization design, such as the high degree of nonlinearity and the recessiveness of constraint conditions, but also make up for deficiencies which exist in the basic genetic algorithm, such as the weak local searching ability, the nonhigh optimization accuracy, and premature convergence. The GA consumes less time and has high efficiency.(2)The comparative analysis results between the initial design and optimization show that, after optimization, the dam slope became steeper, the downstream rock-fill area increased, and the quantities of projects and project costs were reduced.(3)This paper conducts a finite element calculation of stress deformation and dam slope stability on Jishixia CFRD. Results show that, after optimization, both the dam stability and stress deformation conform with the general rule of CFRD, and the maximum stresses meet the requirements. These results confirm that the optimal result is safe, reliable, economic, and reasonable.

COMPETING INTERESTS

The authors declare that they have no competing interests.

ACKNOWLEDGMENTS

This research is financially supported by the CRSRI Open Research Program (Program SN: CKWV2014208/KY), the National Natural Science Foundation of China (nos. 51309190, 51409206, and 51579207), Program 2013KCT-15 for Shaanxi Provincial Key Innovative Research Team, China Postdoctoral Science Foundation (no. 2014M552470), Open Research Fund Program of State Key Laboratory of Water Resources and Hydropower Engineering Science (2014SGG03), and the Innovative Research Team of Institute of Water Resources and Hydroelectric Engineering, Xi'an University of Technology.

REFERENCES

1. M. Zhou and S. Sun, Genetic Algorithms: Theory and Applications, National Defense Industry Press, Beijing, China, 1999.
2. C. Xin, Structure Analysis and Optimization Design of Concrete-Faced Rock-Fill Dam, China Water Power Press, Beijing, China, 2005.
3. J. H. Holland, Adaptation in Natural and Artificial Systems, University of Michigan Press, Ann Arbor, Mich, USA, 1975.
4. D. H. Ackley, A Connectionist Machine for Genetic Hillclimbing, Kluwer Academic Publishers, Boston, Mass, USA, 1987.
5. D. Whitley, K. Mathias, and P. Fitzhorn, "Delta coding: an iterative search strategy for genetic algorithms," in Proceedings of the 4th International Conference on Genetic Algorithms, pp. 77–84, San Diego, Calif, USA, 1991.
6. J. Dzubera and D. Whitley, "Advanced correlation analysis of operators for the traveling salesman problem," in Parallel Problem Solving from Nature—PPSN III: International Conference on Evolutionary Computation The Third Conference on Parallel Problem Solving from Nature Jerusalem, Israel, October 9–14, 1994 Proceedings, vol. 866 of Lecture Notes in Computer Science, pp. 68–77, Springer, Berlin, Germany, 1994.
7. H. Bersini and G. Seront, "In search of a good evolution-optimization crossover," in Proceedings of the 2nd Conference on Parallel Problem Solving from Nature, pp. 479–488, North Holland, Amsterdam, Netherlands, 1992.
8. X.-M. Dai, Z.-G. Chen, R. Feng, X.-F. Mao, and H.-H. Shao, "Improved algorithm of pattern extraction based mutation approach to genetic algorithm," Journal of Shanghai Jiaotong University, vol. 36, no. 8, pp. 1158–1160, 2002.
9. H.-L. Zhao, X.-H. Pang, and Z.-M. Wu, "A building block coded parallel genetic algorithm and its application in TSP," Journal of Shanghai Jiaotong University, vol. 38, no. 2, pp. 213–217, 2004.
10. L. Jiang, "Study of elastic TSP based on parallel genetic algorithm," Microelectronics, vol. 22, no. 8, pp. 130–133, 2005.
11. L. Zhang, Y. Liu, B. Li, G. Zhang, and S. Zhang, "Study on real-time simulation analysis and inverse analysis system for temperature and stress of concrete dam," Mathematical Problems in Engineering, vol. 2015, Article ID 306165, 8 pages, 2015.
12. H. Gu, Z. Wu, X. Huang, and J. Song, "Zoning modulus inversion method for concrete dams based on chaos genetic optimization algorithm," Mathematical Problems in Engineering, vol. 2015, Article ID 817241, 9 pages, 2015.
13. Z. Li, C. Gu, and Z. Wu, "Nonparametric change point diagnosis method of concrete dam crack behavior abnormality," Mathematical Problems in Engineering, vol. 2013, Article ID 969021, 13 pages, 2013.
14. M. Rezaiee-Pajand and F. H. Tavakoli, "Crack detection in concrete gravity dams using a genetic algorithm," Proceedings of the Institution of Civil Engineers: Structures and Buildings, vol. 168, no. 3, pp. 192–209, 2015.
15. X. Zongben and L. Guo, "The simulated-biology-like algorithms for global optimization, part I: simulated evolutionary aslgorithms," Chinese Journal of Operations Research, vol. 14, no. 2, pp. 1–13, 1995.
16. D. B. Fogel, "Introduction to simulated evolutionary optimization," IEEE Transactions on Neural Networks, vol. 5, no. 1, pp. 3–14, 1994.
17. D. E. Goldberg and P. Segrest, "Finite Markov chain analysis of genetic algorithms," in Proceedings of the 2nd International Conference on Genetic Algorithms, pp. 1–8, Cambridge, Mass, USA, July 1987.
18. H. Asoh and H. Mühlenbein, "On the mean convergence time of evolutionary algorithms without selection and mutation," in Parallel Problem Solving from Nature-PPSN III, vol. 866 of Lecture Notes in Computer Science, pp. 88–97, Springer, Berlin, Germany, 1994.

19. C. A. Ankenbrandt, "An extension to the theory of convergence and a proof the time complexity of genetic algorithms," Foundations of Genetic Algorithms, pp. 53–58, 1990.

20. G. Xuejing and W. Pu, "An adaptive genetic algorithm based on real coded and its application," Journal of Beijing University of Technology, vol. 33, no. 2, pp. 144–148, 2007.

CHAPTER 5

HEDGING RULES FOR WATER SUPPLY RESERVOIR BASED ON THE MODEL OF SIMULATION AND OPTIMIZATION

Yi Ji, Xiaohui Lei, Siyu Cai and Xu Wang

Keywords: Hedging rule, two-period model, acceptable damage depth, simulation and optimization model

INTRODUCTION

Reservoirs, as projects to regulate runoff, play an important role in making water resources more suitable for social development and environmental protection. Making reasonable regulation rules is significant for reservoirs to play roles in water resources allocation [1,2,3,4]. Based on the research of reservoir release, starting standard, ending condition, and the determination of periods supplying water, different kinds of water supply rules have been put forward, such as Pack Rule (PR) [5], Space Rule (SP) [6], New York City Rule (NYC) [7], Linear Decision Rule (LDR) [8] and Standard Operation Policy (SOP) [5], Hedging Rules (HR) [9,10,11,12,13], and so on. Among the rules, the SOP is widely applied in reservoir operation. The maximum water supply and minimum water shortage of the water supply system are chosen by SOP [10]. The reservoir water is fully released under SOP when the available water is smaller than the water demand. During the dry season, water is not reserved in advance under SOP, which will cause severe damage for water supply in the later period. HR is obtained on the basis of optimizing the operation rules of SOP in dry season, which is an effective measure for reservoir regulation [14]. In order to avoid severe water deficit of a few periods in the dry season, HR mostly chooses several periods to decrease the water supply so that water can be reserved for the later period [15]. The loss caused by a severe water deficit of a few periods is more serious than that caused by the slight water shortage of many periods. HR determines the starting water availability according to the judgements of reservoir storage or available water in facing period [10]. Bayazit *et al.* [16] proposed the starting water availability and ending water availability of hedging rules based on the judging condition of available water at the beginning or end of the period, and analyzed the influence of the available water threshold on reservoir release performance. Srinivasan *et al.* [17] added a hedging factor (HF) index on the basis of the available water threshold index at the beginning or end of the period by Bayazit*et al.* [16], and analyzed the influence of the three indices on reliability, resiliency, and vulnerability of reservoir release.

Currently, in the studies of hedging rules, the research of the two-period model is a hot issue. The two-period model divides operation periods into a facing period and remaining period. This study focuses on the operation rules through analyzing the water supply of a facing period and water storage target of the next period. The operation rules contact the water supply of the facing period, starting water availability (SWA), and ending water availability (EWA). The study mainly focuses on two aspects: one is the determination of the three elements of hedging rules, including the starting water availability (SWA), ending water availability (EWA), and hedging factors (HF); another is the influence of inflow uncertainty on operation rules [18]. Draper *et al.* [10] and Shiau [15] determined starting water availability according to marginal costs of reservoir storage and marginal benefits of reservoir release based on the marginal benefit principle of economics. According to different reservoir hedging thresholds, SWA and EWA are given by the analytical expression of the period water supply. You *et al.* [12] and Zhao *et al.* [19,20] discussed effects of inflow uncertainty and reservoir operation constraint conditions on optimal conditions and operation rules of the two-period reservoir operation model.

In the two-period model, the weighting factor and water storage target are important parameters for determining

Citation: Yi Ji, Xiaohui Lei , Siyu Cai and Xu Wang, Hedging Rules for Water Supply Reservoir Based on the Model of Simulation and Optimization, doi:10.3390/w8060249.

hedging rules. Due to short-term defects of the two-period model, it is difficult to determine the weighting factor and water storage target. The two-period model chose the upper-basic operation curve as the storage target which could not guarantee the optimal release benefits in the long term. Therefore, this study proposes a hedging model that consists of the two-period model and parameter simulation and an optimization model. The relationship between the different operation periods is established by the optimization model of weighting factors and the release target, so it makes up the deficiency of the two-period model. This paper proposes a hedging rule which is composed of the two-period model and the parameter simulation optimization model. The two-period model is adopted for simulating the process of the water supply. The parameter simulation optimization model, with a particle swarm optimization algorithm, is adopted for optimizing the parameter. Heiquan reservoir is employed as a case study to verify the reasonability and efficiency of the hedging rule.

METHODOLOGY

Hedging Rules Based on the Two-Period Model

The Optimal Reservoir Release

For a water supply reservoir, the water availability at time t is the sum of initial water stored in a reservoir at time $t$plus the current inflow and minus the evaporation loss at this time period, *i.e.*,

$$WA_t = S_t + I_t - E_t \tag{1}$$

where WA_t, is the water availability of the reservoir during time period t; S_t, I_t, and E_t are, respectively, the initial water stored in a reservoir at the beginning of time period t, inflow into reservoir at time t, evaporation and seepage loss at time t.

By applying the principle of water balance, the initial water stored in a reservoir at the beginning of time period t+1 is equal to the sum of the initial water stored in a reservoir at time t plus the current inflow and minus the sum of evaporation loss and reservoir release at time period t. That is:

$$S_{t+1} = S_t + I_t - E_t - R_t \tag{2}$$

where R_t is the reservoir release at time t, which is used for projected demand D_t and constrained by $0 < R_t \leq D_t$when considering breakage depth.

Equations (1) and (2) simplify to Equation (3):

$$WA_t = R_t + S_{t+1} \tag{3}$$

Equation (3) implies that water availability during time period t includes the reservoir release of the current periodR_t, along with water storage capacity S_{t+1} at the beginning time t+1. The benefits of a water supply reservoir during a period can be separated into two parts, including the benefit of release and benefit of storage. The benefit of release is measured by water deficit loss at time t, while the benefit of storage is measured by the difference between storage at the beginning of the next period and storage targets. The optimization framework of reservoir operation is to minimize the sum of the water supply benefit function $B(R_t)$ and the carryover storage benefit function $C(S_{t+1})$ [19].$B(R_t)$ and $C(S_{t+1})$ is formulated as Equations (5) and (6):

$$\min z = B(R_t) + C(S_{t+1}) \tag{4}$$

$$B(R_t) = \left(\frac{D_t - R_t}{D_t}\right)^m \tag{5}$$

$$C(S_{t+1}) = (1-\omega)\left(\frac{S_{t+1}^T - S_{t+1}}{S_{t+1}^T}\right)^m \tag{6}$$

where ω is the weighting factor assigned to the loss function of reservoir release target, m is an exponent to define the shape of the loss functions (m should exceed 1 in order to have a convex function that can be minimized).

The key finding of Draper and Lund [10] was the derivation of the optimal hedging condition for a water supply reservoir, which states that the marginal benefit of release is equal to the marginal benefit of storage. That is:

$$\frac{\partial B(R)}{\partial R} = \frac{\partial C(S)}{\partial S} \tag{7}$$

Using the Equations (5)–(7), the optimal reservoir release R∗t can be deduced, which is formulated as Equation (8):

$$R_t^* = \frac{(S_{t+1}^T)^{m/m-1} D_t + \left(\frac{1-w}{w}\right)^{1/m-1} (D_t)^{m/m-1} (WA_t - S_{t+1}^T)}{\left(S_{t+1}^T\right)^{m/m-1} + \left(\frac{1-w}{w}\right)^{1/m-1} (D_t)^{m/m-1}} \tag{8}$$

Starting Water Availability (SWA) and Ending Water Availability (EWA) of Hedging Rules

A factor ηt, which is called the inverse-weighted target ratio (IWTR), is developed by Shiau [19]. IWTR is defined as:

$$\eta_t = \left(\frac{1-w}{w}\right)\left(\frac{D_t}{S_{t+1}^T}\right) \tag{9}$$

By putting Equation (9) into Equation (8), the optimal reservoir release R∗t can be simplified using ηt, which is formulated as Equation (10):

$$R_t^* = \frac{S_{t+1}^T + \eta_t^{1/m-1} (WA_t - S_{t+1}^T)}{\frac{S_{t+1}^T}{D_t} + \eta_t^{1/m-1}} \tag{10}$$

Under the hedging rules, when m = 2, the relationship between the optimal reservoir release and water availability (WA) has a linear relationship with a slope of$D_t / (S_{t+1}^T/\eta + D_t)$. As demonstrated in Figure 1, O-S1-D, O-D, and O-S2-D represent hedging rules under three different situations, where S1 and S2 are SWA and D is the EWA. In the figure, D is a fixed point, because $EWA_t = S_{t+1}^T + D_t$.

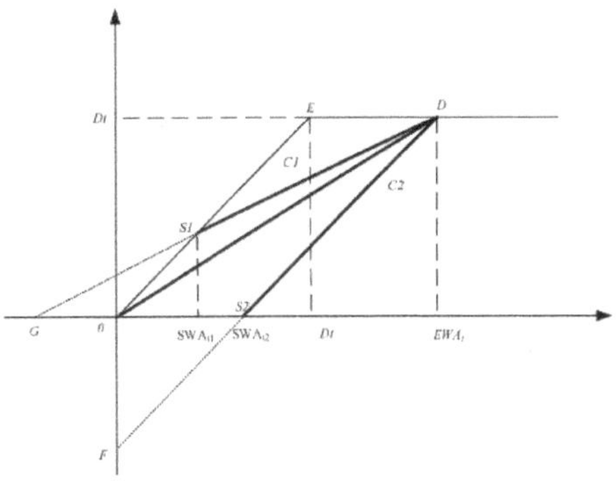

Figure 1. The hedging rules and the standard operating policy.

The linear optimal hedging of ηt=1 is line OD, the slope of line OD is $D_t/(S_{t+1}^T + D_t)$. When ηt=1, Equation (10) can be simplified as:

$$R_t^* = WA \times \frac{D_t}{S_{t+1}^T + D_t} \qquad (11)$$

It is shown from Equation (11) that WA is allocated according to the ratio of demand target and storage target, and the two targets get the same satisfaction degree in line OD. Line OD separates area O-S1-D-S2-O into two parts. As D is a fixed point, the slope of any line DS_1 in the upper area of line OD is smaller than the slope of line OD. The slope of line DS_1 is $D_t/(S_{t+1}^T/\eta + D_t)$, so $D_t/(S_{t+1}^T/\eta + D_t) < D_t/(S_{t+1}^T + D_t)$, then ηt<1. As a result, when *WA*remains constant, R_t^* is larger, and the satisfaction degree of the demand target is larger than that of the storage target. On the other hand, the slope of any line DS_2 in the lower area of line OD is larger than the slope of line OD. The slope of line OS_2 is $D_t/(S_{t+1}^T/\eta + D_t)$, so $D_t/(S_{t+1}^T/\eta + D_t) > D_t/(S_{t+1}^T + D_t)$, then ηt>1. As a result, when WA remains constant, R_t^* is smaller, and the satisfaction degree of the demand target is smaller than that of the storage target.

The Expression Form of Hedging Rules

Damage depth is an important index for the ability of water supply of reservoir. This index for evaluating the reservoir with low inflow in dry season is particularly important. The acceptable damage depth (ADD) is used to express the vulnerability of water supply reservoir, generally defined as the proportion of the minimum water supply to the water demand in a single period for water-supply area. The minimum water supply is obtained by the minimum demand in different departments. The supply water is less than the minimum water supply, which can bring a devastating effect on the local economy. After considering ADD, hedging rules can be expressed in three situations, which is shown in Table 1.

Table 1: The three situations of the hedging rules.

Scenario		Water Availability	Reservoir Release	Rules
$\eta_t \leqslant 1$	(I) $SWA_t \leqslant DWA_{t1}$	$0 < WA_t < \alpha D_t$	WA_t	O-A
		$\alpha D_t < WA_t < DWA_{t1}$	αD_t	A-B
		$DWA_{t1} \leqslant WA_t < EWA_{t1}$	R^*	B-C
		$WA_t \geqslant EWA_{t1}$	D_t	C-D
	(II) $SWA_t \geqslant DWA_{t2}$	$0 < WA_t < SWA_{t2}$	WA_t	O-B
		$SWA_{t2} \leqslant WA_t < EWA_{t2}$	R^*	B-C
		$WA_t \geqslant EWA_{t2}$	D_t	C-D
$\eta_t \geqslant 1$	(III) $SWA_t \leqslant DWA_t$	$0 < WA_t < \alpha D_t$	WA_t	O-A
		$WA_t < DWA_t$	αD_t	A-B
		$DWA_t \leqslant WA_t < EWA_t$	R^*	B-C
		$WA_t \geqslant EWA_t$	D_t	C-D

Note: Where α is ADD.

Situation I: Point E is defined as SWA by the traditional model. Point A of ADD is higher than Point E (Figure 2). Water is supplied under the guide of the SOP rule in line EA. Point A is chosen as the SWA point, then the quantity of the water supply is kept fixed at an allowable damage depth in line AB, and point C is defined as the EWA point. In hedging rules: $SWA_{t1} = \alpha D_t, DWA_{t1} = \partial D_t + \frac{(\partial + \eta - 1)}{\eta} S_{t+1}^T, EWA_{t1} = S_{t+1}^T + D_t$

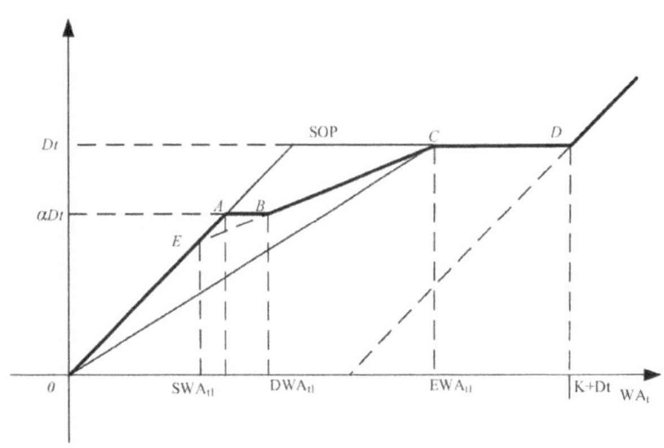

Figure 2: The hedging rules considering an allowable damage depth of Situation I.

Situation II: Point B is defined as the SWA by the traditional model. Point A of ADD is lower than Point B (Figure 3). Point B is chosen as the starting point and point C is chosen as the water-supply hedging ending point. In hedging rules: $SWA_{t2} = D_t(1-\eta)$, $DWA_{t2} = \alpha D_t$, $EWA_{t2} = S^T_{t+1} + D_t$.

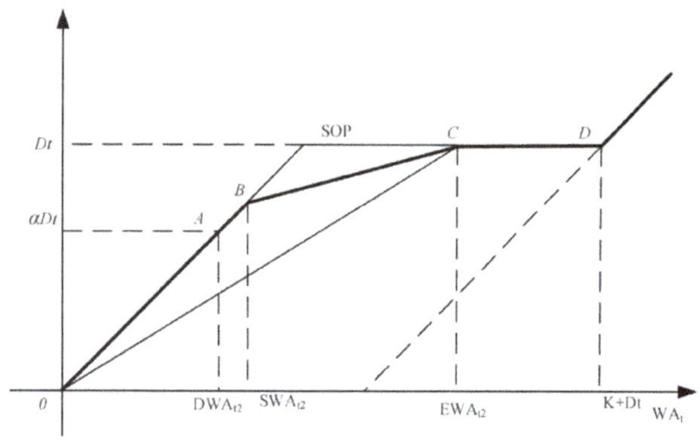

Figure 3: The hedging rules considering allowable damage depth of Situation II.

Situation III: Point A of ADD is higher than the water-supply hedging starting point E, which is defined as the SWA by the traditional model (Figure 4). Point A is chosen as the starting point and point C is chosen as the water-supply hedging ending point, water is supplied under the guide of the SOP rule in line OA and the quantity of water supply is kept fixed at an allowable damage depth in line AB. In hedging rules: $SWA_{t3} = \alpha D_t$, $DWA_{t3} = \partial D_t + \frac{(\partial+\eta-1)}{\eta} S^T_{t+1}$, $EWA_{t3} = S^T_{t+1} + D_t$

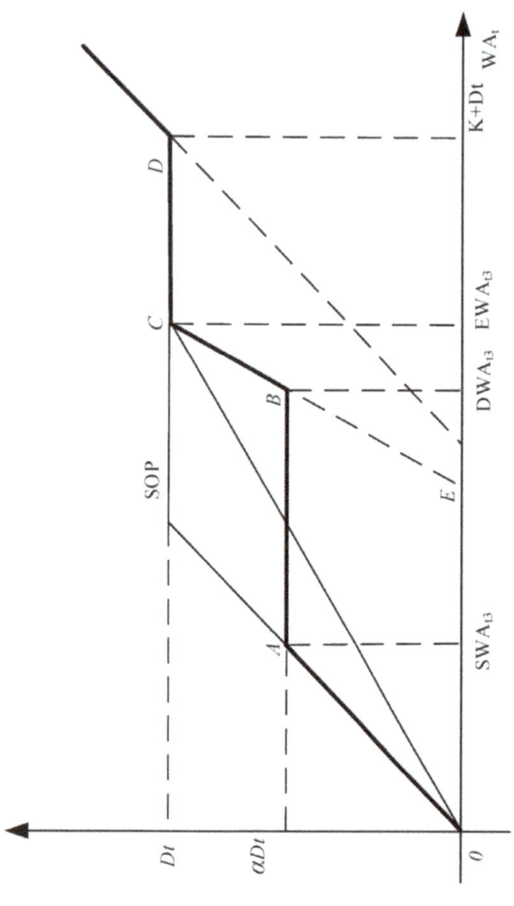

Figure 4: The hedging rules considering allowable damage depth of Situation III.

Parameter Simulation and Optimization Model in Hedging Rules

Water supply decision in a certain period can change the reservoir storage condition at the beginning of the next period, which affects the water supply decision in the next period and eventually influences the overall benefit of the reservoir in the long-series release. In order to ensure maximum long-term benefits, weighting factors and storage targets as the parameters need to be optimized by the simulation and optimization models. The structure of the models include state variable, decision variable, bound condition, and object function. The state variable is the reservoir storage and the decision variable is the reservoir release. The specific steps are as follows: (1) a set of parameters is randomly selected in the feasible region, the water supply process of the long-series is simulated through the two-period model, and the value of objective function is obtained; (2) according to the value of objective function, the parameters are adjusted by the optimization algorithm; (3) the water supply process of the long-series is simulated again after the readjustment of the parameters; and (4) repeat from step (1) to (3) again until a satisfactory value of the objective function is obtained. In this paper, one regulation cycle is separated into 12 regulation periods; the chosen parameters are the water storage target and the water supply weighting factor for each period w, which are 24 in total.

Objective Function

In order to allocate limited water resources between different regulation periods and avoid severe damage during some periods, the minimum shortage index (SI) is chosen as the objective function:

$$\min f = \sum_{i=1}^{m}\sum_{j=1}^{n}\left(\frac{D_j - R_{i,j}}{D_j}\right)^2 \tag{12}$$

System Constraint

(1) Water balance constraint

$$S_{t+1} = S_t + I_t - E_t - R_t \tag{13}$$

(2) Water supply constraint

$$0_t \leqslant R_t \leqslant D_t \tag{14}$$

(3) Reservoir capacity constraint

$$S_d \leqslant S_{t+1} \leqslant S_u \tag{15}$$

where, S_d is the dead storage capacity and S_u is the storage capacity of the reservoir.

Model Calculation

Parameter simulation and optimization models of hedging rules firstly determine a group of decision variables by use of the particle swarm optimization algorithm (PSO). Hedging rules can guide reservoir operation by this group of decision variables. The process of water supply is simulated in each period. The final statistical index uses the SI as the fitness value, and decision variables are readjusted by the PSO. Parameter simulations and the optimization model repeat the above steps again until the objective satisfies the requirements. The process is shown in Figure 5.

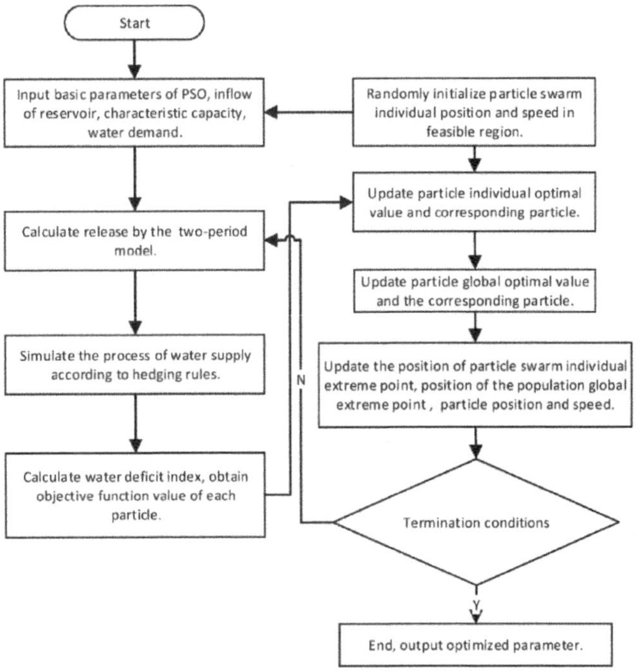

Figure 5: The flowchart of the solving simulation and optimization model for hedging rules using PSO.

CASE STUDY

The developed model is applied to the Heiquan reservoir. The Heiquan reservoir is the most important water resource for Xining City, which is located in the Qinghai Province of China. The Heiquan reservoir was built in 1996 with a normal storage capacity of 124 million m^3 and a dead storage of 17 million m^3. Water demand has shown an upward trend along with the rapid development of the economy, which leads to the severe shortages of the Heiquan reservoir. To evaluate the HR rule, 55 years of monthly inflow data of the Heiquan reservoir is used from 1956 to 2010, resulting in a total of 660 months. The input data include reservoir properties, evaporation, and seepage of the reservoir. In order to verify the feasibility of HR, which is composed of the two-period model and the parameter simulation optimization model, three kinds of comparable operation models are established, including standard operation policy (SOP), operation chart, and dynamic programming. The decision interval of solution process of the different operation model is one month in this study. The water shortage index (SI) and water supply damage recovery index (RI) are chosen as indexes, in order to evaluate scheduling results of the operation schemes. RI refers to average frequency of recovering from a water-deficiency condition during the operation period. RI is usually measured by the maximum number of successive water-deficiency periods in which the water supply degree is less than ADD for a risk analysis. The number of successive water-deficiency periods is fewer, the recovery ability is stronger. Conversely, the recovery ability is weaker.

Reasonable Analysis of Optimized Parameter

Impact Analysis of Parameters of HR

(1) Impact analysis of ω on HR

Equation (10) shows that the optimal reservoir release at time $t\,(R_t^*)$ is a function of η_t, when the carryover storage target at time $t\,(S_{t+1}^T)$ is fixed. The derivative of the Equation (10) is calculated with respect to η_t, the result of derivation is as follows:

$$(R_t^*)' = \frac{S_{t+1}^T \times D_t}{\left(S_{t+1}^T + D_t \times \eta\right)^2}\left(WA_t - S_{t+1}^T - D_t\right) \tag{16}$$

During the hedging period, WA cannot completely meet the benefit of storage and release, namely $WA_t - S_{t+1}^T - D_t \leqslant 0$. According to the Equation (16), it shows that function R_t^* about η_t is monotonically decreasing. While the result that η_t is monotonically decreasing with ω in the range of 0 to 1 can be determined from Equation (9). As a result, R_t^* is monotonically increasing with ω during the hedging period, which indicates that ω is an important parameter for measuring current water supply.

Table 1 shows that ω does not have impact on EWA and SWA of HR in situation I and III, only influences that in situation II. The SWA of situation II is WAt=Dt(1−ηt), so WAt decreases with η_t during the hedging period. However, η_t decreases with ω in the range of 0 to 1. So, WAt is positively associated with ω, namely, WAt increases with ω during the hedging period.

(2) Impact analysis of storage target on HR

Equation (8) shows that the optimal reservoir release at time $t\,(R_t^*)$ is a function of the carryover storage target at time $t\,(S_{t+1}^T)$, when ω is fixed. The derivative of the Equation (8) is calculated with respect to S_{t+1}^T, the result of derivation is as follows:

$$(R_t^*)' = \frac{\left(\frac{1-\omega}{\omega}\right) \times D_t}{\left(S_{t+1}^T + D_t \times \frac{1-\omega}{\omega}\right)^2}\left(D_t - \left(\frac{1-\omega}{\omega}\right) \times D_t - WA_t\right) \tag{17}$$

Equation (17) shows that R_t^* decreases with the increase of S_{t+1}^T when $WA_t \geqslant D_t\left(\frac{2\omega-1}{\omega}\right)$ during the hedging period. R_t^* increases with the increase of S_{t+1}^T when $WA_t \leqslant D_t\left(\frac{2\omega-1}{\omega}\right)$ during the hedging period. When ω≤0.5, R∗tmonotonically decreases with the increase of S_{t+1}^T, WA_t has no effect on R_t^*.

The carryover storage target at time $t\,(S_{t+1}^T)$ mainly influences the EWA of the HR. The larger S_{t+1}^T is, the larger the hedging interval will be; otherwise, the hedging interval will be smaller. Table 1 shows that S_{t+1}^T does not have impact on SWA of HR.

Rationality Analysis of Results of Optimization Parameters

The optimal values of ω and η_t are shown in Figure 6. It shows that ω is negatively correlated with η_t. Both ω and η_t show large fluctuations in scheduling period. Figure 7 shows the change of the storage target in scheduling period. The upper bound and the lower bound represent operation lines of the operation chart. The upper bound and the lower bound are selected as the reference values for the rationality analysis of the storage target. In order to make a concrete analysis, the operation process is divided into three periods.

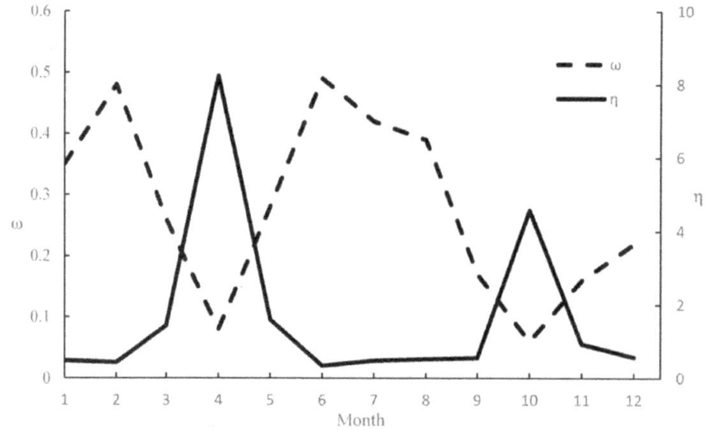

Figure 6: Optimization result of the weighting factor(ω) and IWTR(η).

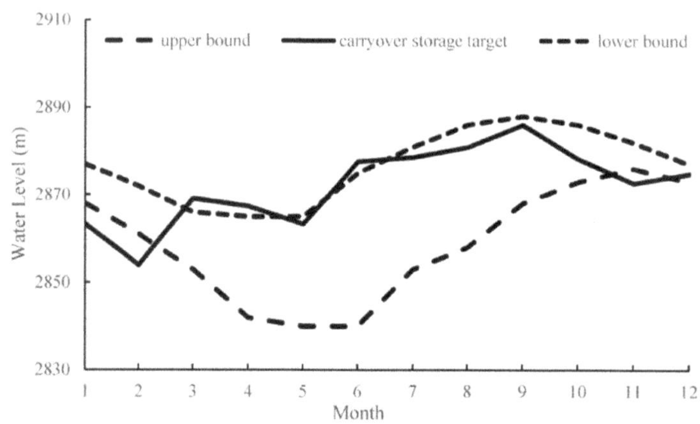

Figure 7: Optimization result of the carryover storage target.

While the value of ω has a gradual increasing trend, the storage target changes smoothly from Novermber to Feburary in the following year. In this period, ω is smaller than 0.5, η_t is smaller than 1, and the storage target is mostly under the lower bound. It shows that the satisfaction degree of the release target is larger than that of the storage target in this period. In this period both the reservoir inflow and water demand is less, because this period is in the dry season, and there is no water supply for agriculture. Thus, water stored in the flood season can be used to avoid water supply damage in the beginning of this period. Consequently, the carryover storage target should be kept at a lower level to reduce the hedging water supply interval. With the decrease of stored water in the flood season, it is necessary to increase the water supply by increasing the value of ω during the hedging period.

There has been a high level of water demand from March to August; this period is called the constant water supply period (CWSP). The reservoir inflow also increases during the CWSP. The storage target decreases gradually, but it is higher than the upper bound in the first half of the CWSP (from March to May). The value of ω is smaller than that in the last period and η_t is larger than 1. Although reservoir inflow is increasing, water demand is higher than reservoir inflow in this period. Constant water supply can lead to the decrease of reservoir storage. Therefore, in order to avoid water supply damage, it is necessary to reduce the hedging water supply interval by decreasing the storage target. However, the satisfaction degree of release target is smaller than that of the storage target by decreasing the value of ω in this period, so that water is stored to avoid severe damage in the next period in advance.

The storage target increases gradually in the second half of the CWSP (from June to August). The value of ω is larger than that in the first half of the CWSP and η_t is smaller than 1. The flood season is the period from June to August, and the reservoir inflow is higher than water demand in this period. The increase of storage target is beneficial for water supply in the next period. Water demand needs to be satisfied as much as possible in this period. Increasing the value of ω can make the satisfaction degree of the release target higher than that of the storage target.

September and October are two special months. In September, inflow is higher, and water demand is lower. On the contrary, inflow is lower, and water demand is higher in October. The storage target is highest in September, η_t is smaller than 1. This allows for water to be stored in September, and the satisfaction degree of the release target is larger than that of the storage target. Oppositely, the storage target decreases in October, but still remains at a higher level, and η_t is larger than 1. Water is stored to avoid severe damage at the beginning of dry season.

Rationality Analysis of Hedging Rules

Table 2 and Table 3 show the statistical results of water supply index of the four operation schemes. The statistical results show that the index result of dynamic pragramming is the most optimal. Dynamic pragramming is one optimum model; it divides the reservoir storage process into several parts, utilizes a step-by-step inducing principle to make decisions on every part, and then gets the optimistic operation performance of the total problem. The inflow condition is not taken into account under the SOP rule. The operation performance under the SOP is the worst with SI of 0.81 and RI of 7. The SI of the SOP scheme is the largest among the four schemes, which shows that the average degree of water supply is lower than that of other schemes. Meanwhile, there are 101 periods in which the water supply degree is less than ADD, accounting for 15.3% of the total periods. The proportion of the successive water-deficiency periods (≥2ΔT) is the largest during these periods, accounting for 85.1% of the water-deficiency periods. This water supply policy is not suitable for the water supply in the dry season.

Table 2: Water deficit index of different schemes.

	SOP	Operation Graph	Hedging Rules	Dynamic Programming
SI	0.82	0.74	0.49	0.31

Table 3: Water supply damage recovery ability of different schemes.

Operation Schemes	Length Distribution of Water-Deficiency Periods							RI	Water-Deficiency Periods
	7ΔT	6ΔT	5ΔT	4ΔT	3ΔT	2ΔT	1ΔT		
SOP	1	1	2	7	3	13	15	7	101
Operation graph	0	1	1	6	7	11	10	6	88
Hedging rules	0	1	0	0	3	5	13	6	38
Dynamic programming	0	0	1	0	2	5	9	5	30

The operation performance under the operation chart is better than that under SOP with SI of 0.74 and RI of 6. The operation curves is obtained on the basis of analysis of the long-series inflow in the operation chart scheme. Consequently, operation curves can guide the reservoir to take measures to decrease water supply before the low water period. The water can be stored in advance to avoid water-deficiency, it proves that the SWA of the operation chart scheme is more reasonable. However, due to the operation feature of the operation chart scheme, the hedging coefficient of this model is fixed, and the hedging release is not optimal. There are 88 periods in which the water supply degree is less than ADD, accounting for 13.3%. That is a significant difference between the operation performance under the operation chart and that under dynamic prigramming.

The minimum SI is chosen as the objective function in the parameter simulation and optimization model, so the value of SI under HR is close to the value of SI under DP. The main advantage of the HR scheme is that the number of water-deficiency periods under HR is very close to that under DP. The results show that HR scheme is optimal except the DP scheme.

To analyze the reason for water deficiency and the operating features of the different operation schemes in the extreme dry season, this study chose two successive water-deficiency periods in the SOP scheme (October 1991–April 1992, January 1978–June 1978) for research.

The longest successive water-deficiency periods occurred in Period 430–436 (October 1991–April 1992). The changing process of reservoir storage in period 421–444 (January 1991–December 1992) is shown in Figure 8. Water is impounded in advance in Period 425–429 under the scheduling chart and HR before the successive water-deficiency period. The inflow level is far below the average, the total inflow of this period is 39.42 million m^3, and the average inflow is 80.88 million m^3. Water stored in advance fails to meet the water demand in this successive water-deficiency periods.

A successive water-deficiency period from January 1978 to June 1978 lasted six months under the SOP rule, lasted one month under scheduling chart, and zero months under the hedging rule. As a result, selecting the period from January 1977 to December 1978 (Period 253–276) for research, the changing process of reservoir storage during this period is shown in Figure 9. Before the successive water-deficiency period, water is impounded in advance in Period 256–262 under the scheduling chart and HR. The reservoir stores water only at the end of flood season (Period 260–262) under the SOP rule. Thus, the storage water under the operation chart and HR is more than that under SOP at the beginning of the successive water-deficiency period (Period 263). In addition, the water supply hedging coefficient is fixed under the scheduling chart, so water supply cannot be adjusted according to the dry season. Water stored in advance cannot satisfy water supply requirement in the successive water-deficiency period, leading to deep damage. HR can guide the reservoir to carry out different levels of hedging the water supply at different periods by optimal reservoir release and avoid the deep damage better.

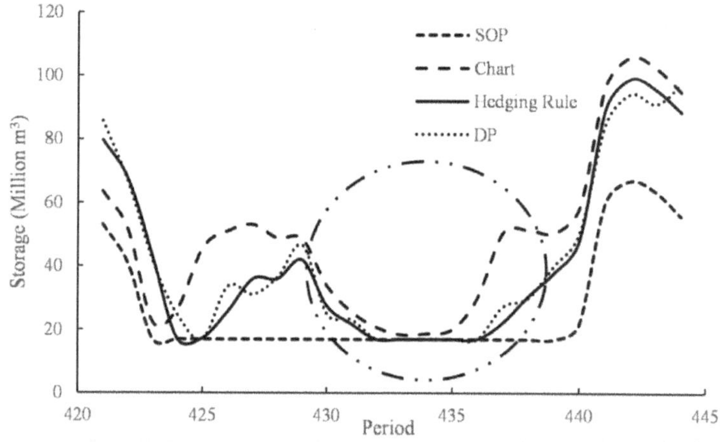

Figure 8: Scheduling results for different scheduling scheme in Period 421–444.

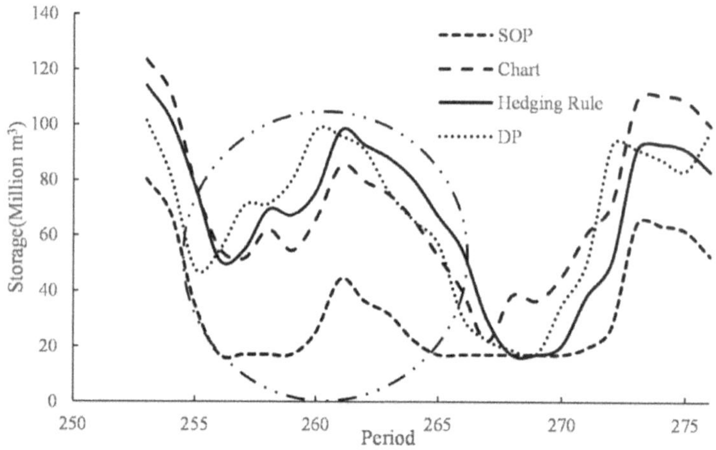

Figure 9: Scheduling results for different scheduling scheme in Period 253–276.

CONCLUSIONS

It is essential to develop the reasonable HR model for water supply reservoirs. This study proposed the hedging rule model which is composed of the two-period model and the parameter simulation and optimization model. The HR model contains two types of undetermined parameters. The undetermined parameters are optimized by the parameter simulation and optimization model. In order to verify the effectiveness of the HR model, the Heiquan reservoir is employed as a case study. The kinds of comparable operation models are developed, including standard operation policy (SOP), operation chart, and dynamic programming. Through comparing the operation performance of the models, the results are summarized below:

(1) The values of the optimal parameters have a larger fluctuation in a scheduling cycle, including the weighting factor and the storage target. The optimal parameters are proved reasonable by analyzing the inflow and water demand and can be used to guide the water supply; and

(2) The result shows that the operation performance of HR is close to that of DP and is better than that of the operation chart and SOP. The operation curves are obtained on the basis of analysis of the long-series inflow under the operation chart, so the result of the operation chart is superior to that of SOP. Due to the fixed hedging coefficient under the operation chart, the hedging release is non-optimized. The SWA, EWA, and hedging release are optimized in HR schemes, so the operation performances of HR is close to the optimal schemes

ACKNOWLEDGMENTS

This paper was jointly supported by National Key Technology R&D Program of the Ministry of Science and Technology (2013BAB05B05), National 973 project (2013CB036406).

AUTHOR CONTRIBUTIONS

Yi Ji designed the experiments and wrote the manuscript; Xiaohui Lei provided suggestions on the data analysis and manuscript preparation, Siyu Cai and Xu Wang revised the manuscript.

CONFLICTS OF INTEREST

The authors declare no conflict of interest.

REFERENCES

1. Zeng, X.; Hu, T.; Guo, X.; Li, X. Water transfer triggering mechanism for multi-reservoir operation in inter-basin water transfer-supply project. *Water Resour. Manag.*

2014, *28*, 1293–1308.

2. Peng, Y.; Chu, J.; Peng, A.; Zhou, H. Optimization operation model coupled with improving water-transfer rules and hedging rules for inter-basin water transfer-supply systems. *Water Resour. Manag.* 2015, *29*, 3787–3806.
3. Birhanu, K.; Alamirew, T.; Dinka, M.O.; Ayalew, S.; Aklog, D. Optimizing reservoir operation policy using chance constraint nonlinear programming for Koga irrigation Dam, Ethiopia. *Water Resour. Manag.* 2014, *28*, 4957–4970.
4. Jothiprakash, V.; Shanthi, G. Single reservoir operating policies using genetic algorithm. *Water Resour. Manag.*2006, *20*, 917–929.
5. Maass, A.; Hufschmidt, M.M.; Dorfman, R.; Thomas, H.A., Jr.; Marglin, S.A.; Fair, G.M. Design of water resource system. *Soil Sci.* 1962, *94*, 135.
6. Bower, B.T.; Hufschmidt, M.M.; Reedy, W.W. Operating procedures: Their role in the design of water-resource systems by simulation analyses. In *Design of Water Resource System*; Maass, A., Hufschmidt, M.M., Dorfman, R., Thomas, H.A., Jr., Marglin, S.A., Fair, G.M., Eds.; Harvard University Press: Cambridge, MA, USA, 1962; pp. 443–458.
7. Clark, E.J. Impounding reservoirs. *J. Am. Water Work. Assoc.* 1956, *48*, 349–354.
8. Revelle, C.; Joeres, E.; Kirby, W. The linear decision rule in reservoir management and design: 1, Development of the stochastic model. *Water Resour. Res.* 1969, *5*, 767–777.
9. Tu, M.-Y.; Hsu, N.-S.; Tsai, F.T.-C.; Yeh, W.W.-G. Optimization of hedging rules for reservoir operations. *J. Water Resour. Plan. Manag.* 2008, *134*, 3–13.
10. Draper, A.J.; Lund, J.R. Optimal hedging and carryover storage value. *J. Water Resour. Plan. Manag.* 2004, *130*, 83–87.
11. Tu, M.-Y.; Hsu, N.-S.; Yeh, W.W.-G. Optimization of reservoir management and operation with hedging rules. *J. Water Resour. Plan. Manag.* 2003, *129*, 86–97.
12. You, J.Y.; Cai, X. Hedging rule for reservoir operations: 1. A theoretical analysis. *Water Resour. Res.* 2008, *44*, W01415.
13. You, J.Y.; Cai, X. Hedging rule for reservoir operations: 2. A numerical model. *Water Resour. Res.* 2008, *44*, W01416.
14. Celeste, A.B.; Billib, M. Evaluation of stochastic reservoir operation optimization models. *Adv. Water Resour.*2009, *32*, 1429–1443.
15. Shiau, J.T. Analytical optimal hedging with explicit incorporation of reservoir release and carryover storage targets. *Water Resour. Res.* 2011, *47*.
16. Bayazit, M.; Ünal, N. Effects of hedging on reservoir performance. *Water Resour. Res.* 1990, *26*, 713–719.
17. Srinivasan, K.; Philipose, M. Effect of hedging on over-year reservoir performance. *Water Resour. Manag.* 1998,*12*, 95–120.
18. Shiau, J.; Lee, H. Derivation of optimal hedging rules for a water-supply reservoir through compromise programming. *Water Resour. Manag.* 2005, *19*, 111–132.
19. Zhao, T.; Zhao, J.; Lund, J.R.; Yang, D. Optimal hedging rules for reservoir flood operation from forecast uncertainties. *J. Water Resour. Plan. Manag.* 2014, *140*, 04014041.
20. Zhao, T.; Zhao, J. Optimizing operation of water supply reservoir: The role of constraints. *Math. Probl. Eng.*2014, *2014*, 853186.

CHAPTER 6

FUZZY DECISION SUPPORT SYSTEM TO DETERMINE SWELL/SHRINK FACTOR AFFECTING EARTHWORK OPTIMIZATION OF HIGHWAYS

A. Burak Göktepe, A. Hilmi Lav, Selim Altun and Gökhan Altıntaş

Keywords: Highway, swelling, shrinkage, fuzzy logic, earthwork optimization

INTRODUCTION

Highway alignment, in general, is a three-dimensional optimization problem requiring both vertical and horizontal considerations of possible design alternatives. This complex optimization problem, which is subject to economical, topographical, environmental, sociological, and geometrical constraints, can be solved by an independent direct searching technique [1-4]. However, numerous researchers reduced highway alignment problem to two-dimensions–either horizontal or vertical, and focused on mathematical methods to obtain optimum solutions. Several horizontal [5-7] and many vertical [8-12] alignment models were used in previous studies. Various optimization techniques were employed within this context, i.e. linear, integer, quadratic, and dynamic programming; gradient and random search techniques; state parameterization; and genetic algorithm (GA).

A three-dimensional highway alignment approach was first introduced by Turner and Miles [13]. Their model was based on searching for the shortest path on a calculated grid network characterized by relative costs. Later, Parker [14] developed a two-staged gradient-based methodology for selecting an optimal route candidate from a grid network. Several researchers also employed dynamic programming technique to solve the optimal three-dimensional alignment problem [15]. Chew et al. [16] applied a fundamentally different approach by combining the cubic spline interpolation and state parameterization to find a three-dimensional optimal route candidate. Recently, genetic algorithm (GA) has emerged as a powerful tool for simultaneous optimization of threedimensional highway alignments [2-4, 17]. Several drawbacks existing in previous models, such as the lack of some design constraints, non-smooth alignments, and incapability of backward route bending, are overcome by the GA based highway alignment approach [2].

Goktepe and Lav [18] recently suggested a hypothetical weighted ground elevation concept to balance cut-fill volumes and to minimize total amount of earthwork. In this method, the integration of weighted ground elevations along the centerline defines a hypothetical reference ground line to determine optimum grades for both hand and computer calculations. Later, this method was modified to consider some soil properties essential for an accurate earthwork optimization [19].

The determination of soil properties affecting earthwork optimization, such as the swelling/shrinkage factor, is highly ambiguous due to the complex behavior of soils. It may be necessary, with the help of previous experience, to utilize several additional linguistic expressions for the definition of related soil properties. Therefore, it is necessary to design a decision support system for the determination of soil properties affecting earthwork optimization under uncertain conditions. The fuzzy set theory provides a powerful tool for modeling uncertainty associated with vagueness, imprecision, and lack of information. If the system is less complex and significant amount of data exists, model-free methods, such as artificial neural networks (ANN) or simulated annealing (SA) appear

Citation: A. Burak Göktepe, A. Hilmi Lav, Selim Altun and Gökhan Altıntaş, Fuzzy Decision Support System to Determine Swell/Shrink Factor Affecting Earthwork Optimization of Highways, doi:10.3390/mca13010061.

to be more appropriate choices. Consequently, fuzzy logic provides an efficient way of handling the uncertainty for complex systems without sufficient data or only with vague information [20, 21].

Considering swelling/shrinkage factors of soils, the modification of alignment models provides greater accuracy, if earthwork optimization techniques are used. It is possible to use several linguistic expressions to define these properties with the help of expert experience. Therefore, a decision support system for the determination of soil properties affecting earthwork optimization under uncertain conditions is necessary. In this study, a fuzzy rule-based decision support system is developed to determine the swelling/shrinkage factor influencing highway earthwork optimization. Successful results encourage the potential use of fuzzy approach for any highway alignment problem.

FUZZY LOGIC AND FUZZY INFERENCE METHODOLOGY

Fuzzy logic is a kind of multi-valued logic utilizing fuzzy sets to perform approximate reasoning. Additionally, a fuzzy rule-based system is a methodology for the interpretation of natural language, which is essential for linguistic expressions. Fuzzy rules and fuzzy reasoning are the fundamentals of fuzzy inference processes that are utilized to derive meaningful conclusions from ambiguous information [21].

In this context, fuzzy inference systems (FIS), also known as fuzzy rule-based systems, are well-known tools for the simulation of nonlinear behaviors with the help of fuzzy logic and linguistic fuzzy rules. There are currently several popular inference techniques developed for fuzzy systems, such as Mamdani [22], Sugeno [23], Tsukamoto [24]. Mamdani FIS was selected to use in this study. In the Mamdani FIS, inputs and outputs are represented by fuzzy relational equations in a canonical rulebased form. These linguistic IF-THEN rules are associated with logical connectives, namely AND, OR, ELSE. For example, in the following expression the conjunctive (AND) is used as connectives in a fuzzy IF-THEN rule:

$$\text{IF } x \text{ is } \mathbf{A}^1 \text{ AND } \mathbf{A}^2 \ldots \text{ AND } \mathbf{A}^N \text{ THEN } y \text{ is } \mathbf{B}^1 \quad (1)$$

$$\mu_{B^1}(x) = min\left[\mu_{A^1}(x), \mu_{A^2}(x), \ldots, \mu_{A^N}(x)\right] \quad (2)$$

Analogously, disjunctive connectives are employed as follows:

$$\text{IF } x \text{ is } \mathbf{A}^1 \text{ OR } \mathbf{A}^2 \ldots \text{ OR } \mathbf{A}^N \text{ THEN } y \text{ is } \mathbf{B}^1 \quad (3)$$

and obtained membership function is given by a maximization procedure as:

$$\mu_{B^1}(x) = max\left[\mu_{A^1}(x), \mu_{A^2}(x), \ldots, \mu_{A^N}(x)\right] \quad (4)$$

Another important point that should be explained about fuzzy rule-based systems is how the aggregation of fuzzy rules is performed. It is necessary to obtain an overall conclusion through a consideration of results from each rule. The combination of entire outcomes in a rule-base is referred as the aggregation of fuzzy rules. Similar to the association of fuzzy variables, there are two cases used in the aggregation process, namely conjunctive and disjunctive systems of rules [20, 21]. A graphical representation of a Mamdani inference system with two rules and two crisp inputs is shown in Figure 1. The Mamdani fuzzy inference process gives a two-dimensional solution area, as can be seen in Figure 1. But it is necessary to obtain a single value instead of a region to reach a decision; therefore, the solution should be defuzzified to get a crisp outcome. There are several methods developed for defuzzification process, such as centeroid, weighted average, and center of sums. According to the centeroid defuzzification method chosen in this research, a single output (x^*) can be calculated as follows:

$$x^* = \frac{\int \mu_A(x) \times x \, dx}{\int \mu_A(x) \, dx} \quad (5)$$

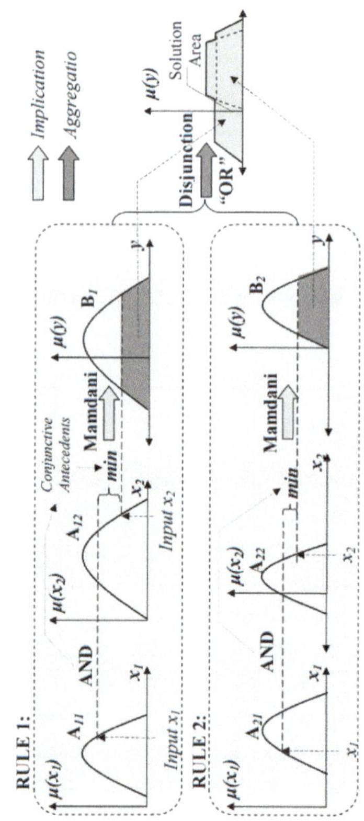

Figure 1: Graphical illustration of Mamdani inference methodology (for two rules and two inputs)

SWELLING AND COMPACTIBILITY OF SOILS

Although earthwork optimization is related with both swelling and compaction behavior of fill material, it is possible to combine these characteristics by a unique swelling/shrinkage ratio that accounts for field densities measured before excavation and after compaction. Compaction is a soil densification process achieved by the application of mechanical energy and improves several engineering properties of soils. Commonly, it is essential to control certain compaction parameters, namely, dry density and water content, with field tests conducted throughout the earthwork construction. It is desirable that fill material has a field unit weight as close as possible to the maximum dry unit weight obtained by the laboratory Proctor test. The measure of the closeness is defined as the relative compaction (C_R), which is required to be higher than a threshold value determined by the project specifications. C_R is calculated as follows:

$$C_R = \frac{\gamma_{df}}{\gamma_{d\max}} \quad (6)$$

where, γ_{df} denotes field dry unit weight and γ_{dmax} is maximum dry unit weight obtained from laboratory Proctor test. It is thereby possible to determine a shrinkage/swelling factor (F_S) using field compaction parameters, i.e. γ_{df}, γ_{dmax}, and C_R. Consequently, for known field and maximum dry unit weights, F_S can be calculated by:

$$F_S = \left(\frac{C_R \times \gamma_{d\max}}{\gamma_{df}} - 1 \right) \quad (7)$$

In order to determine the swelling/shrinkage behavior of a material, field and laboratory tests should be performed to measure field dry unit weight and maximum dry unit weight. Swelling/shrinkage parameters can then be calculated using these test results based on the project compaction criterion and the construction equipment being used. However, soil behavior is inherently ambiguous and the actual compaction control process is usually carried out while earthwork construction is continuing. Therefore, for most of the highway designs, swelling/shrinkage factors are selected from predetermined tables according to specific soil types being considered [25-30].

The swelling/shrinkage behavior of soils can also be characterized based on their particle size classifications (either fine or coarse grained based on the amount passing No. 200 sieve). In this context, gradation (well or poor) determined by the coefficient of curvature (C_c) and coefficient of uniformity (C_u) parameters, can be taken into consideration for coarse grained soils, whereas plasticity index (PI) is the primary distinguishing variable for expressing the swelling/shrinkage behavior of fine grained soils (silts and clays). Natural water content (w_n) is also a significant factor influencing the shrinkage/swelling potential of both fine and coarse grained soils. For fine-grained soils, an increase in PI reduces the swelling/shrinkage potential. At a certain applied energy level, the dry unit weight of a soil reaches to the maximum level for optimum water content. Therefore, the natural water content (either at wet or dry of optimum) should also be considered to characterize swelling/shrinkage behavior [25-29].

FUZZY DECISION SUPPORT MODEL FOR THE SELECTION OF SWELLING/SHRINKAGE FACTOR

In this study, two different procedures using the fuzzy inference methodology were developed to determine the swelling/shrinkage factor (F_S). In the first fuzzy decision support system, swelling/shrinkage factor (F_S) is approximated by field density, maximum dry density, and relative compaction fuzzy variables. In the second procedure, F_S parameter is considered by utilizing the sieve analysis and liquid-plastic limit test results. The aim of using two different procedures with different parameters is to present a choice to the decision-maker in accordance with the data at hand.

It should be noted for the first procedure that the estimation of C_R ratio and soil heterogeneity along roadway play important roles in the dimension of uncertainty. Fuzzy partitionings of input parameters (γ_{df} and C_R) and the output variable (F_S) are obtained. Additionally, Mamdani type inference methodology and centeroid defuzzification technique are preferred for the development of the decision support system. The grid partitioning technique was applied with 288 established fuzzy IFTHEN rules. Detailed information on the fuzzy decision support system 1 (FDSS-1) is given in Table 1. Processing scheme of FDSS-1 is depicted in Figure 2. A representative fuzzy IF-THEN rule is:

$$\text{IF } C_R \text{ is } \mathbf{C_R}^1 \text{ AND } \gamma_{df} \text{ is } \boldsymbol{\gamma_{df}}^1 \text{ AND } \gamma_{dmax} \text{ is } \boldsymbol{\gamma_{dmax}}^1 \text{ THEN } F_S \text{ is } \mathbf{F_S}^9 \quad (8)$$

Table 1: Details of FDSS-1

Fuzzy Decision Support System-1										
Classification	Fuzzy input variable						Fuzzy output		Rule-Base	
	C_R		γ_{df}		γ_{dmax}		F_S			
	Range	Part.	Range	Part.	Range	Part.	Range	Part.	ID	Rule #
All Soils	[0.90,0.96]	6	[1.2,2.6]	8	[1.6,2.6]	6	[-20,60]	9	1	288

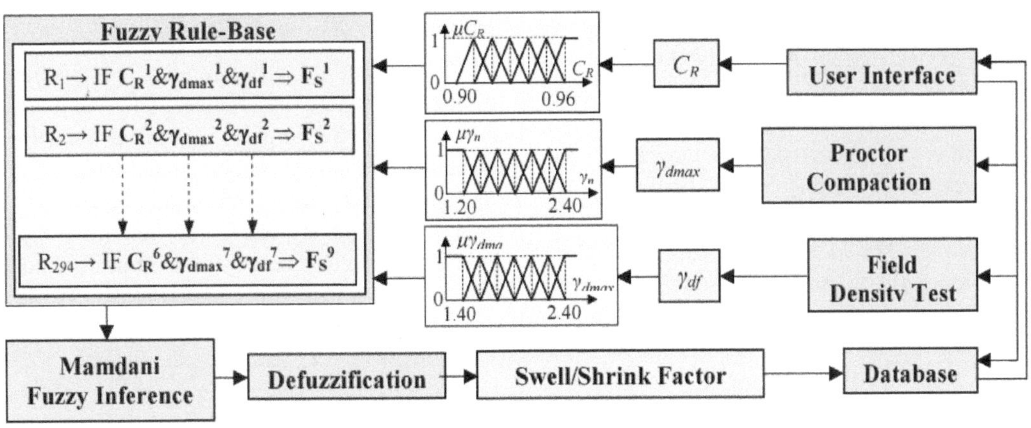

Figure 2: Processing scheme of FDSS-1 (for site-based analysis)

For the second fuzzy decision support system (FDSS-2), which requires laboratorybased information to derive a conclusion, swelling/shrinkage behavior is characterized by soil classification and several index properties, i.e. C_C and C_U that are necessary for coarse-grained soils to characterize C_R. On the other hand, PI is selected as distinguishing design variable for fine-grained soils. Natural water content (w_n) is the last fuzzy variable used for both fine and coarse-grained soils. A sample from the rule of the second procedure for coarse-grained soils is given as:

$$\text{IF } C_u \text{ is } \mathbf{C_u}^3 \text{ AND } C_c \text{ is } \mathbf{C_c}^2 \text{ AND } w \text{ is } \mathbf{w}^4 \text{ THEN } F_S \text{ is } \mathbf{F_S}^9 \quad (9)$$

Partitionings of related input variables, PI and w_n for silts and clays and C_U, C_C, and w_n for sands and gravels, are obtained. Although no direct tests are performed to measure the actual field compaction value, the source of the ambiguity is limited by the possible C_R ranges and index properties influencing the compactibility. Processing scheme of FDSS-2 is given in Figure 3. Analogous to the previous procedure, Mamdani inference methodology and centeroid defuzzification technique were adopted. Contrary to the first fuzzy decision methodology, there are two sub-procedures for fine and coarse-grained soils in FDSS-2, namely FDSS-2.a and FDSS-2.b. Further details on these sub-procedures (FDSS-2.a and FDSS-2.b) are presented in Tables 2 and 3, respectively. As can be derived from the ranges in the tables, the partitionings of fuzzy input variables are different and there are nine fuzzy rule-bases (No.2 to 10) utilized in the second procedure. There are 60 to 168 rules in the rule-bases because of the grid partitioning technique applied for input space characterizations. Namely, there are 168 rules in Clay rule-base which is a part of FDSS-2.a, as well as there are 60 rules in sandy clay and silty sand rule-bases of FDSS-2.b. One of the fuzzy rules used in the second procedure for fine-grained soils is as follows:

$$\text{IF } PI \text{ is } \mathbf{PI}^7 \text{ AND } w \text{ is } \mathbf{w}^4 \text{ THEN } F_S \text{ is } \mathbf{F_S}^9 \quad (9)$$

Table 2: Details of FDSS-2.a

Fuzzy Decision Support System-2.a									
Group	Classification	Fuzzy input				Fuzzy output		Rule-base	
		PI		w		F_S			
		Range	Part.	Range	Part.	Range	Part.	ID	Rule #
Fine grained	Clay	[5,30]	7	[6,25]	6	[-13,-7]	4	2	168
	Silt	[0,5]	3	[12,18]	4	[-20,-10]	6	3	72
	Clayey silt	[4,10]	4	[10,18]	5	[-17,-9]	4	4	80
	Silty clay	[5,12]	4	[8,20]	6	[-15,-8]	4	5	96
	Sandy clay	[5,15]	5	[7,15]	5	[-12,-9]	3	6	75

Table 3: Details of FDSS-2.b

Fuzzy Decision Support System-2.b											
Group	Classification	Fuzzy input						Fuzzy output		Rule-base	
		Cu		Cc		w		F_S			
		Range	Part.	Range	Part.	Range	Part.	Range	Part.	ID	Rule #
Coarse grained	Gravel	[1,6]	4	[0,5]	3	[6,32]	7	[-14,-4]	5	7	84
	Sand	[1,6]	4	[0,5]	3	[6,20]	6	[-13,-7]	4	7	72
	Sandy clay	[1,6]	4	[0,5]	3	[7,17]	5	[-13,-7]	4	8	60
	Gravelly sand	[1,6]	4	[0,5]	3	[6,25]	6	[-13,-5]	4	9	72
	Silty sand	[1,6]	4	[0,5]	3	[10,18]	5	[-15,-9]	4	10	60

In the second procedure, soils are classified according to the Unified Soil Classification system. There are ten different categories considered in this study, i.e. clay, silt, sand, gravel, clayey silt, silty clay, gravelly sand, silty sand, sandy silt, and sandy clay. It should be noted that, organic soil and peat classifications are neglected since they commonly are not used as fill materials. Further information on Unified Soil Classification system and material appropriateness can be found elsewhere [25, 29]. In this study, several sources available in the literature [25-30] and the personal experiences of the authors are used to characterize swelling/shrinkage behavior and to determine possible ranges of all soil types. Comprehensive illustration of flow charts for the second fuzzy decision support procedure is given in Figure 3.

Users need to input relative compaction values based on their personal experiences and valid circumstances as shown in Figure 2 (project information, machinery capacities, specifications, environmental parameters). In addition, field dry unit density and maximum dry density values can be entered into the system after field density and laboratory Proctor compaction tests have been performed. For the second decision support procedure (Figure 3), laboratory sieve analysis and liquid-plastic limit tests should be carried out to obtain PI, w_n, C_C, C_U, and w_n parameters.

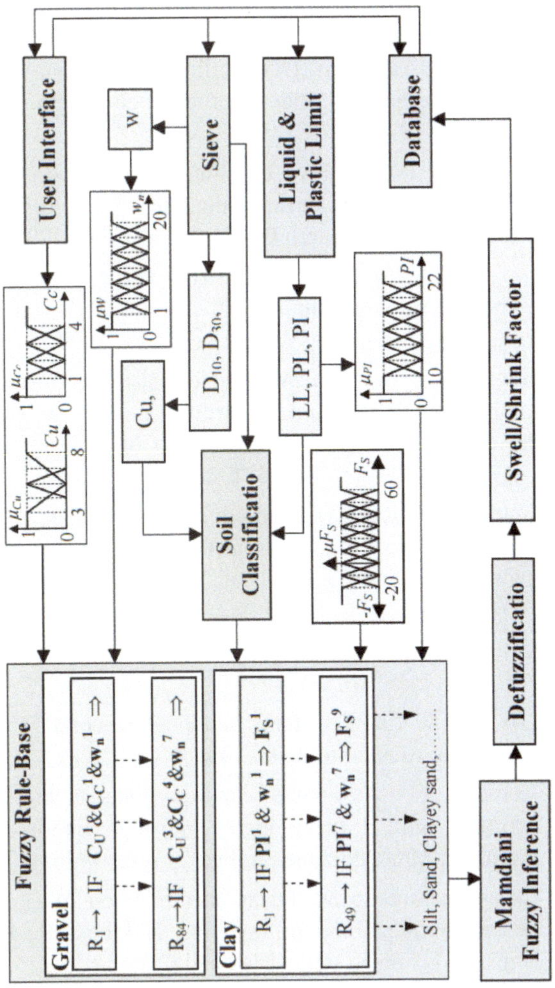

Figure 3: Processing scheme of FDSS-2.a and FDSS-2.b (for laboratory-based analysis)

VALIDATION OF THE METHOD

The following application is for validation and evaluation of the purposed model. In this context, the soil data, consists of 55 experimental data patterns collected from 14 different locations located in several parts of Turkey. The data patterns comprise C_U, C_C, w_n, and F_S information on the collected granular soil samples. Accordingly, FS values were obtained by field density tests conducted in accordance with ASTM D 698, as well as C_U, C_C, and w_n, parameters were calculated by conventional oven and grainsize analyses due to ASTM 2216, ASTM D 422, and ASTM D 3282 procedures [31- 34]. In this regard, data necessary to evaluate the mapping established by the fuzzy decision support system (FDSS 2.b) using the Rule-base #7 for gravels was obtained. The swell/shrink factors are given in road design manuals for each type of soil [25, 28]; however, the ranges of the factors are considerably large in the manuals. Therefore, the final decision is made by experienced design engineers in situ. In other words, exact FS values are determined by project engineers in the classical method.

Scatter plots between (a) project engineers' decisions, (b) FDSS 2.b's inferences and in-situ measurements (actual swell/shrink factors) are given in Figure 6. As can be seen from Figure 6, namely the scatter plots and the coefficients of determination (R^2), the proposed method is superior than the classical method in practice. Because the fuzzy decision support system accounts for the input variables affecting the swell-shrink behaviors and makes non-crisp inference without certain assumptions (such as F^S value should have a value for wet granular soils, and etc). Besides, the existing uncertainty in the problem domain can also be handled by fuzzy reasoning using the partitionings (regions) to define input and output spaces instead of a singleton. For that reason, fuzzy approach makes an inference in a different way, which is appropriate for such a problem involving uncertain information and requiring linguistic (human-like) knowledge. It must be noted that it is possible to estimate a correct F^S value individually by chance and to make in-situ experiments before the construction. However, it is neither rational nor feasible to drill boreholes and to carry out a large amount of experiments in order to estimate the F^S value in the design phase. Additionally, as mentioned before, soil behaviors are highly ambiguous, resulting the variations in FS values; in this regard, it is uncertain how well an experimental value characterizes an entire road sector. Consequently, regarding the heterogeneity and uncertainty existing in soil parameters, proposed fuzzy inference methodology is a powerful way for use with the swelling/shrinkage factor decision process.

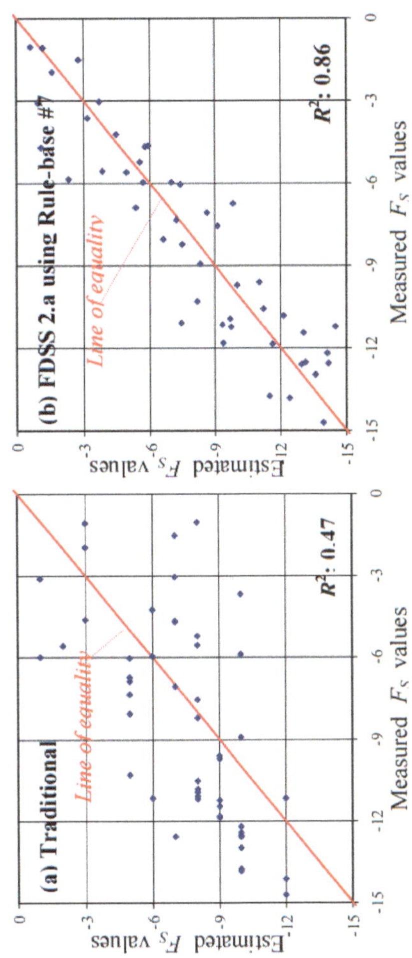

Figure 6: Scatter plots between measured and estimated FS values for (a) traditional and (b) fuzzy methods

CONCLUSIONS

The purpose of this study is to present a method that aides in determining the swelling/shrinkage factor based on soil type, site conditions, and project criterion for an earthwork optimization process of a highway project. This system can also be accompanied with any other problem that may require the consideration of swelling/shrinkage for earthwork optimization. Within the context of this study, fuzzy logic based decision support methodology with high degree of uncertainty is incorporated to the existing estimation method of swelling/shrinkage factor. This enables a designer to consider any ambiguity related to the problem domain. As a result, this approach improves previous highway alignment optimization methods by supporting to the determination of more reliable shrinkage/swelling factors.

This study focused on soils as a fill material to be placed along roadway construction. Nevertheless, it is possible to design similar fuzzy decision support systems for rock materials. Moreover, this approach is not only applicable to highway applications, but also to any other civil engineering designs involving earthwork constructions.

REFERENCES

1. Schoon JG. Geometric design projects for highways: an introduction. New York: ASCE Publication; 2001.
2. Jong J-C. Optimizing highway alignments with genetic algorithms PhD Dissertation, Dept. of Civil And Environmental Engineering 1998
3. Jong J-C and Schonfeld P. An evolutionary model for simultaneously optimizing three-dimensional highway alignments. Transportation Research Part B, 37, 107-128, 2003.
4. Jha MK. Criteria-based support system for selecting highway alignments. ASCE Journal of Transportation. Engineering, 129 (1), 33-41, 2003.
5. Howard BE, Bramnick Z, Shaw JFB. Optimum curvature principle in highway routing. ASCE Journal of Highway Division, 94 (1), 61-82, 1968
6. Shaw JFB, Howard BE. Expressway route optimization by OCP. ASCE Journal of Transportation Engineering, 10 (8), 227-243, 1982
7. Trietsch NR. Comprehensive design of highway networks. Transportation Science, 21(1),26-35, 1987
8. Hayman RW. Optimization of vertical alignment for highways through mathematical programming. Highway Research Record, 306, 1-9, 1970.
9. Easa SM. Selection of roadway grades that minimize earthwork cost using linear programming. Transportation Res. Part A, (22A) 2, 121-136, 1988.
10. Fwa TF. Highway vertical alignment analysis by dynamic programming. Transportation Research Board, 1239, 1-9, 1989.
11. Moreb AM. Linear programming model for finding optimal roadway grades that minimize earthwork cost. European Journal of Operational Research, 93, 148-154,1996.
12. Fwa TF, Cha WT, Sim YP. Optimal Vertical Alignment for Highway Design. ASCE Journal of Transportation Engineering, 128 (5), 395-402, 2002.
13. Turner AK, Miles RD. A computer-assisted method of regional route location. Highway Research Record, 348, 1-15, 1971.
14. Parker NA. Rural highway route corridor selection. Transportation Planing Technology, 3, 247-256, 1977.
15. Nicholson AJ, Elms DG, Williman A. A variational approach to optimal route location. Highway Engineers, 23, 22-25, 1976.
16. Chew EP, Goh CJ, Fwa TF. Simultaneous optimization of horizontal and vertical alignments for highways. Transportation Research Part B, 23B (5), 315-329, 1989.
17. Jong J-C, Schonfeld P. Optimizing vertical alignments with genetic algorithms. 78th Annual TRB Meeting Proceedings, 99-0719, Washington, D.C, 1999.
18. Goktepe AB, Lav AH. "Method for balancing cut-fill and minimizing the amount of earthwork in the geometric design of highways. ASCE Journal of Transportation Engineering, 129(5), 564-571, 2003.
19. Goktepe AB, Lav AH. Method for optimizing earthwork considering soil properties in the geometric design of highways. ASCE Journal of Surveying Engineering 2004, in press.
20. Ross TJ. Fuzzy logic with engineering applications. New York: McGraw Hill Co. 1995.
21. Cox E. The fuzzy systems handbook Second Edition. California: Academic Press, 1999
22. Mamdani EH, Assilian S. An experiment in linguistic synthesis with a fuzzy logic controller. International Journal of Man-Machine Studies, 7, 1-13, 1975.
23. Takagi T, Sugeno M. Fuzzy identification of systems and its applications to modeling and control. IEEE Transactions on Systems Man and Cybernetics, 15,116-132, 1985.
24. Tsukamoto Y. An approach to reasoning method. In: Gupta M, Ragade R.K., Yager R.R., editors. Advances in Fuzzy Set Theory and Applications, Amsterdam, 137-149 1979.
25. Holtz RD, Kovacs WD. An introduction to geotechnical engineering. New Jersey: Prentice Hall, 1981.
26. Rodriguez AR, Castillo H, Sowers GF. Soil Mechanics in Highway Engineering, Germany: Trans Tech Publications, 1988.
27. FLH. Federal Lands Highway Project Development and Design Manual. U.S. Dept. of Transportation, Federal Highway Administration, Pub. No.FHWA-DF-88-003, D.C. 1996.
28. FDOT. Investigation of shrinkage and swell factors for soils used in FDOT construction FDOT, Special Report No. WPI 0510796, Gainesville, Florida, 1999.
29. VDOT RDM. . Road design manual. Vol. 2. Rev. 06/2001. Virginia: Virginia Department of Transportation, Location and Design Division, 2001.
30. NYSDOT HDM. Highway design manual. Rev. 1. New York: New York State Department of Transportation, 2002.
31. ASTM D. Moisture density relations of soils and soil aggregate mixtures (5.5 lb hammer, 12 in drop). West Conshohocken: American Society for Testing & Materials, 698,2000.
32. ASTM D. Standard Test Methods for Laboratory

Determination of Water (Moisture) Content of Soil and Rock by Mass. West Conshohocken: American Society for Testing & Materials, 2216, 2005.

33. ASTM D. Test method for particle-size analysis of soils. West Conshohocken: American Society for Testing & Materials, 422, 1998.

34. ASTM D. Standard practice for classification of soils and soil-aggregate mixtures for highway construction purposes. West Conshohocken: American Society for Testing & Materials, 3282,2004.

CHAPTER 7

LIFE CYCLE ASSESSMENT AND OPTIMIZATION-BASED DECISION ANALYSIS OF CONSTRUCTION WASTE RECYCLING FOR A LEED-CERTIFIED UNIVERSITY BUILDING

Murat Kucukvar , Gokhan Egilmez , and Omer Tatari

Keywords: Multi-criteria decision analysis, economic input-output analysis, life cycle assessment, construction waste management, LEED

INTRODUCTION

Residential and commercial buildings generate a significant amount of construction and debris (C&D) waste in the United States. The estimated total amount of building-related C&D materials is approximately 170 million tons [1]. Based on the U.S. Environmental Protection Agency's (EPA) waste report, 39% of these wastes are residential and 61% are from commercial buildings [1]. Recycling or appropriate treatment of these C&D wastes not only reduces the amount of waste land-filled or incinerated, but additionally minimizes the environmental impacts associated with producing new materials from virgin resources. In this context, one of the barriers for effective policy making towards shifting to a more sustainable C&D waste management is that C&D generation statistics are not rigorously collected [2]. Even though the statistics vary significantly, a recent report indicates that recycling could create credible benefits as a sustainable solution [2]. For instance, according to the same report, in 2012, the estimated magnitude of GHG emissions offset corresponded to taking 4.7 million passenger cars off the road for an entire year. The green building movement has adopted several strategies to reduce C&D waste. Among the green building initiatives, the LEED (Leadership in Energy and Environmental Design) rating system, which was established by the U.S. Green Building Council (USGBC), has gained wide acceptance and has been adopted by several federal and state agencies for evaluating their building designs. LEED green building certification systems employ a simplified checklist that is mainly used in the design process [3]. To obtain LEED certification, a building must first satisfy certain prerequisites and then obtain points for credits related to sustainable sites, water efficiency, energy and atmosphere, materials and resources, indoor environmental quality and design process. In general, LEED can be applied to new constructions, major renovations, existing buildings, commercial interiors, core and shell, schools, retail, healthcare, homes and neighborhood development [4]. LEED has two main construction waste material diversion credits, which are as follows [3]:

Credit 2.1 (one point): recycle and/or salvage at least 50 percent of construction, demolition and land-clearing waste.

Credit 2.2 (one point): recycle and/or salvage an additional 25 percent (75 percent total) of construction, demolition and land-clearing waste.

The credits are proposed to divert construction, renovation and demolition debris from landfill areas and redirect recyclable materials back to the manufacturing process. To accomplish this goal, it is necessary to develop a detailed waste management plan for recycling of various construction waste materials, such as cardboard, metal, brick, mineral fiber

Citation: Murat Kucukvar , Gokhan Egilmez and Omer Tatari, Life Cycle Assessment and Optimization-Based Decision Analysis of Construction Waste Recycling for a LEED-Certified University Building, doi:10.3390/su8010089.

panel, concrete, asphalt, plastic, clean wood, glass, gypsum wallboard, carpet and insulation materials. However, setting a goal, such as "50 percent or 75 percent of construction waste must be recycled" without defining the possible economic and environmental impacts associated with recycling of each C&D waste material may lead to a misrepresented understanding of the comprehensive impact that recycling of that nature may cause. Therefore, recycling goals should be supported by robust decision making models considering environmental and economic impacts, simultaneously [5].

Studying the construction materials from a life cycle point of view is critical for an overall understanding about the sustainability impacts of processes associated with the entire life cycle of buildings. In fact, the literature is abundant with works that focus on the process life cycle of construction materials based on case studies. Several construction materials are analyzed from a life cycle perspective to quantify the environmental impacts; for example, wood and concrete [6], wood and alternative materials [7], generic *vs.* product specific comparisons [8], socio-economic aspects of life cycle impacts [9] and the case of multiple life cycles [10,11,12,13]; for recent comprehensive reviews, see [14].

Novelty and Organization of the Research

Even though a significant amount of work related to LCA-based sustainability assessment of construction materials and buildings has been done, the current literature lacks a comprehensive LCA of the recycling of construction materials that considers both process and supply chain-related impacts as a whole. Additionally, a decision support framework has not been also addressed in any work, which provides a quantifiable understanding about the potential savings and implications associated with recycling of such construction materials from the life cycle perspective. The goal of this research is to develop a comprehensive framework that aids in quantifying and minimizing the environmental impact of C&D waste, while maximizing the economic value added to the solid waste industry. To realize this goal, the following tasks were undertaken: (1) assess the net environmental impacts of C&D waste management strategies using the proposed hybrid LCA model; (2) optimize C&D waste recycling strategies using a multi-criteria optimization model considering all direct and indirect environmental and economic impacts of C&D waste management alternatives; and (3) optimize the sustainable waste recycling strategies by taking LEED requirements into consideration. Hence, this research aims at integrating the solid waste requirements of the LEED green building rating system in order to devise the most environmentally-friendly C&D waste treatment strategies. For a general research framework, please see Figure 1.

The rest of the paper is organized as follows: First, the case study is explained. Then, the proposed methodology that consists of hybrid LCA and optimization models is presented. Next, the LCA results of the certified green building and the results of the optimization model are presented. Finally, the conclusion and the future work are pointed out.

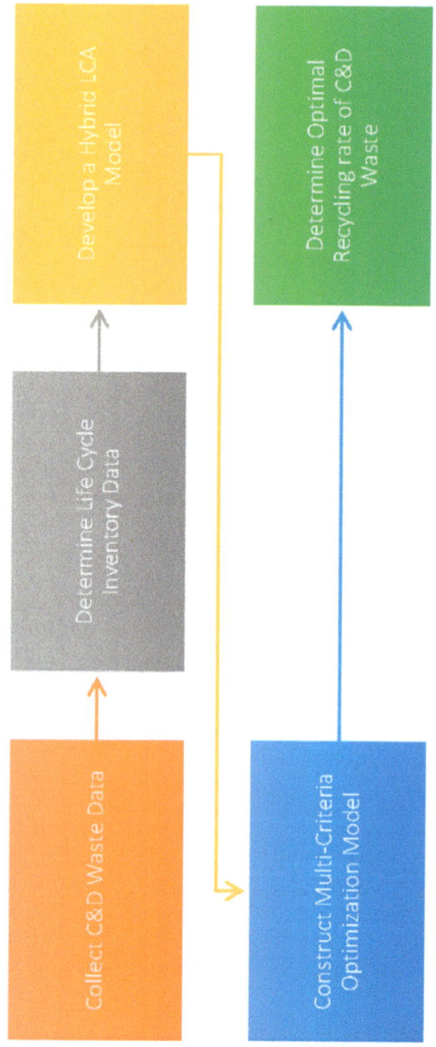

Figure 1: General research framework. C&D, construction and debris.

CASE STUDY

In this paper, the LEED-certified Physical Science Building at the University of Central Florida was chosen as a case study for this research. In this analysis, the amount of C&D waste and its composition data were gathered from the LEED waste documentation, which is publicly available through the University of Central Florida (UCF) Office of Sustainability for the Physical Science Building [15]. The percentage of the waste composition of C&D materials of the building is included, such as asphalt, concrete, wood, non-ferrous and ferrous metals, cardboard, plastic, glass and cardboard. Total waste composition shows that concrete and asphalt have the largest share (over 60% of the total) among the C&D waste materials. The composition of all other building waste materials is presented in Figure 2, which shows the % shares of C&D waste of different materials related to the case study. Although the total composition of these waste materials is important to know, the life cycle impacts related to treatment of these wastes are critical. Hence, the net greenhouse gas (GHG) emissions, energy consumption and water withdrawal associated with recycling, land filling and incineration of C&D materials were quantified using the economic input-output (EIO)-based hybrid LCA methodology. Later, an optimization model is developed to optimize the construction waste recycling strategies.

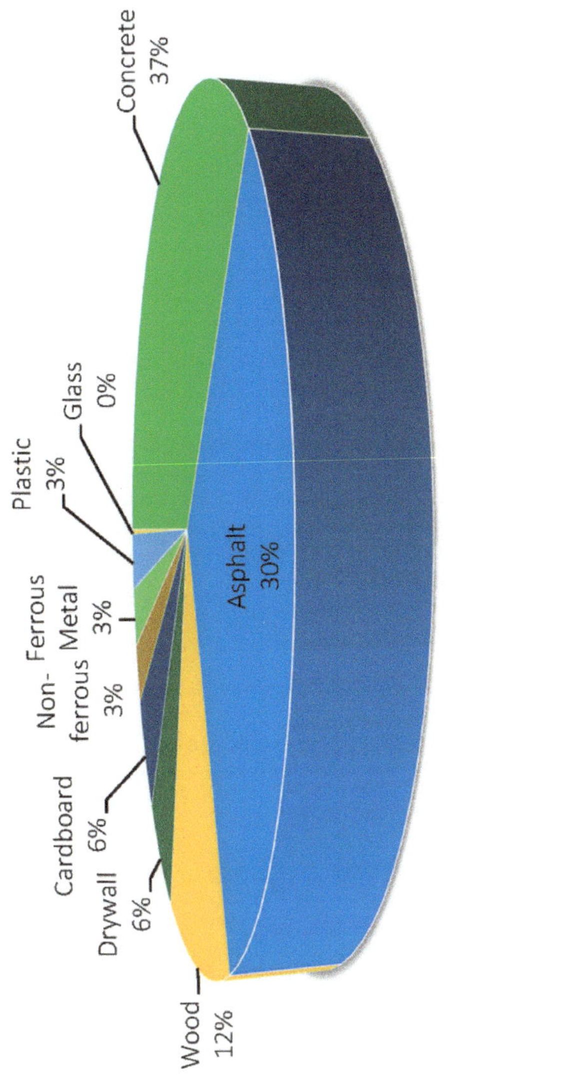

Figure 2: Percentage distribution of C&D wastes for the LEED-certified university building.

METHODOLOGY

Several important life cycle phases, including production of building materials from virgin and recycled resources, transportation, material recovery, incineration and conventional land filling, were analyzed for nine C&D waste materials, namely: asphalt, concrete, wood, non-ferrous and ferrous metals, cardboard, plastic, glass and cardboard. For the recycling, collection and transportation of wastes, material recovery process and producing new products from recycled materials were holistically investigated under the scope of this research. For the incineration, transportation of waste to the incineration facility, processing of waste and energy recovery from combustible waste are analyzed. In addition, the environmental impacts related to a conventional land filling process are quantified for the transportation of C&D to landfill and land filling of each C&D waste material, respectively.

Hybrid LCA Model

The proposed hybrid LCA model consists of process-based LCA (P-LCA) and economic input-output LCA (EIO-LCA). P-LCA is a commonly-used method to analyze the life cycle impacts of solid waste management systems. This LCA approach provides a detailed view of the processes and impacts involved in the management of waste. The EIO-LCA model, augmented with sector-level water use, energy consumption and GHG emissions vectors, has been used in this LCA study. In general, EIO analysis tackles the sector-level interdependencies and represents sectoral direct requirements, which are represented by the matrix A [16,17]. This matrix includes the dollar value of inputs required from other sectors to produce one dollar of output. The total output of a sector in this economic model with a final demand of f can be written as [18,19]:

$$x = (I-A)^{-1} \times f \quad (1)$$

where x is the total output vector, I represents the diagonal identity matrix and f refers to the final demand vector representing the change in the final demand of the desired sector. After the EIO model has been established, the total environmental impacts can be calculated by multiplying the economic output of each industrial sector by the environmental impacts associated with per dollar of output. A vector of environmental outputs can be expressed as [20]:

$$R_i = E_i \times X = E_i \times (I-A)^{-1} \times f \quad (2)$$

where R_i is the total environmental output vector for the environmental impact category of i and E_i represents a diagonal matrix, which consists primarily of the environmental impacts per dollar of output for each industrial sector. In this research, a hybrid EIO-LCA model is built to consider the environmental impacts associated with different waste management scenarios. This LCA model quantifies the total environmental burdens associated with the waste management system, which is presented in the following equation [20]:

$$K_i = E_i \times (I-A)^{-1} \times f + Q_i \times e_i \quad (3)$$

where K_i denotes the total environmental impact defined as the summation of environmental burdens associated with the production of resource inputs (by tracing all supply chains) and the direct environmental impacts related to waste treatment processes. Q_i is the total input requirement for a process, and e_i is the unit environmental impact factor associated with the consumption of Q_i. For example, the production of reinforced steel, which is widely used in residential and commercial buildings, has high GHG emissions. During its production process, electricity is consumed as an energy source for steel manufacturing. In our hybrid LCA model, to quantify the direct and indirect GHG emissions considering the whole supply chain of electricity production, we used Equation (2). In addition, Q_i represents the amount of electricity used in the steel production process, and e_i represents the emission factor related to electricity generation. In this way, Equation (3) presents the total carbon emission related to the indirect supply chains of electricity production ($E_i \times (I-A)^{-1} \times f$) and onsite electricity production processes ($Q_i \times e_i$).

Multi-Criteria Optimization Model

After quantifying the total environmental impacts of different waste management strategies, the next challenge is selecting the best recycling strategy considering environmental and economic benefits, simultaneously. As mentioned earlier, several environmental impact categories, such as energy consumption, GHG emissions and water withdrawals, are quantified using a hybrid LCA model. In addition, the developed LCA model quantified the direct and indirect economic value added associated with the recycling of the analyzed C&D materials. At this point, a multi-criteria optimization model will be critical for finding a feasible alternative that yields the most preferred amount of recycling for each building waste material. Hence, a compromise programming model was developed. This approach is widely used for optimally solving multi-objective linear, non-linear or integer programming problems [21,22,23,24]. The compromise programming model measures the distance based on the L_a metric. The L_a metric defines the distance between two points, such as $Z_k^*(X)$ and $Z_k(X)$. As can been seen from Equation (4), compromise programming uses a distance-based function in order to minimize the difference between ideal and compromise solutions. The formulation of the L_a metric is presented as follows:

$$L_a = \text{Min}\left(\sum \pi_k{}^a (Z_k^*(X) - Z_k(X))^a\right)^{\frac{1}{a}} \tag{4}$$

Due to each objective function having different units, normalization is needed before the optimization analysis is performed. The values after normalization will be confined to a given range, such as zero to one. The normalization function Z is presented in Equation (5):

$$Z = \frac{Z_k^*(X) - Z_k(X)}{Z_k^*(X)} \tag{5}$$

After completing the normalization procedure, the distance-based compromise programming formulation can be written as [25]:

$$\text{MinL}_a = \text{Min}\left(\sum \pi_k{}^a \left(\frac{Z_k^*(X) - Z_k(X)}{Z_k^*(X)}\right)^a\right)^{\frac{1}{a}} \tag{6}$$

Subject to:

$$\sum_{k=1}^{p} \pi_k{}^a = 1 \tag{7}$$

$$1 \leq a \leq \infty \tag{8}$$

In this formulation, Z_k^* represents the ideal solution for objective k. The parameter p represents the total number of objectives, and $\pi_k{}^a$ refers to the corresponding weight associated with each objective. Since we give an equal importance for our economic and environmental objectives, $\pi_k{}^a$ is assumed to be equal for each objective function. In general, three points of the compromise set are calculated for decision analysis, such as a = 1, 2 and ∞. After presenting the theoretical background of the compromise programming, this model has been used for selecting the best recycling strategy. As mentioned earlier, we have the following four primary objectives: maximizing economic value added, maximizing GHG savings, maximizing the net reductions in energy and water consumption. Based on these objectives, the following equations are solved using a multi-objective optimization approach, which is presented as follows:

Notation:

Index:

i: Material index

Parameters:

Ci: economic value added per ton recycled waste for material *i*

GHGi: GHG emission savings per ton recycled waste for material *i*

Wi: water savings per ton recycled waste for material *i*

Ei: energy savings per ton recycled waste for material *i*

Qi: total amount of waste generated by the LEED-certified building for material *i*

LEEDrf: recycling factor

Decision variable:

Xi: optimal amount of recycled waste allocated for material *i*

Objective function:

$$Max\ Z_1(X_i) = \sum_{i=1}^{M} (C_i \times X_i) \tag{9}$$

$$Max\ Z_2(X_i) = \sum_{i=1}^{M} (GHG_i \times X_i) \tag{10}$$

$$Max\ Z_3(X_i) = \sum_{i=1}^{M} (W_i \times X_i) \tag{11}$$

$$Max\ Z_4(X_i) = \sum_{i=1}^{M} (E_i \times X_i) \tag{12}$$

Subject to:

$$\sum_{i=1}^{M} X_i \leq LEED_{rf} \times \sum_{i=1}^{M} Q_i \tag{13}$$

$$X_i \leq Q_i\ for\ i = 1, 2, \ldots, M \tag{14}$$

$$\forall Xi \geq 0 \tag{15}$$

The first objective is to maximize the total economic value added (Equation (10)). The second objective maximizes the GHG emission-based savings (Equation (10)). The water savings are addressed by the third objective, as shown in Equation (11). The energy savings objective function is also represented in Equation (12). The total of the optimal waste for each material *(i)* is less than or equal to the total recycled waste multiplied by the LEED recycling factor ($LEED_{rf}$) (see Equation (13)). Subsequently, the decision variable (X_i) must be less than or equal to the recycled waste (Q_i), as shown in Equation (14). Finally, all decision variables are greater than or equal to zero (see Equation (15)).

Since our goal is to select the best combination of C&D materials for 50% recycling of overall construction waste, the total waste amount multiplied by 0.5, which is known as the $LEED_{rf}$, is used. Due to being one of the most robust optimization software in the applied optimization field, the LINGO© software package is used for solving the multi-objective optimization model [26]. Since we have four objective functions (Z_1, Z_2, Z_3 and Z_4) and three compromise programming functions ($a = 1$, $a = 2$ and $a = \infty$), the LINGO© program has been run to solve each mathematical model. By using the mathematical optimization model, the optimal recycling amount of each construction material has been calculated for the overall 50% recycling goal for single and multiple objectives.

Data Collection

In this paper, several sources have been used to collect the life cycle inventory data for different C&D waste management alternatives. First, the process data for producing each building material from virgin resources and recycled C&D waste are gathered from Christensen [27]. This detailed data included all electricity, fuel and other resource inputs, as well as atmospheric GHG emissions associated with producing building materials from recycled or virgin materials. Additionally, electricity, fuel consumption and GHG emissions data for the material recovery process were compiled from the LCA study of Denison [28]. The waste reduction model (WARM), which was developed by the U.S. EPA, is utilized for quantifying the emissions related to incineration and landfilling of each C&D waste [29]. The energy production efficiency and electricity generation associated with incineration of cardboard, paper, plastic and wood waste are obtained from the Waste Analysis Software Tool for Environmental Decisions (WASTED) model developed by Diaz and Warith [30].

For transportation of C&D materials from the building construction site to both the material recovery facility and final disposal area, a 50-km transportation distance is assumed for each transfer process. Diesel fuel consumption and emission factors data are also provided by the National Renewable Energy Laboratory (NREL) life cycle inventory database for a diesel-powered single-unit truck [31]. After quantifying the life cycle inventory data for each waste management alternative, the producer prices of each energy and material input are obtained, and the Carnegie Mellon's EIO-LCA software was used for calculating direct plus indirect environmental impacts related to C&D waste management [32].

RESULTS

LCA Results

In this part, environmental impacts analysis results are presented in terms of energy, GHG and water savings.

Energy Savings

The amount of fossil fuel consumption has been quantified in terms of terajoules (TJ) for recycling, landfilling and incineration of per ton building-related C&D waste. Results indicated that recycling of non-ferrous metals and plastics resulted in considerable reductions for the energy consumption among the alternative waste management approaches (see Table 1). This is due to the recycling of these materials reducing the production-related fuel and electricity input requirements. Among C&D materials, recycling of non-ferrous metals and plastics showed the highest potential to reduce the total energy footprint.

On the other hand, recycling of some C&D wastes, such as wood, drywall and cardboard, did not have a significant impact on minimizing the net energy consumption compared to other C&D materials. Although concrete and asphalt have the highest percentage contributions to the total waste amount, their recycling did not significantly reduce the net energy footprint due to high energy consumption during the recycling process. Additionally, landfilling and incineration did not have significant impacts on the overall energy footprint compared to the recycling of waste materials. Therefore, it is likely to conclude that a high recycling of metals and plastics, which accounts for almost 90% of the total energy savings of the nine construction materials, will be critical for reducing the net energy footprint of the C&D management systems.

Table 1: Energy savings of construction materials (TJ).

Material	Recycling	% Share	Landfilling	% Share	Incineration	% Share
Cardboard	0.271	3.6%	−0.004	5.6%	0.000	27.2%
Non-ferrous	5.720	75.5%	−0.002	2.8%	N/A	N/A
Ferrous Metal	0.156	2.1%	−0.002	2.8%	N/A	N/A
Concrete	0.241	3.2%	−0.024	37.0%	N/A	N/A
Plastic	0.978	12.9%	−0.002	2.7%	0.000	13.2%
Wood	0.093	1.2%	−0.008	12.3%	0.000	59.6%
Glass	0.010	0.1%	0.000	0.2%	N/A	N/A
Drywall	0.084	1.1%	−0.004	6.1%	N/A	N/A
Asphalt	0.027	0.4%	−0.020	30.3%	N/A	N/A

GHG Emission Savings

In addition to energy consumption, the current study quantified the amount of GHG emissions and savings related to C&D waste recycling. We utilize the definition of carbon footprint as the total emissions of carbon dioxide or GHGs expressed in terms of CO_2 equivalents related to recycling, landfilling and incineration of per ton C&D waste. Based on the analysis results, recycling of ferrous and non-ferrous metals are found to have significant benefits for reducing total GHG emissions with a total percent share of 79.3% (see Table 2). This is because recycling of these metals reduced the amount of electricity and fuel inputs, which are highly utilized for the production of metal products. Additionally, on-site emissions are decreased when using recycled metals instead of virgin resources. Recycling of other C&D materials, such as paper, glass and cardboard, also contribute to reductions in the net GHG emissions.

Table 2: GHG emission savings of construction materials (Mt CO_2-eqv.).

Material	Recycling	% Share	Landfilling	% Share	Incineration	% Share
Cardboard	9.430	7%	−2.714	6%	-26.718	18%
Non-ferrous	38.142	30%	−1.370	3%	N/A	N/A
Ferrous Metal	61.732	49%	−1.370	3%	N/A	N/A
Concrete	−23.800	-	−17.918	37%	N/A	N/A
Plastic	15.733	12%	−1.318	3%	−32.250	22%
Wood	−7.910	-	−5.955	12%	−90.400	61%
Glass	0.436	0%	−0.108	0%	N/A	N/A
Drywall	0.562	0%	−2.963	6%	N/A	N/A
Asphalt	−8.930	-	−14.663	30%	N/A	N/A

On the contrary, recycling of C&D wastes, such as wood, drywall and asphalt, did not have a significant impact on GHG savings when compared to other C&D materials. Moreover, recycling of asphalt and concrete is not found to be an environmentally-friendly option due to increasing GHG emissions. This is because recycling of these materials requires high fuel consumption during crushing and emitted on-site GHG emissions in this process. When incineration was more closely analyzed, wood, plastic and cardboard resulted in additional GHG emissions. For this reason, combustion of these materials is not found to be an environmentally-friendly waste treatment option due to increased GHG emissions. Therefore, recycling has a positive impact on reduced energy consumption, and a high recycling of metal and plastic materials will have a key importance for decreasing the net carbon footprint of the LEED construction management strategies.

Water Savings

Water footprint analysis results are also presented for recycling, landfilling and incineration of per ton C&D waste. Recycling of non-ferrous metals and asphalt is found to be beneficial due to reductions in overall water consumption (see Table 3). This is because recycling of these materials reduced water consumption for the direct and indirect processes required for the production of these materials.

Table 3: Water savings of construction materials (kgal).

Material	Recycling	% Share	Landfilling	% Share	Incineration	% Share
Cardboard	0.201	0.2%	−1.102	5.6%	−0.048	27.3%
Non-ferrous	5.707	5.2%	−0.556	2.8%	N/A	N/A
Ferrous Metal	0.130	0.1%	−0.556	2.8%	N/A	N/A
Concrete	−0.211	-	−7.276	37.0%	N/A	N/A
Plastic	0.945	0.9%	−0.535	2.7%	−0.023	12.9%
Wood	−0.058	-	−2.418	12.3%	−0.106	59.8%
Glass	0.007	0.0%	−0.044	0.2%	N/A	N/A
Drywall	0.009	0.0%	−1.203	6.1%	N/A	N/A
Asphalt	101.988	93.6%	−5.954	30.3%	N/A	N/A

Among these C&D materials, asphalt showed the highest potential (94% of total share) for reducing the overall water footprint. This is due to recycling of asphalt producing a large amount of natural aggregate, which requires a large amount of water during its mining process. On the other hand, recycling of C&D wastes, such as wood, drywall, concrete and cardboard, did not show a significant contribution to water footprint savings. When compared to recycling, landfilling did not help the environmental sustainability, due to increasing water consumption. It is important to note that landfilling of concrete and asphalt showed a higher water footprint value compared to other materials. This is because the waste composition of concrete and asphalt wastes was found to be the highest among C&D materials, and disposal of these materials through landfilling required a higher amount of water. In conclusion, a high recycling of metals, glass, plastic and paper should be encouraged by policy makers to diminish the net water footprint of the building-related construction wastes.

Optimization Results

Figure 3 presents the optimal recycling percentage of construction wastes with respect to each single objective. Mathematical optimization results show that 100% of cardboard, ferrous and non-ferrous metals, plastic and glass wastes should be recycled for maximizing economic and environmental benefits. For GHG gas reductions and water footprint savings, recycling of concrete waste, which accounts for 37% of total waste, is not found to be a feasible option. However, recycling a small portion of concrete makes a positive contribution to net economic savings and energy minimization. In addition, 100% recycling of asphalt waste is found to be a feasible policy when economic and GHG savings are under consideration. For economic savings, 100% recycling of ferrous and non-ferrous metals, asphalt, plastic and glass is suggested, whereas recycling of drywall is not found to be an economically feasible recycling strategy. On the other hand, approximately 100% of produced drywall should be recycled to maximize energy and GHG emissions savings, as well (see Figure 3).

After generating the output of this single objective optimization, a compromise programming model is used to determine the optimal set of recycling amounts when considering the maximization of all objectives, simultaneously. For this model, three points of the compromise set, such as a = 1, 2 and ∞, are calculated for decision analysis, and the percentage recycling rates are presented in Figure 4 for each waste material. Multi-criteria optimization results revealed that recycling of wood and concrete is not found to be a feasible solution when all objectives are aimed to be maximized. However, 100% recycling ferrous and non-ferrous metals, plastic, glass and cardboard will be a sound policy from economic and environmental perspectives. The recycling of a small portion of wood is suggested for a compromise solution in which *a* is infinity. However, this recycling rate is found to be negligible when compared to the recycling rates of other construction materials. According to optimization results, recycling of over 90% of drywall is suggested for the three points in the set. Consequently, by using the results of the LCA model in conjunction with the multi-objective optimization model, the decision makers will have a better understanding of optimum recycling rates of each building waste material (see Figure 4).

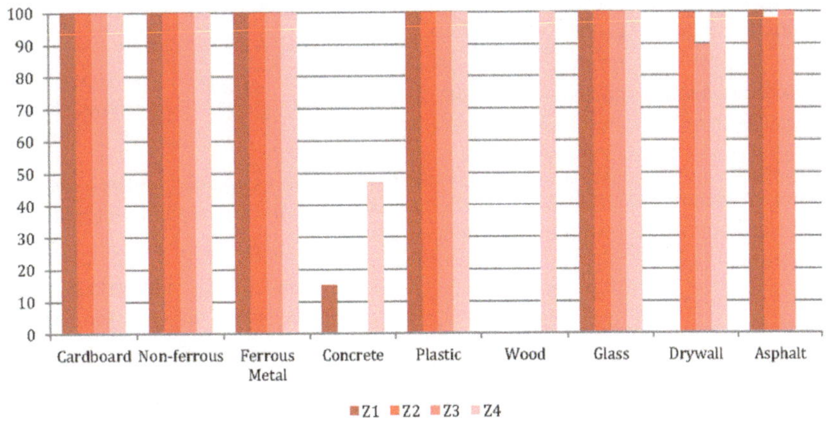

Figure 3: Optimal recycling percentages of C&D wastes for single objectives.

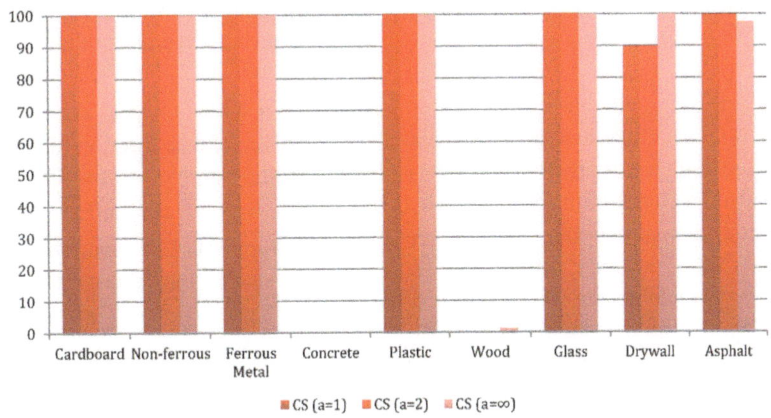

Figure 4: Optimal recycling percentages of C&D wastes for compromise solutions (CS).

CONCLUSIONS, LIMITATIONS AND FUTURE WORK

The overarching goal of this research was to offer a decision making methodology for recycling of building materials, specifically for LEED-certified green buildings. This study is the first attempt to combine the EIO-based LCA model with multi-objective decisions analysis for construction waste management options. First, the EIO-based hybrid LCA model was used for quantifying the environmental impacts of building materials associated with different waste management scenarios. Second, a multi-criteria optimization model was developed to propose sustainable waste management strategies considering both environmental and economic impacts all together. Our analysis shows that recycling of asphalt and concrete increased the net carbon footprints due to high fuel consumption and emissions during the recycling process. Moreover, recycling of these materials has a minimum impact on the net energy footprint reductions. On the contrary, ferrous and non-ferrous metals are critical for reducing the net carbon footprint of waste management systems. In addition to that, 100% recycling of metals, plastic, glass and cardboard will be an economically- and environmentally-sound policy based on multi-objective optimization results. Even though the case study focuses on a LEED-certified university building, the proposed approach can be robustly integrated for other types of buildings as long as comprehensive C&D waste data are readily available.

It is critical to note that this work presents an integrated decision making framework combining optimization and LCA methods. Our results are based on recycling of 50% of total C&D debris for a selected LEED-certified university building in the United States. Based on assumptions made and the collected data, recycling of concrete is not found to be a feasible option when environmental and economic impacts are considered simultaneously. In other words, recycling of other C&D materials, such as ferrous and nonferrous metals, plastic and glass, is found to be a better strategy, and policy makers should give priority to these materials for recycling. On the other hand, our results still showed that recycling of concrete has a positive contribution to the economy and environment in terms of cost and energy savings. Keeping in mind that among different types of C&D wastes, the composition of concrete waste is found to be higher than 80% of the volume of C&D waste in many countries, such as Australia and Japan, to minimize the concrete waste generated from construction activities, recycling of concrete waste is one of the best methods, and many studies propose that recycling concrete as aggregate for new concrete production can provide a cost-effective method for the construction industry and help save the environment [33,34].

It is certain that the net environmental impacts of C&D waste management are not limited to the findings presented in this study. The application domain of the proposed approach can be extended to other types of buildings, which is a horizontal research extension. Additionally, the vertical depth of the methodology can be further extended by adding different weight scenarios for economic *vs.* environmental impact domains for policy making. Ecological impacts due to toxic releases, hazardous waste generation and ground water pollution can also be considered for analyzing the environmental impacts of different waste treatment options. Additionally, a multi-criteria decision making model is used to select the most appropriate building-related waste materials to maximize environmental and economic savings. However, the social impacts of C&D waste recycling are still critical and can be considered for a more comprehensive sustainability analysis. Therefore, we plan to develop a multi-criteria-based decision making model to consider all economic, social and environmental impacts of waste recycling strategies for the future. In this way, it is possible to have an optimized solution for C&D recycling by considering the triple-bottom-line sustainability impacts. In addition, as a future research direction, further analyzing alternative buildings from different universities worldwide could provide a more comprehensive framework that can be used for sensitivity analysis purposes. Consequently, this research provides an important decision making model, which offers vital guidance for policy makers when developing environmentally- and economically-sound waste management policies.

AUTHOR CONTRIBUTIONS

Murat Kucukvar carried out the analyses, including work related to life cycle inventory, data collection and processing. Gokhan Egilmez contributed to the multi-criteria optimization model development and literature review parts. Omer Tatari supervised the research and contributed to the framework of the LCA methodology. All of the authors contributed to preparing and approving the manuscript.

CONFLICTS OF INTEREST

The authors declare no conflict of interest.

REFERENCES

1. USEPA. *Estimating 2003 Building-Related Construction and Demolition Materials Amounts*; USEPA: Washington, DC, USA, 2009.
2. Townsend, T.; Wilson, C.; Beck, B. *The Benefits of Construction and Demolition Materials Recycling in the United States*; University of Florida: Gainesville, FL, USA, 2014.
3. USGBC. *LEED 2009 for New Construction and Major Renovations Rating System*; U.S. Green Building Council: Washington, DC, USA, 2009.
4. Tatari, O.; Kucukvar, M. Cost premium prediction of certified green buildings: A neural network approach. *Build. Environ.* 2010, *46*, 1081–1085.
5. Lave, L.B.; Hendrickson, C.T.; Conway-Schempf, N.M.; McMichael, F.C. Municipal Solid Waste Recycling Issues. *J. Environ. Eng.* 1999, *125*, 1–16.
6. Gustavsson, L.; Sathre, R. Variability in energy and carbon dioxide balances of wood and concrete building materials. *Build. Environ.* 2006, *41*, 940–951.
7. Petersen, A.K.; Solberg, B. Environmental and economic impacts of substitution between wood products and alternative materials: A review of micro-level analyses from Norway and Sweden. *For. Policy Econ.* 2005, *7*, 249–259.
8. Lasvaux, S.; Habert, G.; Peuportier, B.; Chevalier, J. Comparison of generic and product-specific Life Cycle Assessment databases: Application to construction materials used in building LCA studies. *Int. J. Life Cycle Assess.* 2015, *20*, 1473–1490.
9. Dixit, M.K.; Culp, C.H.; Fernandez-Solis, J.L. Embodied energy of construction materials: Integrating human and capital energy into an IO-based hybrid model. *Environ. Sci. Technol.* 2015, *49*, 1936–1945.
10. Pierucci, A. LCA evaluation methodology for multiple life cycles impact assessment of building materials and components. *Tema:Tempo Mater. Arch.* 2015, *1*, 1–6.
11. Cabeza, L.F.; Rincón, L.; Vilariño, V.; Pérez, G.; Castell, A. Life cycle assessment (LCA) and life cycle energy analysis (LCEA) of buildings and the building sector: A review. *Renew. Sustain. Energy Rev.* 2014, *29*, 394–416.
12. Dixit, M.K.; Fernández-Solís, J.L.; Lavy, S.; Culp, C.H. Need for an embodied energy measurement protocol for buildings: A review paper. *Renew. Sustain. Energy Rev.* 2012, *16*, 3730–3743.
13. Van den Heede, P.; de Belie, N. Environmental impact and life cycle assessment (LCA) of traditional and "green" concretes: Literature review and theoretical calculations. *Cem. Concr. Compos.* 2012, *34*, 431–442.
14. Ortiz, O.; Castells, F.; Sonnemann, G. Sustainability in the construction industry: A review of recent developments based on LCA. *Constr. Build. Mater.* 2009, *23*, 28–39.
15. University of Central Florida Sustainability and Energy Management Department. LEED GOLD Physical Sciences Building: Material and Resources Credits. 2012. Available online: http://www.sustainable.ucf.edu/?q=node/108 (accessed on 5 March 2015).

16. Hendrickson, C.; Lave, L.; Matthews, H. *Environmental Life Cycle Assessment of Goods and Services: An Input-Output Approach*; Routledge: London, UK, 2006.
17. Suh, S.; Huppes, G. Methods for Life Cycle Inventory of a product. *J. Clean. Prod.* 2005, *13*, 687–697.
18. Egilmez, G.; Kucukvar, M.; Tatari, O. Sustainability assessment of U.S. manufacturing sectors: An economic input output-based frontier approach. *J. Clean. Prod.* 2013, *53*, 91–102.
19. Egilmez, G.; Kucukvar, M.; Tatari, O. Supply chain sustainability assessment of the U.S. food manufacturing sectors: A life cycle-based frontier approach. *Resour. Conserv. Recycl.* 2014, *82*, 8–20.
20. Kucukvar, M.; Egilmez, G.; Tatari, O. Evaluating environmental impacts of alternative construction waste management approaches using supply-chain-linked life-cycle analysis. *Waste Manag. Res.* 2014, *32*, 500–508.
21. Noori, M.; Kucukvar, M.; Tatari, O. A macro-level decision analysis of wind power as a solution for sustainable energy. *Int. J. Sustain. Energy* 2013, *34*, 629–644.
22. Tatari, O.; Nazzal, M.; Kucukvar, M. Comparative sustainability assessment of warm-mix asphalts: A thermodynamic based hybrid life cycle analysis. *Resour. Conserv. Recycl.* 2012, *58*, 18–24.
23. Kucukvar, M.; Noori, M.; Egilmez, G.; Tatari, O. Stochastic decision modeling for sustainable pavement designs. *Int. J. Life Cycle Assess.* 2014, *19*, 1185–1199.
24. Onat, N.C.; Kucukvar, M.; Tatari, O.; Zheng, Q.P. Combined Application of Multi-Criteria Optimization and Life-Cycle Sustainability Assessment for Optimal Distribution of Alternative Passenger Cars in US. *J. Clean. Prod.* 2015, *30*, 1–17.
25. Chang, N.-B. *Systems Analysis for Sustainable Engineering*; McGraw-Hill: New York, NY, USA, 2011.
26. Lindo Inc. LINDO Systems—Optimization Software: Integer Programming, Linear Programming, Nonlinear Programming, Stochastic Programming, Global Optimization. 2012. Available online: http://www.lindo.com/(accessed on 10 May 2012).
27. Christensen, T.H.; Bhander, G.; Lindvall, H.; Larsen, A.W.; Fruergaard, T.; Damgaard, A.; Manfredi, S. Experience with the use of LCA-modelling (EASEWASTE) in waste management. *Waste Manag. Res.* 2007, *25*, 257–262.
28. Denison, R.A. Environmental life-cycle comparisons of recycling, landfilling, and incineration: a review of recent studies. *Annu. Rev. Energy Environ.* 1996, *21*, 191–237.
29. USEPA. Waste Reduction Model (WARM). 2010. Available online: http://www.epa.gov/climatechange/wycd/waste/calculators/Warm_Form.html (accessed on 3 March 2015).
30. Diaz, R.; Warith, M. Life-cycle assessment of municipal solid wastes: Development of the WASTED model. *Waste Manag.* 2006, *26*, 886–901.
31. NREL. *U.S. Life Cycle Inventory Database: Transport, Train, Diesel Powered*; NREL: Washington, DC, USA, 2010.
32. Carnegie Mellon University Green Design Institute. Economic Input-Output Life Cycle Assessment (EIO-LCA), US 2002 Industry Benchmark Model. 2012. Available online: http://www.eiolca.net/ (accessed on 5 March 2015).
33. Tam, V.W. Economic comparison of concrete recycling: A case study approach. *Resour. Conserv. Recycl.* 2008, *52*, 821–828.
34. Tam, V.W. Comparing the implementation of concrete recycling in the Australian and Japanese construction industries. *J. Clean. Prod.* 2009, *17*, 688–702.

CHAPTER 8

AN OPTIMIZATION SYSTEM FOR CONCRETE LIFE CYCLE COST AND RELATED CO_2 EMISSIONS

Tae Hyoung Kim, Sung Ho Tae, Sung Joon Suk, George Ford and Keun Hyek Yang

Keywords: Concrete, life cycle assessment, CO_2 emission, cost

INTRODUCTION

The Korean government plans to reduce its greenhouse gas emissions by 37% by 2030. However, carbon dioxide (CO_2) emissions per capita in Korea have actually increased by 113% since 1990, the highest for any of the Organization for Economic Co-operation and Development (OECD) countries.

Because the CO_2 emissions created by the Korean construction industry account for 40% of the nation's total, it is essential to reduce the amount of CO_2 generated by construction activities if Korea is to attain its ambitious greenhouse gas reduction goal [1]. Concrete is known to emit particularly high amounts of environmentally damaging waste over its life cycle of production, construction, maintenance and, eventually, demolition. The industry is well aware of this; the amount of CO_2 emissions produced is specified in a ready-mixed concrete (RMC) report that is provided whenever concrete is sold commercially in Korea. Although this has led to many studies addressing the quantitative evaluation and reduction of the environmental effects of concrete, little research has considered the development of optimization systems that would enable concrete production companies to reduce the CO_2 emissions associated with concrete. This study therefore focused on developing an optimization system, dubbed the concrete life cycle assessment system (CLAS), to recommend options to help minimize CO_2 emissions and/or the costs incurred at every stage of the concrete production process.

Citation: Tae Hyoung Kim, Sung Ho Tae, Sung Joon Suk , George Ford and Keun Hyek Yang, An Optimization System for Concrete Life Cycle Cost and Related CO_2 Emissions, doi:10.3390/su8040361.

REVIEW OF LIFE CYCLE ASSESSMENT PROGRAMS

A number of software programs have been developed to help those seeking to perform life cycle assessment (LCA) calculations to assess the environmental impact of different construction materials. The TOTAL (Tool for Type labeling and LCA) program was developed by researchers at the Korean Ministry of the Environment to define how product labels must comply with the data format required by the country's environmental performance declaration system [2].

An additional program, Product Assessment for Sustainable Solutions (PASS), was developed by staff at the Korean Ministry of Knowledge Economy to perform LCA calculations. The energy section of the associated Life Cycle Index database includes fuels such as diesel, gasoline and coal, while the construction-related materials section includes items such as cement, concrete and different types of glass [3]. Another useful tool is the COOL program, which was exclusively designed to help companies create accurate carbon footprint labels and was developed by the Korean Environmental Industry & Technology Institute to assist companies applying for carbon footprint label certification [4].

The effort to identify environmental impacts is international. For example, the Building for Environmental and Economic Sustainability (BEES) program was developed by the National Institute of Standards and Technology in the United States to help designers make informed materials decisions by integrating the environmental impacts and costs for buildings and materials into a single tool. BEES helps architects and engineers select a suitable product for their desired application that balances environmental considerations with economic performance [5]. The ATHENA Impact Estimator was developed by Canadian researchers to facilitate building design by modelling the environmental impact of changing the shape, design, or material makeup of a building, allowing designers to optimize the operating energy effects over the complete building life cycle [6].

In Europe, GaBi software was developed to manage sustainability through LCA and to assess environmental design and energy efficiency. The associated database is managed by the GaBi database manager and the program provides detailed classification of substances and production processes [7]. In the United Kingdom, the building LCA program ENVEST2 was developed by the Building Research Establishment to assess building structures. It not only simplifies the multiple processes of determining life cycle costs and environmental impacts when designing buildings, but also identifies the elements with the greatest influence on a building's environmental impact and life cycle cost, and shows the effects of selecting different materials [8]. Also in the UK, the program LCA in Sustainable Architecture (LISA) was developed by Newcastle University and BHP Research Laboratories with a convenient interface consisting of a simple input–output form to facilitate its use [9]. Also in Europe, the BECOST building LCA program was developed by the VTT Technical Research Centre of Finland for use in developing carbon emission reduction technologies based on the environmental impact data deduced from the life cycle assessment of buildings, including their design, construction, maintenance, and demolition [10]. Elsewhere, Eco-Quantum is Australia's leading life cycle assessment program, providing LCA and greenhouse gas and carbon assessment services based on the life cycle data for various products or services. Eco-Quantum is the first building life cycle assessment program to assess environmental effects based on the energy consumption incurred in building structures [11].

DEVELOPMENT OF AN OPTIMIZATION SYSTEM

Characteristics of the Optimization System

The CLAS optimization system developed for this study is designed to evaluate the CO_2 emissions and costs of concrete from an LCA perspective and suggest methods to reduce both. The system includes a method and database that allows users to evaluate CO_2 emissions and costs relatively simply. The first step in the development of the new system was to establish a system boundary for the life cycle CO_2 emission and cost evaluations for concrete that would be used in the program (Figure 1). The product stage of concrete (the so-called "cradle to gate" for the product, up to the point at which it leaves the manufacturer's premises), based on ISO 14044 [12] and ISO 21930 [13], was selected as a convenient system boundary. The product stages for concrete were divided into three stages: raw material, transportation, and manufacturing.

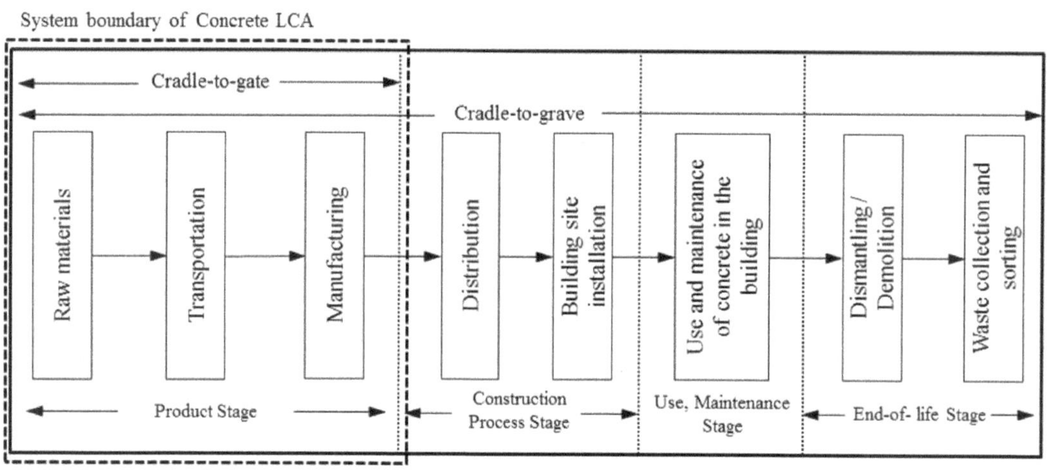

Figure 1: System Boundary for Concrete Life Cycle Assessment (LCA).

	Raw material stage	Transportation stage	Manufacturing stage
Optimization Technique	Evolution Algorithm	Trigonometric Method	Estimation Model
Input	▪ Strength, Admixture, ▪ Evaluation condition (CO_2, COST)	▪ Selection of mean of transportation	▪ Using Energy of factory
Output	▪ Mix design	▪ Supplier	▪ Energy consumption during concrete production
Database	▪ Mix Designs	▪ Location information ▪ CO_2 / Cost information	▪ Standard energy consumption

Figure 2: Evaluation Process for the Optimization System.

The raw material stage includes the CO_2 emissions created during the production of the major components of concrete, namely cement, aggregate, and water, while the transportation stage looks at the CO_2 emissions due to the transportation of raw materials to the RMC manufacturing plant and the manufacturing stage considers the CO_2emissions caused by the generation of electricity and the use of fossil fuels in the concrete batch plant. CLAS assesses the CO_2 emissions and costs when the optimization technique is applied during the product stages of concrete, and then identifies the specific mix design and/or supplier that will minimize either the CO_2 emissions or costs. The methods applied at each stage, as well as the database and input/output items, are shown in Figure 2. A mechanism to cope with fluctuations in the basic unit of CO_2 emission (kg-CO_2/kg) and the costs of concrete components such as cement and aggregate was incorporated into CLAS; its database structure was designed to allow the basic units of CO_2 emission and costs to be easily updated and changed. The updated database is then accessed by the program when calculating the optimal result.

Raw Material Stage

CO_2 emission from concrete production was calculated as the sum of the quantity of each ingredient used for producing 1 m^3 of concrete and the CO_2 emission base units. The base units of the CO_2 emission for cement, aggregate, and water were based on the Korea LCI (Life Cycle Inventory) database (DB). In addition, blast furnace slag, fly ash, and chemical admixtures which are not database units in Korea, were applied to the overseas LCI database [13,14]. Equation (1) is used for calculating of CO_2 emission during the production of the raw material required for manufacturing 1 m^3 of concrete. Table 1 lists the CO_2 emission reference of each ingredient.

$$\text{CO}_2\,\text{M} = \sum(\text{M}(i) \times \text{CO}_2\,\text{emission factor M}) \quad (i = 1:\text{cement}, 2:\text{aggregate}, 3:\text{admixture}, 4:\text{water}) \tag{1}$$

$$\text{Cost M} = \sum(\text{M}(i) \times \text{Unit price M}) \quad (i = 1:\text{cement}, 2:\text{aggregate}, 3:\text{admixture}, 4:\text{water}) \tag{2}$$

Table 1: LCI (Life Cycle Inventory) database (DB) of raw materials.

Material	Unit	Reference Basis
Ordinary Portland Cement	kg	National LCI Database (Korea)
Coarse aggregate	kg	National LCI Database (Korea)
Fine aggregate	kg	National LCI Database (Korea)
Blast furnace slag powder	kg	Overseas LCI DB (ecoinvent)
Fly ash	kg	Overseas LCI DB (ecoinvent)
Water	kg	National LCI Database (Korea)
Chemical admixture compound	kg	Overseas LCI DB (ecoinvent)

CO_2 M is the CO_2 emission at the raw material stage for the production of 1 m^3 concrete (kg-CO_2/m^3); CO_2emission factor M is the CO_2 emission factor for each material (kg-CO_2/kg); Cost M is the cost at the raw material stage for the production of 1m^3 concrete (\$/m^3); Unit price M is the cost for each material (\$/kg); and M(i) is the amount of material used of concrete (kg/m^3).

Transportation Stage

For assessing the CO_2 emission due to transportation, the total quantity used and the load for each component were measured for calculating the vehicle number required for transportation. The calculated number of vehicles, distance, and fuel efficiency of each vehicle were used for assessing the CO_2 emissions. In this study, a truck's speed and traffic weren't considered. Equation (3) gives the quantity of CO_2 emission equation for the transportation stage.Table 2 lists the CO_2 emission reference of each transportation method.

$$\text{CO}_2\,\text{T} = \sum[(\text{M}(i)/\text{Lt}) \times (\text{d/e}) \times \text{CO}_2\,\text{emission factor T}] \quad (i = 1:\text{cement}, 2:\text{coarse aggregate}, 3:\text{fine aggregate}, 4:\text{admixture}) \tag{3}$$

$$\text{Cost T} = \sum[(\text{M}(i)/\text{Lt}) \times (\text{d/e}) \times \text{Unit price T}] \quad (i = 1:\text{cement}, 2:\text{coarse aggregate}, 3:\text{fine aggregate}, 4:\text{admixture}) \tag{4}$$

Table 2: LCI DB of transportation method.

Transportation Equipment	Unit	Reference Basis
Truck	km	National LCI Database (Korea)
Train	km	National LCI Database (Korea)

CO_2 T is the CO_2 emission at the transportation stage for the production of 1 m^3 concrete (kg-CO_2/m^3); CO_2emission factor T is the CO_2 emission factor of the energy resource (kg-CO_2/kg); Cost T is the cost at the transportation stage for the production of 1m^3 concrete (\$/m^3); Unit price T is the cost of the energy resource (oil) (\$/L); M(i) is the amount of material used of concrete (kg/m^3); Lt is the transportation load (tons), d is the transportation distance (km); and e is the fuel efficiency (km/L).

Manufacturing Stage

The CO_2 emitted from concrete manufacturing can be calculated using the amount of energy consumed by the manufacturing equipment for producing 1 m^3 of concrete and converting that to CO_2. For following this approach, the concrete manufacturing process must be divided and the consumed energy must be calculated. The concrete manufacturing process can be divided into the following five stages: loading; storage; transportation; measurement for mixing; and mixing. The equipment required and the data related to the power and fossil fuel energy consumed in each stage are examined; then, by analyzing the ratio between the capacity of each piece of equipment and the total amount of electricity used, the energy consumed for manufacturing 1 m^3 of concrete can be calculated. Equation (5) is used for calculating the quantity of CO_2 emitted during the manufacturing process, and Table 3 lists the CO_2emission reference of energy source.

$$CO_2\,F = \sum[(E(i)/R) \times CO_2\,\text{emission factor F}]$$
$$(i = 1: \text{electricity usage}, 2: \text{oil usage}, 3: \text{water usage}) \quad (5)$$

$$\text{Cost F} = \sum[(E(i)/R) \times \text{Unit price F}]$$
$$(i = 1: \text{electricity usage}, 2: \text{oil usage}, 3: \text{water usage}) \quad (6)$$

Table 3: LCI DB of energy source.

Division		Reference basis
Energy	Electric	National LCI Database (Korea)
	Diesel	National LCI Database (Korea)

CO_2 F is the CO_2 emission at the manufacturing stage for production of 1 m^3 concrete (kg-CO_2/m^3); CO_2 emission factor F is the CO_2 emission factor (kg-CO_2/kwh, L, kg)] of an energy resource; Cost F is the cost at the manufacturing stage for production of 1 m^3 concrete (kg-CO_2/m^3); Unit price F is the cost of an energy resource (\$/kwh, L, kg), R denotes the annual RMC production (m^3/year)]; and E(i) denotes the annual energy usage (unit/year).

Application of Optimization Techniques

Evolution Algorithm for the Materials Stage

The evolution algorithm applied in this stage generates a probable variable from the first parent group. The initial variable is set to be close to the preset objective value in order to generate the next-generation group [14,15]. The variable most suited to the objective can then be obtained through a mutation and reproduction process [16]. The parent group's selection is critical, because the reactions of concrete mix designs depend on the amount of each raw material. However, the evolution algorithm can reduce the error in the next-generation group because it selects the parent group nearest to the target to derive the mix design [17,18,19,20,21,22].

(1) Mix Design by Applying the Evolution Algorithm

To minimize CO_2 emissions, the maximum possible amount of admixture (granulated blast furnace slag/fly ash (GGBS/FA)) is included in the mix design. This means that the mix design can be derived by using the evolution algorithm and inputting the strength, mix ratio (%), and admixture type without the need for basic information, such as the water/binder ratio, slump, air content, coarse aggregate size, or specific gravity. We can also establish an objective function by analyzing the amounts of the materials involved and their costs for the mix, the admixture types and CO_2 emissions, and the admixture mix ratios and CO_2 emissions [23].

(2) Process of Mix Design Deduction

Figure 3 shows the process of deriving the concrete mix design using the evolution algorithm. After information regarding the concrete strength, admixture types, and mix ratio has been entered, it is immediately transferred to the established database of mix designs, where mix designs that are consistent with the input data (for example, strength, admixture type, and mix ratio) are identified and established as the initial group. These processes are performed to generate the initial group needed to apply the evolution algorithm [17].

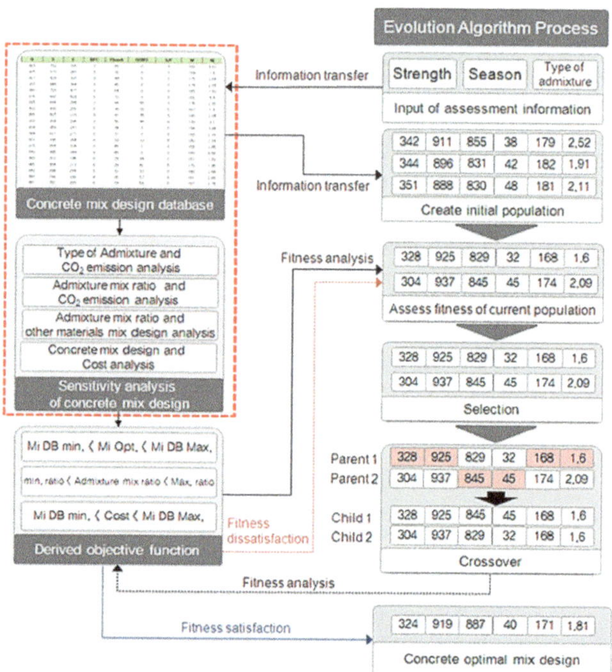

Figure 3: Process Used to Derive the Mix Design Based on the Evolution Algorithm.

A test of fit is performed on the mix designs that constitute the generated initial group. This test is performed to establish the maximum and minimum range of CO_2 emissions, costs, and mix amounts of materials by analyzing the CO_2 emissions and the cost of each mixture design. A test fit is also conducted to evaluate the fit of the mix designs derived from the initial mix design group and any recombination options.

(3) Analysis for the Mix Design Selection Process

Among the initial mix design groups, those mix designs that satisfy the fit test are selected to generate a new group of mix designs according to the approach proposed by Kim *et al.* [17].

$$M(i)\ \text{DB min. mix volume} < M(i)\ \text{mix volume} < M(i)\ \text{DB max. mix volume}$$
$$(i = \text{mixing materials})$$

$$M(i)\ \text{DB min. ratio} < M(i)\ \text{admixture mix ratio} < M(i)\ \text{DB max. ratio}$$

$$M(i)\ CO_2\ \text{emission} < M(i)\ \text{DB min.}\ CO_2\ \text{emission}$$

$$M(i)\ \text{DB min. cost} < M(i)\ \text{cost} < M(i)\ \text{DB max. cost}$$
$$(i = \text{compressive strength})$$

Intersection and combination processes are also performed on the mix designs to derive the new mix design, which is again analyzed using a fit test. These processes are performed repeatedly until the fit criteria are satisfied.

As shown in Figure 4, when the mix types are classified into plain, GGBS mixtures, fly ash mixtures, and GGBS/fly ash mixtures in mix designs with a strength of 24 MPa, the minimum CO_2 emissions identified using this procedure are 319 [kg/m^3], 295 [kg/m^3], 271 [kg/m^3], and 254 [kg/m^3], and the maximum CO_2 emissions are 348 [kg/m^3], 317 [kg/m^3], 307 [kg/m^3], and 282 [kg/m^3], respectively. These results therefore establish the maximum and minimum ranges of CO_2 emissions based on the mix ratios of the admixtures for a given strength of concrete. This allows the quantity of CO_2 emissions to be analyzed for each of the admixtures based on the designs for nominal strengths of 18, 21, and 24 MPa. The mix designs with strengths of 21 and 24 MPa, in which both fly ash and granulated blast-furnace slag are mixed, result in the lowest CO_2 emissions for this example. CO_2 emissions are reduced by approximately 29% and 26%, respectively, relative to those of the plain mix design that did not include any admixture [17].

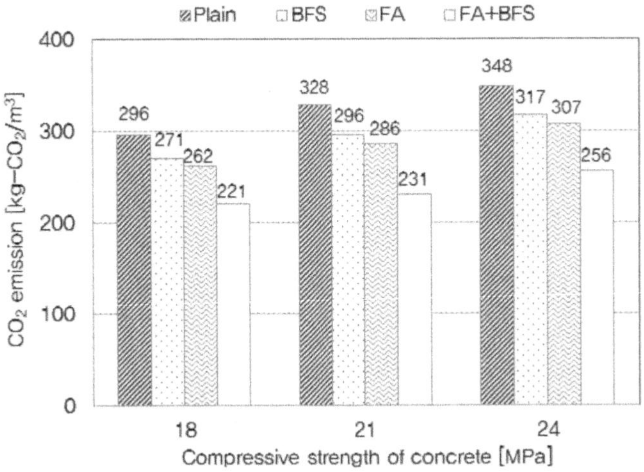

Figure 4: Carbon dioxide (CO_2) Emission Analysis of Concrete by Type of Admixture.

As shown in Figure 5, the CO_2 emissions of the 18 MPa strength concrete vary between 251 [kg/m^3] and 310 [kg/m^3] when the mix ratio of the admixtures is increased from 10% to 30%. This analysis demonstrates that it is indeed possible to determine the maximum and minimum range of the CO_2 emissions for specific mix ratios of the admixtures for a given concrete strength. The CO_2 emissions decrease as the mix ratios of admixtures such as fly ash and granulated blast-furnace slag increase. Kim *et al.* [17] analyzed 18, 21, and 24 MPa strength concretes and demonstrated that the concrete mix designs with admixtures of 30% showed CO_2 emission that were reduced by 27%, 29%, and 32%, respectively, compared to those with admixtures of 10%.

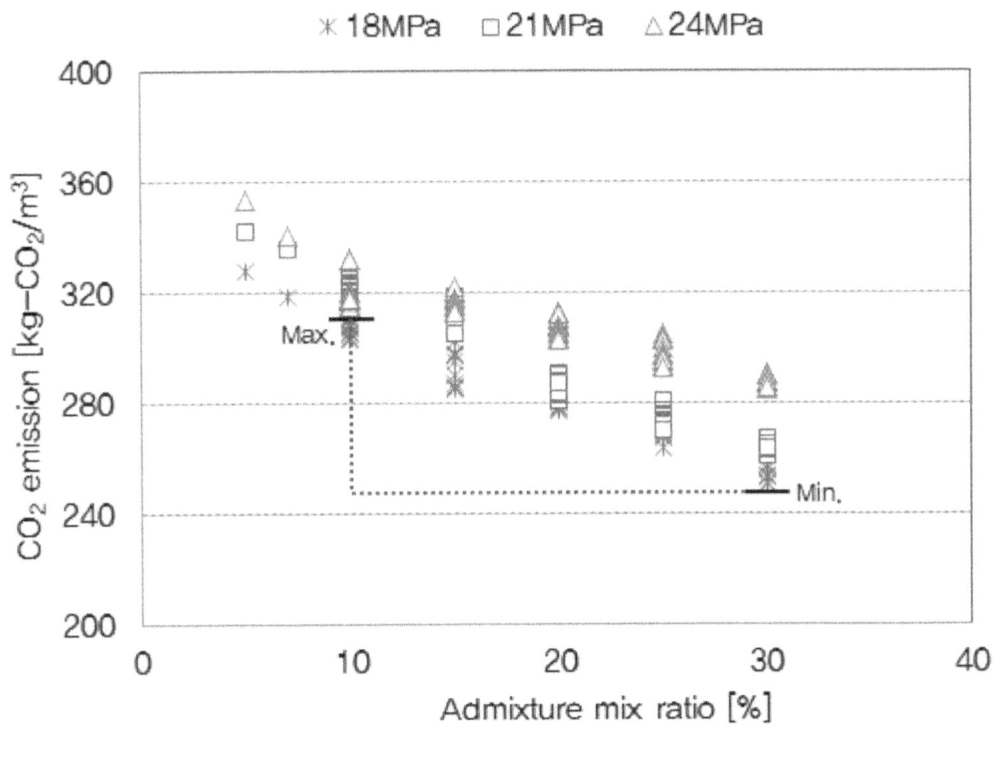

Figure 5: Analysis of the CO_2 Emissions for Various Admixture Mix Ratio.

Looking at the admixture mixing ratios identified in the study by Kim *et al.* [17], the amount of coarse aggregate ranged from a minimum of 865 kg/m^3 to a maximum of 997 kg/m^3; the amount of fine aggregate ranged from 781 kg/m^3 to 982 kg/m^3; the amount of cement was between a minimum of 158 kg/m^3 and a maximum of 389 kg/m^3; and the amount of water was between 105 kg/m^3 and 185 kg/m^3, as shown in Figure 6. Based on these results, the maximum and minimum mix amounts of cement, aggregate, and water for each mix ratio were established.

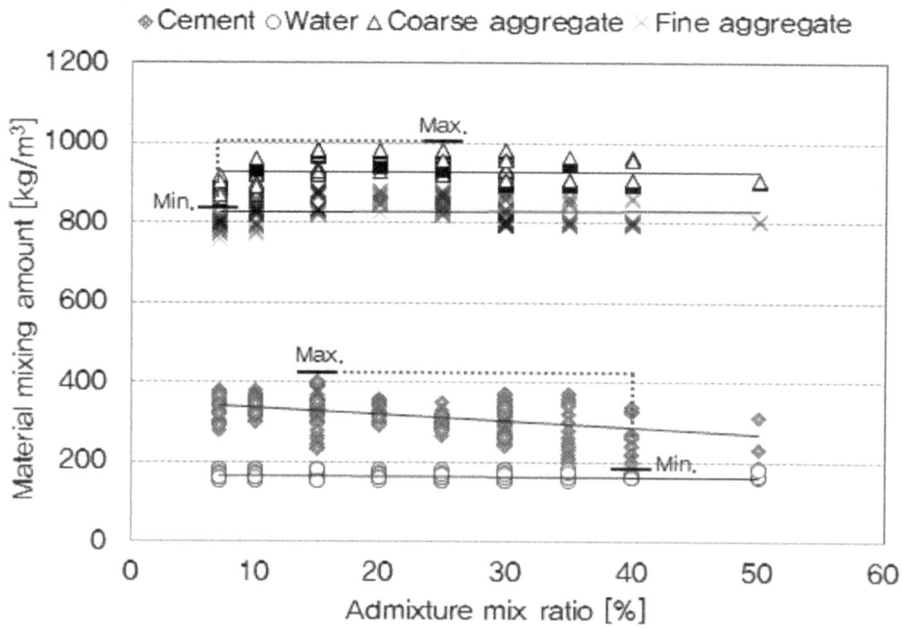

Figure 6: Analysis of Admixture Mix Ratio and Other Materials.

The cost analysis performed for the mix design process utilizes information from Korea's price database to determine the product cost [24]. As shown in Figure 7, the cost of concrete with a strength of 24 MPa ranged between US\$59.20 and 70.60/m^3, and that of concrete with a strength of 30 MPa ranged between US\$63.20 and 73.70/m^3. As a result of this analysis, the maximum and minimum costs as a function of concrete strength were established.

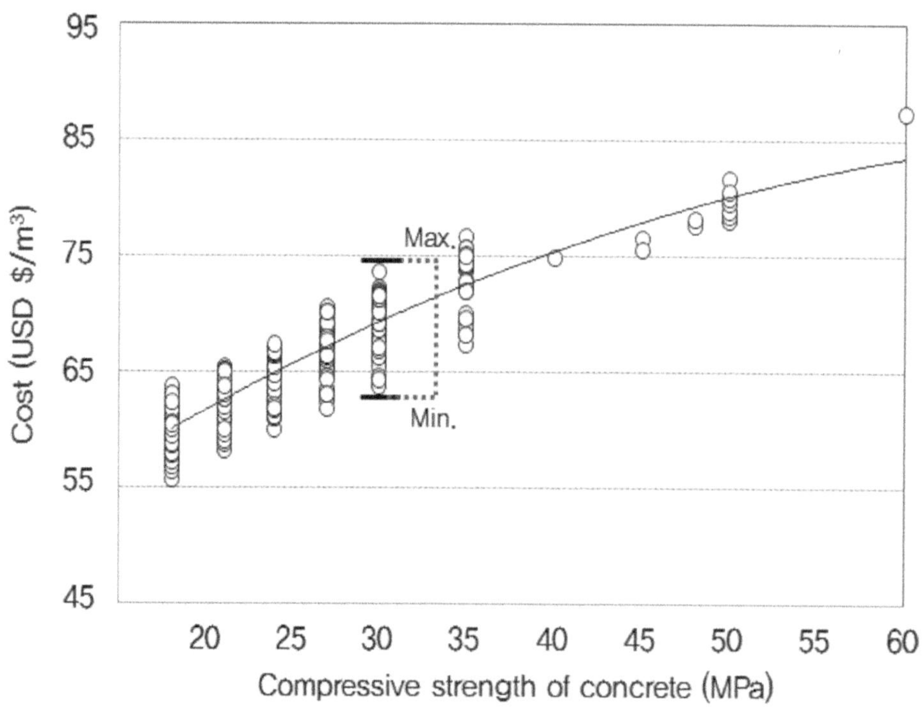

Figure 7: Cost Analysis Based on the Compressive Strength of Concrete.

The largest contributing factor to the cost per m^3 of concrete in the example given here was the mix proportion of cement and aggregate. Not surprisingly, the analysis also showed that the cost increased with strength because the mix ratio for the most expensive cement (in US\$/kg) increased with strength. Adjusting the amount of the coarse aggregate and fine aggregate mixed in to the less expensive cement had less effect on the overall concrete cost.

Trigonometric Function Method for the Transportation Stage

This step determines which supplier to select to minimize CO_2 emissions and costs when the raw materials for the concrete are transported to a ready mix concrete (RMC) production plant. Looking at the production of each of the possible raw material plants, the optimum supplier is identified by analyzing the CO_2 emissions, the unit cost of production and the distance to the concrete production plant. As shown in Figure 8, the distance between the plants can be measured by finding the interval angle (°) between the two points using latitude and longitude coordinates. This can then be multiplied by the radius of the spherical surface, with the surface of the Earth being assumed to be spherical and the distance calculated using the equation below [25,26,27,28]:

$$R = T\cos(\sin(X) \times \sin(X') + \cos(X) \times \cos(X') \times \cos(M))$$

$$D = R \times E_d$$

where R is the angle between two points (°); X is the latitudinal value of the initial point in radians (°); X′ is the latitudinal value of the terminal point in radians (°); M is the difference between the longitudinal values of the initial point and the terminal point in radians (°); T is the radius of the earth, D is the distance between the two points to be measured (km); and E_d is the radius of the earth (km).

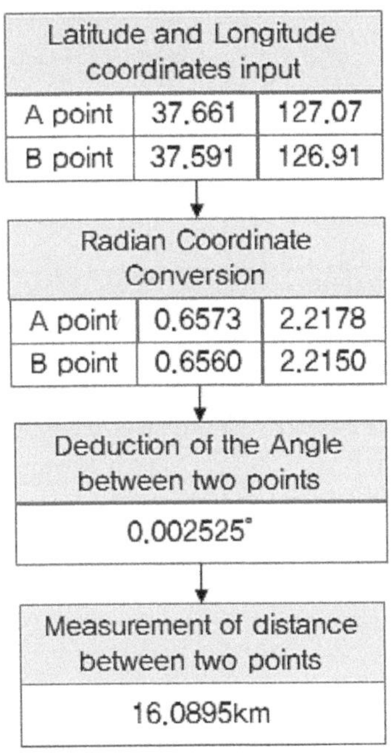

Figure 8: Process of Measuring the Distance Between Latitude and Longitude Coordinates.

As indicated in Table 4, the location information must be collected for the various plants producing cement, aggregate, and admixtures and transformed into latitude and longitude coordinates. These can then be used to create a database to determine the transportation distance. Within Korea, this location information was collected for 100 cement production plants, 390 aggregate production plants, 15 fly ash production companies, five granulated blast-furnace slag production plants, and 15 chemical admixture production plants. The means of transportation for each component material of concrete was limited to either rail or truck, which were classified into cement freight trains and trucks with capacities of 1, 2.5, 3.5, 5, 8, 15, 18, and 25.5 tons, allowing the number of transportation vehicles for each material to be calculated on the basis of the total quantity (kg) of each component material of the concrete and the capacity (in tons) of the various transport options. The number of transportation vehicles was used to evaluate CO_2 emissions and costs by applying the transportation distance, basic unit of CO_2 emissions, and fuel cost [29,30,31,32]. Figure 9 depicts the optimal support algorithm for concrete.

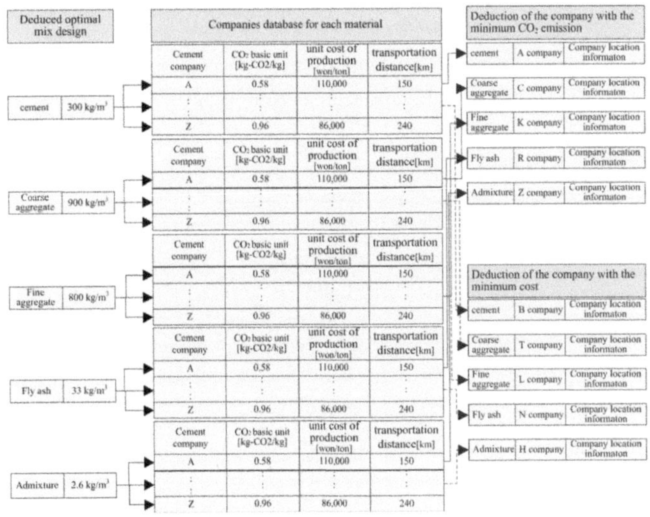

Figure 9: Optimal Support Algorithm for Concrete.

Table 4: Example Database of Contractor Information for Each Material.

Material	Company Address	Latitude-Longitude		CO_2 Emission Basic Unit (kg-CO_2/ton)	COST ($/ton)
		X Coordinate	Y Coordinate		
C	Samcheok, Gangwon	129.1768	37.4286	760	120
G	Gwangmyeong, Gyeonggi	126.8846	37.4236	0.005	8.2
S	Jung-gu, Incheon	126.5082	37.4844	0.005	7.6
B/C	Jung-gu, Incheon	126.6145	37.4410	0.8	117
F/A	Taean, Chungcheong	126.1664	36.9226	0.0196	48
GGBS	Dong-gu, Incheon	126.6231	37.4905	0.0265	54
AE	Asan, Chungcheong	127.0678	36.9301	0.22	880
R/G	Goyang, Gyeonggi	126.8133	37.6896	0.004	2.9
W	running water				

C: cement; G: Coarse Aggregate; S: Fine Aggregate; B/C: Blast Furnace Cement; F/A: Fly ash; GGBS: Granulated Blast Furnace Slag; AD: Admixture; W: Water; R/G: Recycled Aggregate.

Stochastic Model for the Manufacturing Stage

The CO_2 emissions and costs incurred during the manufacturing stage are determined using manufacturing equipment selection and stochastic model methods. The process begins by gathering data on the actual processes and the equipment capacity of the RMC plant. The energy consumption of the manufacturing equipment is calculated using data from the daily energy consumption of a batch plant comprising a material storage silo and blending mixer.

To identify the amount of energy used to produce 1 m^3 of concrete, an analysis was conducted by classifying the batch plant manufacturing equipment into the various types of equipment used for material loading, storage, and mixing (Table 5). Other equipment, such as dust collectors and boilers, must also be classified appropriately. For this step, as indicated in Table 6, a database was established to guide this process by analyzing the annual consumption of electricity, oil, and water at five major concrete production companies in Korea. The analysis revealed that the amounts of energy used and concrete produced varies with the season; there is a significant difference between the summer (April to October) and winter (November to March) figures.

Table 5: Equipment Survey for Manufacturing Process.

Classification		Equipment Types			
Material Loading and Unloading	Unloading equipment	Horizontal Conveyor Belt	Middle Screen	Conveyor Belt	Shuttle Belt
	Heavy Equipment	Wheel Loader		Middle Screen	
Material Storage		Cement Silo	Fly ash silo	Blast Furnace Slag Silo	
Material Transportation	Aggregate	Rotary feeder	Conveyor belt	Gauge	Bucket Elevator
	Cement	Brewer pump	Dust collector	Rotary Feeder	
	GGBS	Brewer Pump	Dust Collector	Rotary Feeder	
	Fly Ash	Brewer Pump	Dust Collector	Rotary Feeder	
	Others	Admixture Pump	Water Pump	Recycled Water Pump	
Material Mixing		Concrete Mixer			
Other Equipment		Dust Collector	Compressor	Boiler	Cooling Chiller
Accessory Equipment	Equipment Wash	Pump			
	Recycle Water Facility	Pump	Aggregate Screen	Sand Screen	Whisk
	Office	Lighting		Power Outlet	

Table 6: Yearly Energy Consumption and Concrete Production.

Company/Season		Production (m^3/year)	Yearly Consumption by Energy Source			
			Electric Power (kwh/year)	Diesel (L/year)	Kerosene (L/year)	Running Water (ton/year)
A	Summer	514,662	775,405	187,953	1056	16,203
	Winter	404,297	783,534	130,132	15,526	17,195
B	Summer	430,271	739,440	30,809	3726	16,203
	Winter	442,059	827,395	31,827	52,774	17,195
C	Summer	430,271	792,704	54,581	895	34,080
	Winter	442,059	946,350	53,561	10,700	26,650
D	Summer	158,792	436,505	38,179	694	34,080
	Winter	137,970	467,193	36,578	7457	26,650
E	Summer	359,644	729,676	187,953	1837	64,189
	Winter	321,316	774,282	66,068	24,703	62,016

Verification of the Optimization Techniques

Mix Design

Actual concrete mix designs were compared and analyzed to assess the reliability of the mix designs derived using the evolution algorithm. The results confirmed that the actual mix design and the mix design derived using the evolution algorithm were indeed similar, as shown in Figure 10, regardless of the concrete strength. The mix design error rates for the cement, water, and aggregate, which determine the physical properties of the concrete, were consistently within 5%. We therefore concluded that this mixture algorithm could be used in the development of the new optimization system, CLAS [33,34,35].

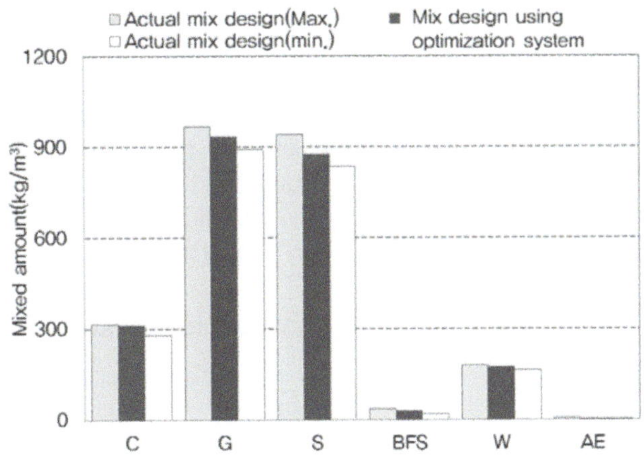

Figure 10: Comparison of mix design for concrete.

Investigation of the Manufacturing Equipment

To establish an appropriate database, a survey requesting information on aspects such as batch plant facilities, production capacities and operation times was distributed to 51 Korean RMC companies. Based on our analysis of the questionnaire survey responses, 80% (43 of 51) of the companies surveyed were determined to be suitable within the terms of the production process and batch plant facility status proposed in this study.

DEVELOPMENT OF CLAS

CLAS, which assesses CO_2 emissions and the costs associated with concrete made using the optimization method, was developed using Visual Basic, as shown in Figure 11. An evaluation of the projected CO_2 emissions and costs can be performed on the basis of the information input by the user [36,37,38].

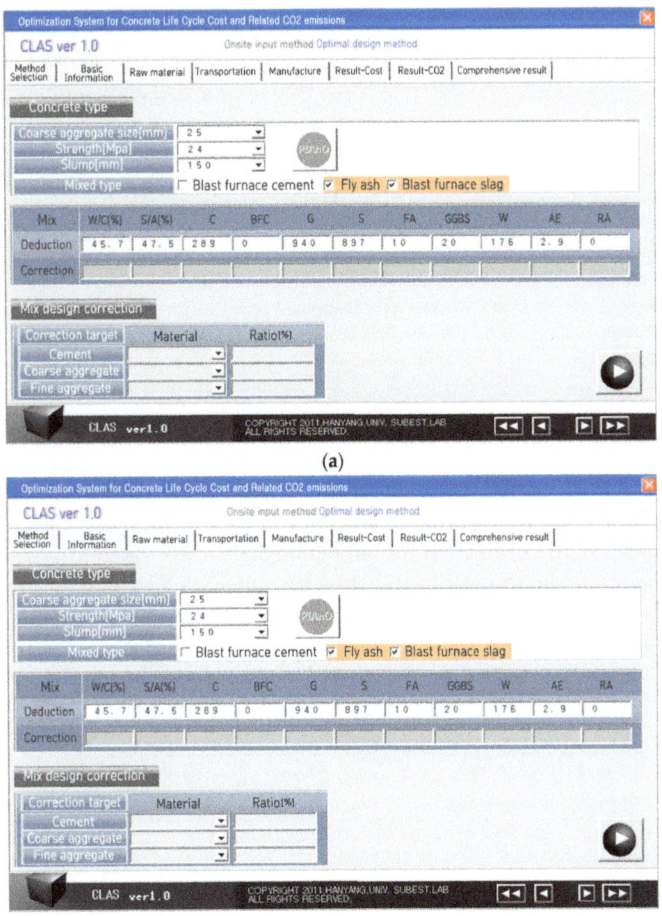

(a)

(b)

Figure 11: Screenshots of the New Optimization System (Concrete Life Cycle Assessment System (CLAS)). (**a**) Basic Information Input Sheet; (**b**) Materials Stage Input Sheet.

Basic Information Input Sheet

The basic information input sheet for CLAS asks for the RMC manufacturer, the evaluation date, and the evaluator. The evaluation company can be selected from the database previously established, and the coordinates deduced by directly entering the address for sites that are not yet in the database. The expected production season (summer or winter), volume, and priority (CO_2 emissions or cost) must also be entered in turn.

Raw Materials Stage Input Sheet

In the raw materials stage, the nominal strength, admixture types, and mix ratios (%) of the admixtures are selected. The mix design is derived from the input data, and the evaluator considers the calibration. In terms of calibrating the mix design, cement can be replaced with blast-furnace cement, ground granulated blast-furnace slag, or fly ash; the natural aggregate can be replaced with recycled aggregate [39].

Transportation Stage Input Sheet

When a means of transportation is selected, the corresponding load capacity and fuel efficiency are automatically assigned. As explained earlier, cement is usually transported by either rail or bulk truck and users will select an option from those offered on a drop-down menu.

Manufacturing Stage Input Sheet

To calculate the energy consumption, either the equipment selection method or the estimation model method can be selected. When the manufacturing and accessory equipment used in the storage, transportation, and mixing processes are selected as the input values in the equipment selection method, the energy consumption of the evaluation company is automatically computed. For the input values of the estimation model method, the yearly concrete output and the oil, running water, and electrical consumption of the company being evaluated are entered.

Evaluation Results Sheet

The evaluation result stage determines the mix design and supplier by evaluating the CO_2 emissions and the cost of the concrete based on the basic information and the evaluation priority selected. The system then calculates the results automatically based on the data provided and the results of the calculation are displayed as both a table and graph.

CASE ANALYSIS

Method

Applying CLAS, the amount of concrete (m^3) with a nominal strength of 24 MPa that was actually produced by the RMC plant in May (summer season) was evaluated in terms of the CO_2 emissions and cost by applying both the input method currently used and the new optimization system. The conventional input method is applied to the actual mix design and transportation distance, energy consumption, while for the optimization system the priority is set as either minimizing the CO_2 emissions or minimizing the costs. The methods for the two approaches are compared inTable 7.

Table 7: Evaluation Method.

Classification	Conventional Input Method	Optimization System Method
Standard	25-24-150, Mixed Materials (Fly ash)	
Quantity	1 m^3	
Materials Stage	Actual Mix Design	Mix Design using CLAS
Transportation Stage	Actual Transportation Distance	Transportation Distance using CLAS
Manufacturing Stage	Estimation Model	Estimation Model

Results

As the data in Table 8 show, the CO_2 emissions generated at the raw materials stage accounted for approximately 96% of the overall emissions and the cost for this stage was approximately 93% of the overall total. The CO_2emissions and costs incurred in the transportation and manufacturing stages accounted for only an insignificant fraction. The CO_2 emissions per m^3 were 340.9 kg-CO_2/m^3 for the conventional input method, significantly more than the 225.8 kg-CO_2/m^3 obtained when minimizing the CO_2 emissions was prioritized using the new optimization system. These results indicate that a potential reduction in CO_2 emissions of 34% can be obtained compared to those obtained using the conventional input method for the example shown here.

Table 8: Evaluation Results.

Classification			Materials	Transportation	Manufacture	TOTAL
CO_2 emission (kg-CO_2/m^3)	Conventional input method		332.4	5.7	2.8	340.9
	Optimization System Method	CO_2 Prioritized	221.8	1.2	2.8	225.8
COST (US$/$m^3$)	Conventional input method		46.30	2.30	0.70	49.30
	Optimization System Method	COST Prioritized	44.20	0.60	0.70	45.50

During concrete production, the cost per m³ was determined to be US$49.30/m³ when using the conventional input method; this value was 1% higher than the value obtained when minimizing the cost was prioritized using the new optimization system (US$45.50/m³).

Materials Stage

The mix design used in both the conventional method and the new optimization system for this case analysis is shown in Table 9. The results in Table 9 indicate clear differences between the two methods for the cement, fine aggregate, and admixtures, but the coarse aggregate, water, and other compounds are very similar. For the raw materials stage, the CO_2 emissions are 221.8 kg-CO_2/m³ when the priority is to minimize the CO_2 emissions for the optimization system calculation. These values are 33% less than those predicted for the conventional input method (332.4 kg-CO_2/m³).

Table 9: Mix Designs Applied for Evaluation Method.

Classification	W/B (%)	S/a (%)	Unit Mixed Amount (kg/m³)					
			C	G	S	F/A	W	AE
Conventional Input Method	48.1	47	314	917	902	45	173	2
Optimization System Method	47.2	47.3	330	924	833	34	172	2.6

C = cement; G = coarse aggregate; S = Fine aggregate; F/A = Fly ash; W = water; AE = other additives.
W/B = water-binder(cement + fly ash) ratio; S/a = Sand(fine) aggregate ratio.

A further reduction is found for cement production when minimizing the CO_2 emissions is prioritized by the optimization system, with the CO_2 emissions dropping to 178.2 kg-CO_2/m³, well below the 228.2 kg-CO_2/m³ obtained by the conventional method at the raw materials stage. This is because the CO_2 emissions due to the cement production processes are significantly greater than those for the aggregate and admixtures. When minimizing the cost is prioritized by the optimization system, the cost drops to US$44.20/m³, which is 6% lower than the value obtained with the conventional input method (US$46.30/m³) because more cement, which has a higher unit production cost, is used in the actual mix design. The unit costs for the production of the aggregate and admixtures have only a relatively insignificant effect.

Transportation Stage

The results shown in Table 10 reveal that the quantity of CO_2 emissions is 1.2 kg-CO_2/m³ when minimizing the CO_2 emissions is prioritized by the optimization system. This is 70% less than that obtained with the typical input method (5.7 kg-CO_2/m³) because minimizing CO_2 emissions assumes the raw material supplier incurring the shortest transportation distance will be selected to minimize fuel consumption when transporting the raw materials to the RMC plants.

Table 10: Location of Raw Material Suppliers.

Classification		Supplier Plant Selection				
		C	G	S	F/A	AE
Conventional Input Method		D company	I company	G company	T company	P company
Optimization System Method	CO_2 Prioritized	U company	G company	G company	D company	A company
Optimization System Method	COST Prioritized	J company	Y company	Y company	T company	P company

When minimizing the cost is prioritized by the optimization system, the cost is US$0.60/m³, 70% below that obtained using the conventional input method (US$2.30/m³). This is because the fuel costs vary according to the transportation distance for the different raw materials suppliers. The supplier that incurs the lowest production unit cost for the raw materials, regardless of the transportation distance, will be selected when minimizing the cost is prioritized.

Manufacture Stage

As indicated in Table 11, both the conventional input method and the new optimization system applied the same estimation model to calculate the cost of the energy and water required for the manufacturing process. In this case, the evaluation is performed using the data (*i.e.*, the annual amount of concrete produced and the energy consumption in terms of the electricity, oil, and water) of the target concrete production company. The results indicate that the CO_2emissions are 2.8 kg-CO_2/m^3 and the cost US\$0.70/m^3 for both methods.

Table 11: Energy Output Applied for Evaluation Method.

Classification	Consumption by Energy Source		
	Electric (kwh/m^3)	Oil (L/m^3)	Water (ton/m^3)
Conventional Input Method Optimization System Method	3.21	1.32	0.29

DISCUSSION AND LIMITATION

This study aimed to develop an optimum design system (concrete life cycle assessment system: CLAS) which can minimizes CO_2 emissions and maximize economic efficiency at the life cycle assessment (LCA) of concrete. The CLAS which assesses the CO_2 emissions and economic efficiency of concrete to which optimum design technique is applied is a program that finds the concrete mix design (kg/m^3) which satisfies minimum CO_2 emissions or costs and selects a raw material supplier.

However, this study has the following limitations:

First, there has been an advancement of studies on the capture and storage of CO_2 emitted by the consumption of fossil fuels in cement manufacturing facilities. Among construction materials, cement is a material with high CO_2emissions so that carbon dioxide capture and storage (CCS) are important. However, this study covers CO_2 emissions and costs which occur during the production of concrete.

Regarding the CO_2 emissions of cement, therefore, the conventional national life cycle database (LCI DB) and each cement manufacturer's database were applied. For cement costs as well, Korea price information and each cement manufacturer's database were applied. It is unknown if the results were derived after reflecting the CCS in the national LCI DB and each cement manufacturer's database. This study did not separately consider CO_2 emissions and costs after the CCS during the production of cement. However, CCS is so important that it would be considered in future studies.

Second, in this study, the Euclidean distance between two latitude and longitude points was applied. Currently, the coordinates-based Euclidean distance would differ from the geodesic distance. Therefore, if correction factors are applied, the information would become more reliable. Since the program in this study is already designed in Visual Basic, it failed to consider this information.

CONCLUSIONS

The new CLAS optimization system proposed here, developed using Visual Basic, applies an evolution algorithm to calculate the mix of materials, a trigonometric function method for the raw material transportation method and a stochastic model for the manufacturing process. The reliability of the resulting program was improved through a thorough analysis of the various mix designs and supplier options. CLAS assesses the CO_2 emissions and costs when the optimization technique is applied during the concrete production stage and identifies the concrete mix design and supplier that will minimize either the CO_2 emissions or the cost, depending on the user's requirements.

A case analysis was performed for a specific RMC plant in Korea. The results indicated that the CO_2 emissions and costs per m^3 of RMC were 340.9 kg-CO_2/m^3 and US\$49.70/m^3, respectively, using the conventional input method currently used in the industry. Applying the new CLAS program, the CO_2 emissions could be reduced by 34% to 225.7 kg-CO_2/m^3 when minimizing the CO_2 emissions was prioritized, and the costs dropped by 1% to US\$48.80/m^{3}when minimizing the cost was prioritized. Based on the case analysis, the most significant way to reduce the CO_2emissions and costs associated with the manufacture of ready mix concrete are to select the mix design and raw material supplier based on whether the CO_2 emission or cost are to be minimized.

ACKNOWLEDGMENTS

This research was supported by Basic Science Research Program through the National Research Foundation of Korea (NRF) funded by the Ministry of Science, ICT & Future Planning (No.2015R1A5A1037548) and the National Research Foundation of Korea (NRF) grant funded by the Korea government (MSIP) (No.20110028794).

AUTHOR CONTRIBUTIONS

All authors contributed substantially to all aspects of this article.

CONFLICTS OF INTEREST

The authors declare no conflict of interest.

REFERENCES

1. World Energy Outlook. International Energy Agency: Paris, France, 2015. Available online: http://www.worldenergyoutlook.org/ (accessed on 11 January 2016).
2. Korea Environmental Industry & Technology Institute. Environmental Declaration of Products. Available online: http://www.edp.or.kr/en/ (accessed on 11 January 2015).
3. Korea National Cleaner Production Center. Korea LCI DB Information Network. Available online: http://www.kncpc.or.kr/main/main.asp (accessed on 15 January 2015).
4. Korea Environmental Industry & Technology Institute. Korea LCI DB Information Network. Available online: http://www.edp.or.kr/en/lci/lci_intro.asp (accessed on 11 January 2015).
5. Peuporier, B.; Putzeys, K. Inter-Comparison and Benchmarking of LCA-based Environmental Assessment and Design Tool. Final Report. Practical Recommendations for Sustainable Construction (PRESCO). Available online: http://www.etn-presco.net/library/index.html (accessed on 11 January 2015).
6. ASMI. Athena Sustainable Materials Institute. Available online: http://www.athenasmi.org (accessed on 11 January 2016).
7. thinkstep GaBi. Available online: http://www.gabi-software.com/international/index/ (accessed on 13 January 2016).
8. Building Research Establishment. Envest 2. Available online: http://envest2.bre.co.uk/account.jsp (accessed on 11 January 2016).
9. Erin, T. LISA (LCA In Sustainable Architecture). Available online: http://sustainableresourcedatabase.weebly.com/lca-in-sustainable-architecture.html (accessed on 11 March 2016).
10. VTT Technical Research Center. Available online: http://www.vttresearch.com/ (accessed on 12 March 2016).
11. Scott, M. Eco Quantum Life Cycle and Greenhouse Gas Assessment. Available online: http://ecoquantum.com.au/ (accessed on 11 January 2015).
12. International Standard. *ISO 14044: Life Cycle Assessment (Requirements and Guidelines)*; International Organization for Standardization: Geneva, Switzerland, 2006; Available online: http://www.iso.org/iso/home/store/catalogue_tc/catalogue_detail.htm?csnumber=38498 (accessed on 19 March 2015).
13. International Standard. *ISO 21930: Environmental Declaration of Building Products*; International Organization for Standardization: Geneva, Switzerland, 2007; Available online: http://www.iso.org/iso/home/store/catalogue_tc/catalogue_detail.htm?csnumber=40435 (accessed on 19 March 2015).
14. Peng, C.H.; Yeh, I.C.; Lien, L.C. Modeling strength of high-performance concrete using genetic operation trees with pruning techniques. *Comput. Concr.* 2009, *6*, 203–223.
15. Tae, S.H.; Shin, S.W.; Woo, J.H.; Roh, S.J. The development of environmental load evaluation system of a standard Korean apartment house. *Renew. Sustain. Energy Rev.* 2011, *15*, 1239–1249.
16. Parichatprecha, R.; Nimityongskul, P. An integrated approach for optimum design of HPC mix proportion using genetic algorithm and artificial neural networks. *Comput. Concr.* 2009, *6*, 253–268.
17. Kim, T.H.; Tae, S.H.; Roh, S.J. Assessment of the CO_2 emission and cost reduction performance of a low-carbon-emission concrete mix design using an optimal mix design system. *Renew. Sustain. Energy Rev.* 2013, *25*, 729–741.
18. Kewalramani, M.; Gupta, R. Concrete compressive strength prediction using ultrasonic pulse velocity through artificial neural networks. *Autom. Constr.* 2006, *15*, 374–379.
19. Goldberg, D. *Genetic Algorithms in Search, Optimization and Machine Learning*; Addison-Welsley: Reading, MA, USA, 1989.
20. Habert, G.; Roussel, N. Study of two concrete mix-design strategies to reach carbon mitigation objectives. *Cem. Concr Compos.* 2009, *31*, 397–402.
21. Holland, J.H. *Adaptation in Natural and Artificial Systems*; University of Michigan Press: Ann arbor, MI, USA, 1975.
22. Jeong, M.J. Integrated Support System for Decision Making in Optimization. Ph.D. Thesis, The University of Tokyo, Tokyo, Japan, 2003; pp. 6–14.
23. Thomas, B. *Evolutionary Algorithms in Theory and Practice*; Oxford University Press: Oxford, UK, 1996.
24. Korea Price Information. Available online: http://www.kpi.or.kr/ (accessed on 15 March 2016).
25. Liu, Y.; Xi, N. Low-cost and automated calibration method for joint offset of industrial robot using single-point constraint. *Ind. Robot* 2011, *38*, 577–584.
26. Park, C.S.; Hong, S.E.; Park, J.O.; Jeong, T.S.; Sung, B.Y. A study on converting cadastral coordinate system into global coordinate system. *J. Korean Soc. Geosp. Inf. Sys.* 2010, *18*, 105–114.
27. Badekas, J. *Investigations Related to the Establishment of a World Geodetic System*; Report 124; Department of Geodetic Science, Ohio State University: Columbus, OH, USA, 1969.
28. Bursa, M. The theory for the determination of the non-parallelism of the minor axis of the reference ellipsoid and the inertial polar axis of the earth, and the planes of the initial astronomic and geodetic meridians from observations of artificial earth satellites. *Stud. Geophys. Geod.* 1962, *6*, 209–214.
29. Krakiwsky, E.J.; Thomson, D.B. Mathematical models for the combination of terrestrial and satellite networks. *Can. Surv.* 1974, *28*, 606–615.
30. Molodensky, M.S.; Eremeev, V.F.; Yurkina, M.I. *Methods for Study of the External Gravitational Field and Figure of the Earth*; Israeli Program for the Translation of

Scientific Publications: Jerusalem, Israel, 1962.

31. Ramsey, P. *The State of Open Source GIS*; Refractions Research Inc.: Victoria, BC, Canada, 2004; pp. 3–10.

32. Wolf, H. Geometric connection and re-orientation of three-dimensional triangulation nets. *Bull. Géod.* 1963, *68*, 165–169.

33. Chou, J.; Tsai, C. Concrete compressive strength analysis using a combined classification and regression technique. *Autom. Constr.* 2012, *24*, 52–60.

34. Zain, M.F. An expert system for mix design of high-performance concrete. *Adv. Eng. Softw.* 2005, *36*, 325–337.

35. Oner, A.; Akyuz, A.; Yildiz, R. An experimental study on strength develpoment of concrete containing fly ash and optimum usage of fly ash in concrete. *Cem. Concr Res.* 2005, *35*, 1165–1171.

36. Kim, T.H.; Tae, S.H.; Lee, J.S. A Study on development of a CO_2 assessment program of concrete. *Int. Conf. Sustain. Build. Asia* 2010, *1*, 303–310.

37. Tae, S.H.; Shin, S.W.; Ha, S.G. Life cycle environmental loads and economic efficiencies of apartment buildings with plaster board drywall. *Renew. Sustain. Energy Rev.* 2011, *15*, 4145–4155.

38. Kjellsen, K.O.; Guimaraes, M.; Nilsson, A. *The CO_2 Balance of Concrete in a Life Cycle 430 Perspective*; Nordic Innovation Centre: Stensberggt, Norway, 2005.

39. Ko, J.H.; Kim, G.T.; Kim, D.H.; Kim, H.S. Development of an optimal design program for a triple-band PIFA using the evolutionary strategy. *J. Korean Inst. Electromagn. Eng. Sci.* 2009, *2*, 746–753.

CHAPTER 9

A FULL FACTORIAL DESIGN BASED DESIRABILITY FUNCTION APPROACH FOR OPTIMIZATION OF PROPERTIES OF C 40/50 CONCRETE CLASS

Barış Şimşek, Y. Tansel İç and Emir H.Şimşek

Keywords: Desirability Function Approach, Full Factorial Design, C 40/50 Concrete Class, Optimization

INTRODUCTION

The optimization of a Ready Mixed Concrete (RMC) mixture for determination of the desired quality is an important issue in the field of material and design engineering [1, 2]. In literature several optimization and modelling methods have been proposed on investigating the optimal mixture proportions for many concrete types. For investigating the effects of parameters on concrete, design of experiment is commonly used in literature. Some of those, Sonebi [3] modeled mix proportion parameters of underwater composite cement grouts using a factorial design. Muthukumar et al. [4] optimized the mechanical properties of polymer concrete and recommended the mixdesigned based on design of experiment. Özbay et al. [5] investigated the mix proportions of high strength self compacting concrete by using Taguchi method. Correia et al. [6] assessed of the recycling potential of fresh concrete waste using a factorial design of experiments. Correia et al. [7] practiced a Factorial design used to model the compressive strength of mortars containing recycled rubber. Santilli et al. [8] applied a factorial design study to determine the significant parameters of fresh concrete lateral pressure and initial rate of pressure decay. Alqadi et al. [9] developed a self compacting concrete using contrast constant factorial design.

Citation: Barış Şimşek , Y. Tansel İç and Emir H.Şimşek, A Full Factorial Design Based Desirability Function Approach for Optimization of Properties of C 40/50 Concrete Class, doi:10.3390/mca18030330.

Multiple response problems include three stages: data gathering, modeling and optimization [10]. In optimization phase; FFD is widely practiced with DFA. Some examples of these applications can be given as followings. Paterakis et al. [11] evaluated and optimized some pellets characteristics using a 3^3 factorial design and desirability function. Mukherjee and Ray [12] applied an optimal process design of twostage multiple responses grinding processes using desirability functions and metaheuristic technique. Gottipati and Mishra [13] optimized the process of adsorption of Cr (VI) on activated carbons prepared from plant precursors by a two-level full factorial design.

Within the scope of this study, it is desired to obtain optimal mixture proportions of C50 in Turkey. First of all, the criteria of determining the quality of the concrete, factors and levels that affect these performance criteria were identified to obtain optimal mixture proportions. A 2^4 full factorial design with two replicates was used for the optimization of dual responses such as slump flow and 28th day compressive strength. An analysis of variance (ANOVA) test was used to find out the significance and percentage contribution of each parameter [14]. The mathematical model of quality criteria has been developed using regression analysis as a function of the water to binder materials ratio, coarse aggregate (II) to total aggregate ratio, the percentage of super plasticizer content and fly ash amount. DFA approach was applied to determine optimal mixture proportions of C50.

MATERIALS AND METHODS

Materials

The cement used in this research for the normal weight concrete is a CEM I 42.5 R has a specific gravity of 3.15 and weighs 350 kg. Fly ash used in this research with a specific gravity of 2.46 weighs 80 kg. Chemical composition of the binder materials is given in Table 1. Crushed and which has particle size smaller than 4 mm (I) was used as the fine aggregate. The fine aggregate ratio was fixed at 50 % in all experiments. Aggregate number (II) with a size between 4mm to 11.2 mm and aggregate number (III) with a size between 11.2 mm to 22.4 mm were used as coarse aggregate in the concrete mixtures. The fine and coarse aggregates have specific gravities of 2.75 and 2.77 and mean water absorptions of 1.5% and 0.9 %, respectively. Superplasticizer content is defined as the ratio of superplasticizer amount of 100 kg cement.

Table 1: Chemical composition of cement and fly ash

Chemical analysis	CEM I 42.5 R (%)	FLY ASH (%)
CaO	66.25	4.76
SiO_2	21.79	56.21
Al_2O_3	5.98	23.1
Fe_2O_3	2.51	6.51
SO_3	1.54	0.73
MgO	1.15	2.11
K_2O	0.61	2.53
Na_2O	0.15	0.27
Cl	0.0071	0.0018
Loss of ignition	3.71	2.24

A polycarboxylic type superplasticizer (SP) was used in all concrete mixtures and physical properties of SP are given in Table 2.

Table 2: Properties of the SP at 200C

Properties	Superplasticizer
Chemical description	Polycarboxylic type polymer
Color	Brown
Specific gravity (kg/L)	1.08 - 1.14
Chlorin content % (EN 480-10)	< 0.1

Proposed multi-response optimization and modeling framework

Optimization of the mixture parameters is aimed to increase the performance and productivity of ready-mixed concrete plant. There are 6 flow steps in performance optimization of ready-mixed concrete plant. This flow diagram is given in Figure 1. Also, FFD used in this study provides possibility of creating a model. Models which may be created give us the ability to predict responses for mixture parameters and two quality characteristics. Modeling and optimization of mixture proportions plays an important role on providing competitive advantage and customer satisfaction.

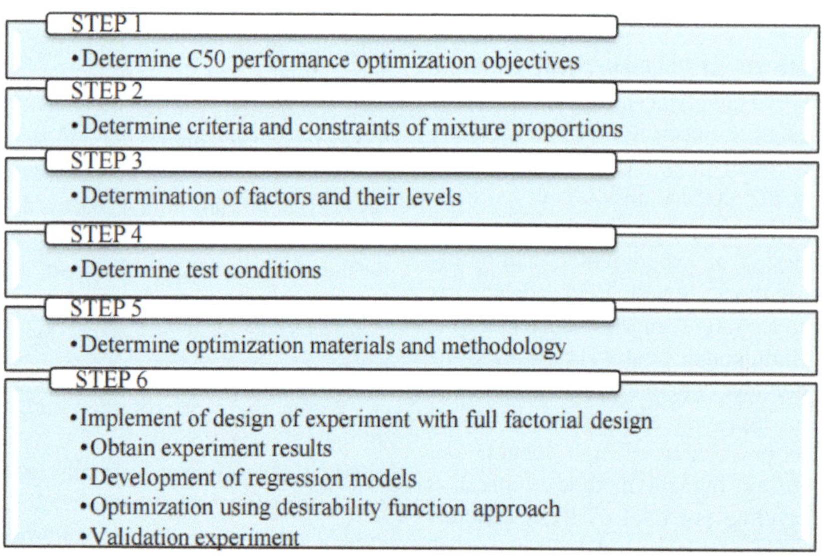

Figure 1: Proposed performance optimization framework

Determining criteria and constraints of optimization and modeling

Two quality characteristics were determined for C50 in the Ready-Mixed concrete plant (Table 3). The viscosity of produced concrete was evaluated through the slump flow test according to TS EN 12350/2 (Turkish standard) [15]. Slump flow range can be 10–220 mm for C50. Slump flow test is a kind of simple and fast implementation experiment both in laboratory and field [16]. The two performance criteria were identified respectively as the slump flow and 28th day compressive strength. Higher slump flow value gives higher workability of concrete. Therefore, the first criterion having effects on concrete quality is the slump flow value which should be maximized. For each concrete mixes, the compressive strength was determined on three 15 cm cubes for 28 days according to TS EN 12390/3 [17]. Each compressive strength experiment was an average of three 150 mm cube specimens. This criterion provides information about concrete durability [5, 18].

Table 3: Quality characteristic and their weights

Quality Characteristic	Symbol	Description	Type of concrete test	Target values	Weights
1	Y_1	Slump flow (cm)	Fresh concrete test	Larger is better	1
2	Y_2	Compressive strength (N/mm^2) 28 days	Hardened concrete test	Larger is better	1

The mixtures were prepared approximately in 4 min using a rotating planetary mixer. The total aggregate mixture weight is 1901 kg/m^3 . The measured concrete temperature varies between $10C^0$ and $16C^0$.

Determination of factors and their levels

Four factors that each has two control levels affects these performance criteria are identified. Water to binder materials ratio, coarse aggregate (II) to total aggregate ratio, the percentage of superplasticizer content and fly ash amount were identified as two level factors. These factors are symbolized X_1, X_2, X_3 and X_4 respectively (Table 4).

Table 4: Levels of factors that affect quality characteristic

Factors	Description	Bounds -1(coded values) First bound	1(coded values) Second bound
X_1	Water to binder materials ratio	0.48	0,52
X_2	Coarse aggregate (II) to total aggregate ratio	0.28	0.32
X_3	Superplasticizer content (kg/m^3)	1.00	1.20
X_4	Fly ash amount (kg/m^3)	60	80

Full factorial design and desirability function

Experimental design is a statistical methodology used to analyze the effect of several factors simultaneously. It makes changes to the independent factors (input) to determine their effect on the dependent response (output). It not only determines the significant factors that affect the response, but also how these factors affect the response [19, 20].

The polynomial regression model which can be considered for the three input factors is given in Eq. (1) [19]. In addition to the main effects of the three factors, interactions among the factors were also included in the regression, as shown in Eq. (1) [20]:

$$Y = \beta_0 + \sum_{i=1}^{k} \beta_i x_i + \sum_i \sum_j \beta_{ij} x_i x_j + \beta_{123} x_1 x_2 x_3 + e, \quad i < j \tag{1}$$

Where in eq.1 Y is the predicted response, β_0 defines the offset term, β_i is the linear effect of factor i, β_{ij} is the two-factor interaction effect between factors i and j and β_{123} represents the three-factor interaction effect.

The desirability function approach transforms an estimated response (e.g., the ith estimated response $\hat{y}_i$) into a scale-free value, called a desirability (denoted as di for $\hat{y}_i$). It is a value between 0 and 1, and increases as the corresponding response value becomes more desirable. The overall desirability D, another value between 0 and 1, is defined by combining the individual desirability values (i.e., di's). Then, the optimal setting is determined by maximizing D [21]. In optimization study, different desirability functions are used depending on the selected criteria (maximum, minimum, target value assignment or have a certain range) for each response [22]. The desirability function for a larger-the-better (LB) - type response is defined as [23].

$$d_i\left(\hat{y}_i\right) = \begin{cases} 0 & \hat{y}_i(x) < L_i \\ \left[\dfrac{\hat{y}_i(x) - L_i}{T_i - L_i}\right]^s & L_i \le \hat{y}_i(x) < T_i \\ 1 & \hat{y}_i(x) > T_i \end{cases} \tag{5}$$

where $d_i(\hat{y}_i(x))$ is the desirability function of $\hat{y}_i(x)$, Li is the lower bounds on the response, Ti is the desired target of the i th response, where Li ≤ Ti and si is the parameters that determine the shape of di($\hat{y}_i$(x)): if $s_i = 1$, the shape is linear; if $s_i > 1$, convex; and if $0 < s_i < 1$, concave. Derringer (1994) proposed a weighted geometric mean as a strategy to aggregate the individual di's:

$$D = \left(d_1^{v_1} \times d_2^{v_1} \times d_3^{v_1} \Lambda \times d_n^{v_1}\right)^{1/\sum v_i} \tag{3}$$

where v_i is the relative weight of the ith response [3, 24].

RESULTS

In Table 5, columns 2–5 represent the four control factors and their levels. In this study, a full factorial design (2^4) was used to implement the experiments and resultsare given in Table 5.

Table 5: Findings were obtained in all experiments

Exp. No	FFD (2^4) uncoded variables X_1	X_2	X_3	X_4	Y_1 mm Replicate 1	Y_1 mm Replicate 2	Y_2 (N/mm^2) Replicate 1	Y_2 (N/mm^2) Replicate 2
1	0.48	0.28	1.0	60	90	100	54.3	54.1
2	0.52	0.28	1.0	60	170	160	52.2	52.0
3	0.48	0.32	1.0	60	70	80	54.0	53.8
4	0.52	0.32	1.0	60	150	150	53.2	53.0
5	0.48	0.28	1.2	60	110	110	53.0	53.0
6	0.52	0.28	1.2	60	200	180	51.0	51.3
7	0.48	0.32	1.2	60	100	100	51.7	51.4
8	0.52	0.32	1.2	60	180	170	51.1	51.5
9	0.48	0.28	1.0	80	100	110	54.5	54.3
10	0.52	0.28	1.0	80	190	190	52.5	52.5
11	0.48	0.32	1.0	80	80	90	54.3	54.6
12	0.52	0.32	1.0	80	170	170	53.8	53.8
13	0.48	0.28	1.2	80	130	130	53.4	53.2
14	0.52	0.28	1.2	80	240	220	51.4	51.2
15	0.48	0.32	1.2	80	120	120	52.0	52.1
16	0.52	0.32	1.2	80	230	210	51.5	51.1

Development of the regression meta-models

Relations among factors and regression equalities were determined by polynomial regression analysis and degree of accuracy by determination coefficient (R^2). MINITAB® Statistical Program Package (version 15.1.1) was used for the data analysis. Regression models obtained with MINITAB® were given in eq. (4) and eq. (5). Estimated effects and coefficients for all criteria were given in Table 6 and 7.

$$Y_1 = 144.375 + 41.875 * X_1 - 7.5 * X_2 + 15 * X_3 + 11.875 * X_4 + 4.375 * X_1 * X_4 + 3.75 * X_3 * X_4 \tag{4}$$

$$Y_2 = 52.7 - 0.64 * X_1 - 0.84 * X_3 + 0.18 * X_4 + 0.34 * X_1 * X_2 - 0.29 * X_2 * X_3 - 0.12 * X_1 * X_3 * X_4 \tag{5}$$

The experimental results are analyzed by ANOVA (ANalysis Of VAriance) procedures. The ANOVA table gives a summary of the main effects and interactions (Tables 6 and 7). MINITAB® 15 displays both the sequential sums of squares (Seq SS) and adjusted sums of squares (AdjSS) [19]. Table 6 and 7 shows the p-values associated with each individual model term. The 'Term' column in Table 6 presents the main effects and all interactions. The second and third column displays the main effects and coefficients of the terms. The fifth and sixth columns display the t-ratios and p-values. The rows of all significant factors are shown in bold in Table 6 and 7 ($p<0.05$). The pvalues lead to following conclusions: (1) Two-way interaction X_1*X_4 ($p=0.005$) and X_3*X_4 ($p=0.012$) for slump flow and X_1*X_2 ($p=0.000$) and X_2*X_3 ($p=0.000$) for 28th day compressive strength is statistically significant, others are not. (2) Main factors X_1, X_3, X_4 ($p=0.000$) are statistically significant for two criteria; X_3, is only statistically significant for slump flow [20].

Table 6: Estimated effects and coefficients for modeling slump flow in coded values

Term	Effect	Coefficient	SE Coef	t	P value
Constant		**144.375**	**1.326**	**108.89**	**0.000**
X_1	**83.750**	**41.875**	**1.326**	**31.58**	**0.000**
X_2	**-15.00**	**-7.500**	**1.326**	**-5.66**	**0.000**
X_3	**30,00**	**15.000**	**1.326**	**11.31**	**0.000**
X_4	**23.750**	**11.875**	**1.326**	**8.96**	**0.000**
X_1*X_2	-0.000	-0.000	1.326	-0.00	1.000
X_1*X_3	5.000	2.500	1.326	1.89	0.078
X_1*X_4	**8.750**	**4.375**	**1.326**	**3.30**	**0.005**
X_2 *X_3	3.750	1.875	1.326	1.41	0.176
X_2 *X_4	0.000	0.000	1.326	0.00	1.000
X_3 *X_4	**7.500**	**3.750**	**1.326**	**2.83**	**0.012**
X_1 *X_2 *X_3	-1.250	-0.625	1.326	-0.47	0.644
X_1 *X_2 *X_4	-0.000	-0.000	1.326	-0.00	1.000
X_1 *X_3 *X_4	2.500	1.250	1.326	0.94	0.360
X_2 *X_3 *X_4	1.250	0.625	1.326	0.47	0.644
X_1 *X_2 *X_3*X_4	1.250	0.625	1.326	0.47	0.644

Table 7: Estimated effects and coefficients for modeling 28th day compressive strength in coded values

Term	Effect	Coefficient	SE Coef	t	P value
Constant		**52.7125**	**0.02932**	**1798.13**	**0.000**
X_1	**-1.2875**	**-0.6437**	**0.02932**	**-21.96**	**0.000**
X_2	-0.0625	-0.0313	0.02932	-1.07	0.302
X_3	**-1.6875**	**-0.8437**	**0.02932**	**-28.78**	**0.000**
X_4	**0.3500**	**0.1750**	**0.02932**	**5.97**	**0.000**
X_1*X_2	**0.675**	**0.3375**	**0.02932**	**11.51**	**0.000**
X_1*X_3	0.0750	0.0375	0.02932	1.28	0.219
X_1*X_4	-0.0375	-0.0188	0.02932	-0.64	0.531
X_2 *X_3	**-0.5750**	**-0.2875**	**0.02932**	**-9.81**	**0.000**
X_2 *X_4	0.0875	0.0437	0.02932	1.49	0.155
X_3 *X_4	-0.1125	-0.0562	0.02932	-1.92	0.073
X_1 *X_2 *X_3	0.0375	0.0188	0.02932	0.64	0.531
X_1 *X_2 *X_4	-0.0500	-0.0250	0.02932	-0.85	0.406
X_1 *X_3 *X_4	**-0.1250**	**-0.0625**	**0.02932**	**-2.13**	**0.049**
X_2 *X_3 *X_4	-0.0750	-0.0375	0.02932	-1.28	0.219
X_1 *X_2 *X_3*X_4	-0.0375	-0.0187	0.02932	-0.64	0.531

Validation of meta-models

The real data of responses versus the predicted responses are given in Figure 2 as the observed and predicted values respectively. The model explains that the experimental range was studied sufficiently. The fitted regression equation shows a good fit to the model [25].

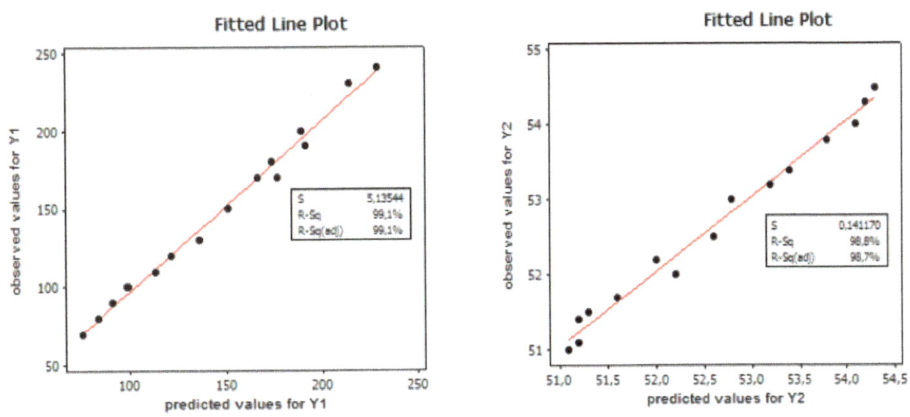

Figure 2: Observed values versus predicted values for criteria

Individual desirability values, the overall desirability D and predicted value were calculated by MINITAB and given in Figure 3. The factors obtained at the maximum points of Y_1 and Y_2 (target: 150 mm and 53 N/mm^2, respectively) were calculated as X_1=0.5090, X_2=0.3123, X_3=1.0012 and X_4=80 which are known as estimated condition (Figure 3).

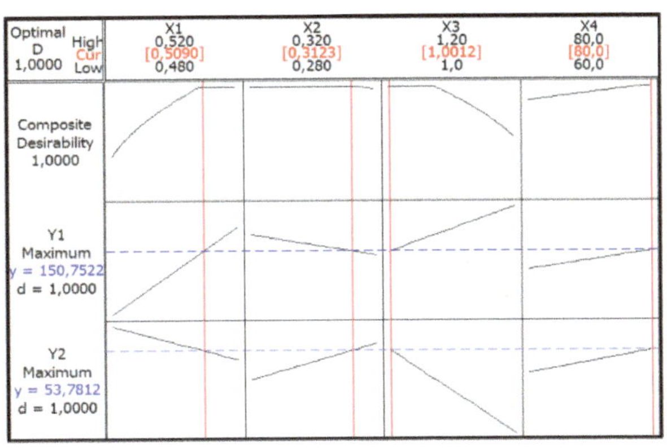

Figure 3: Optimization plot

Verification of optimal parameters

Validation experiments were applied at the estimated condition. The results illustrate that the experimental results are close to the estimated results (Table 8). The accuracy of the results can be taken as a proof for the validity of the estimated models.

Table 8: Test results for the verification of the results of FFD based DFA

Number	Responses	Predicted values	Verification experiment
1	Y_1	150.7522	150
2	Y_2	53.7812	54.0

CONCLUDING REMARKS

In this study, the optimization and the modeling of mixture proportions of C50 were performed by using a FFD based DFA. RMC consists of many conflicting factors; it is critical to use a systematic methodology for determining optimal mixes and modeling quality characteristics of C50 under a set of parameters that make up the concrete. For this reason, a FFD based DFA was used in this study to investigate ranking of the conflicting factor levels and best possible mix proportions of C50. The results showed that the proposed methodology is effective in solving the mixture proportions optimization problem. Also, it can be seen from the results of produced concrete samples, that they satisfied the expected properties of C50. Therefore, this study provides a tool for improvement of system performance for the organizations that have RMC. Outcomes were shared with the Concrete Manufacturer Company and more studies have been initiated to increase the system performance.

REFERENCES

1. S. Yan, H. C. Lin and Y. C. Liu, Optimal schedule adjustments for supplying ready mixed concrete following incidents, Automation in Construction 20, 1041–1050, 2011.
2. C. Y. Chang, R. Huang, P. C. Lee and T. L. Weng, Application of a weighted GreyTaguchi method for optimizing recycled aggregate concrete mixtures, Cement& Concrete Composites 33, 1038–1049, 2011.
3. M. Sonebi, Factorial design modelling of mix proportion parameters of underwater composite cement grouts, Cement& Concrete Composites 31, 1553–1560, 2001.
4. M. Muthukumar and D. Mohan, Studies on polymer concretes based on optimized aggregate mix proportion, European Polymer Journal 40, 2167- 2177, 2004.
5. E. Özbay, A. Öztaş, A. Baykaşoğlu and H. Özbebek, Investigating mix proportions of high strength self compacting concrete by using Taguchi method, Construction and Building Materials 23, 694-702, 2009.
6. S. L. Correia, F. L. Souza, G. Dienstmann and A. M. Segadaes, Assessment of the recycling potential of fresh concrete waste using a factorial design of experiments, Waste Management 29, 2886-2891, 2009.
7. S. L. Correia, T. Partala, F. C. Loch and A. M. Segadaes, Factorial design used to model the compressive strength of mortars containing recycled rubber, Composite Structures 92, 2047-2051, 2010.
8. A. Santilli, I. Puente and M. Tanco, A factorial design study to determine the significant parameters of fresh concrete lateral pressure and initial rate of pressure decay, Construction and Building Materials 25, 1946-1955, 2011.
9. A. N. S. Alqadi, K. N. B. Mustapha, S. Naganathan and Q. N. S. Al-Kadi, Development of self-compacting concrete using contrast constant factorial design, Journal of King Saud University, Engineering Sciences, In press, Available online 2 July 2012.
10. M. Ramezani, M. Bashiri and C.A. Atkinson, A goal programming-TOPSIS approach to multiple response optimization using the concepts of non-dominated solutions and prediction intervals, Expert Systems with Applications 38, 9557-9563, 2011.
11. P. G. Paterakis, E. S. Korakianiti, P. P. Dallas and D. M. Rekkas, Evaluation and simultaneous optimization of some pellets characteristics using a 33 factorial design and the desirability function, International Journal of Pharmaceutics 248, 51-60, 2002.
12. I. Mukherjee and P. K. Ray, Optimal process design of two-stage multiple responses grinding processes using desirability functions and metaheuristic technique, Applied Soft Computing 8, 402-421, 2008.
13. R. Gottipati and S. Mishra, Process optimization of adsorption of Cr(VI) on activated carbons prepared from plant precursors by a two-level full factorial design, Chemical Engineering Journal 160, 99-107, 2010.
14. N. Mandal, B. Doloi, B. Mondal and R. Das, Optimization of flank wear using Zirconia Toughened Alumina (ZTA) cutting tool: Taguchi method and Regression analysis, Measurements 44, 2149–2155, 2011.
15. TS EN 12350/2. Concrete, Fresh concrete experiments, slump flow test, Ankara, Turkish Standardization Institute, 2010 [in Turkish].
16. C. A. Arbeláez Jaramilo, J. Rigueira , W. Victor, J. R. M. Vargas, P. Serna Ros and M. P. Barbosa, Reduced Models Test for The Characterization of the Rheologic Properties of Self Compacting Concrete, in: 3rd International Symposium on SelfCompacting Concrete, Reykjavik, Iceland, 2003.
17. TS EN 12390/3, Testing hardened concrete-Part 3, Compressive strength of test specimens, Ankara, Turkish Standardization Institute; 2010 [in Turkish].
18. Turkish Standard TS EN 206-1. Concrete- Part 1: Specification, performance, production and conformity. Ankara, Turkish Standardization Institute, 2002 [in Turkish].
19. D. C. Montgomery, Design and Analysis of Experiments, 6th ed., New York, USA: John Wiley & Sons; 2005.
20. Y. T. İç, An experimental design approach using TOPSIS method for the selection of computer-integrated manufacturing technologies, Robotics and Computer-

Integrated Manufacturing 28, 245–256, 2012.

21. I. J. Jeong and K.J.Kim, An interactive desirability function method to multiresponse optimization, European Journal of Operational Research 195, 412–426, 2009.

22. G. Derringer and R. Suich, Simultaneous optimization of several response variables, Journal of Quality Technology 12, 214–219, 1980.

23. A. Fitrianto and H. Midi, Multi-Response Optimization via Desirability Function for the Black Liquor DATA, Journal of Science and Technology 4 (1), 2012.

24. G. Derringer, A balancing act: Optimizing a product's properties, Quality Progress 27, 51–57, 1994.

25. H. Cho and D.K. Zoh, Photocatalytic degradation of azo dye (Reactive Red 120) in TiO2/UV system: Optimization and modeling using a response surface methodology (RSM) based on the central composite design, Dyes and Pigments 75, 533-543, 2007.

CHAPTER 10

STUDIES ON PUMICE LIGHTWEIGHT AGGREGATE CONCRETE WITH QUARRY DUST USING MATHEMATICAL MODELING AID OF ACO TECHNIQUES

J. Rex and B. Kameshwari

INTRODUCTION

The aim of the pumice lightweight concrete is to add the quarry dust with the partial replacement of sand for the perfection of strength and the minimization of the cost and time. The High Performance Concrete may be aptly described as that unique brand of concrete, which meets with the distinctive efficiency and reliability constraints that are not capable of habitual estimation usually by employing the traditional materials and the conventional mixing, placing, and curing procedures. The concrete is one of the widely used construction materials which generally toe the line of the conventional bathtub hazard rate function curve [1]. The concrete predominantly boasts of superior compressive strength which is boosted by several qualities like elevated abrasion resistance, stiffness, minimal permeability, superior durability, greater early strength gain, and reduced cost per unit load [2]. The Ordinary Portland Cement (OPC), in fact, is a noteworthy material in the creation of concrete which generally functions as its binder to combine all the collected materials. It is pertinent to note that the OPC invariably requires the burning of huge amount of fuel and decay of limestone [3]. In this regard, the lightweight concrete is extensively employed in place of the usual concrete in view of a large number of improved qualities. The most current vantage point of the lightweight concrete is its unique diminished structural dead weight. The dip in the dead weight goes a long way in the diminution in the building expenditure [4]. The employment of the lightweight aggregates (LWA) in concrete offers a multitude of constructive features. The low-density of the concrete enjoyed by the lightweight aggregates normally paves the way for the cutback in the dead load of edifices, footings size and dimensions of columns, slabs, and beams [5]. The tensile strength of a model concrete habitually exceeds that of the prototype concrete. Hence, it is all the more essential to be aware of the tensile qualities of the model concrete in the whole modeling task [6]. With a view to augmenting the fracture resistance of the cementitious materials, fibers are recurrently supplemented, which leads to the formation of a composite material compressive strength and tensile strength. The compressive strength is individual for the structural applications while the flexural strength is specific for the pavement applications [7]. The fines having less than 150-micron dimension are segregated from the unprocessed crusher dust and the crusher dust devoid of the micro fines is effectively utilized for the overall substitution of sand in masonry and experimented for the fundamental compressive strength [8]. The impact of the crushed stone dust as the fine sand leads to the enhancement in the flexural strength in relation to the concrete with the natural sand and the value goes southward as the percentage of the crusher dust goes upward [9]. In this connection the quarry dust has been kick-started as a viable substitute for the river sand which amazingly adds on additional advantages to the concrete. It is everybody's knowledge that the quarry dust reinforces the strength of the concrete as against the concrete generated out of an equivalent quantity of the river sand, though it paves the way for diminution in the workability of the concrete

Citation: J. Rex and B. Kameshwari, "Studies on Pumice Lightweight Aggregate Concrete with Quarry Dust Using Mathematical Modeling Aid of ACO Techniques," Advances in Materials Science and Engineering, vol. 2016, Article ID 9583757, 12 pages, 2016. doi:10.1155/2016/9583757.

[10]. In an associated expansion, sizeable quantities of the quarry dust are parked in plenty in the region of caliber in the crushed rock sites. The quarry dust is recognized as the highly functional filler in the bituminous concrete [11]. In the highways construction, it is a common procedure to deploy the quarry fines to steady the lateritic materials employed in the road pavements. However, it is unfortunate that the relative benefit is not found to be entirely exploited in the building of structural elements in the edifices and corresponding structures [12]. The most cost-conscious and the simplest method to achieve a viable alternative for the natural sand is attained from the limestone quarries, the lateritic sand, and the crushing natural stone quarries which is called the manufactured sand [13]. Several investigations were carried out on the cubes and beams to assess the compressive, flexural strengths of the concrete made of the Quarry Rock Dust for three diverse proportions and five dissimilar techniques. Durability tests were carried out for the concrete with the Quarry Rock Dust and analyzed and contrasted with the traditional concrete [14]. In view of the fact that their properties are more or less the same as that of the sand, the marble sludge powder and the quarry dust are effectively employed as fine aggregate in the cement concrete [15]. In this regard, the Ordinary Portland Cement (OPC) is somewhat substituted by fly ash, bottom ash, fine aggregate, coarse aggregate, and Light Expanded Clay Aggregate (LECA) by weights of 5%, 10%, 15%, 20%, 25%, 30%, and 35%, respectively. The mathematical modeling based forecast, in turn, is performed with the active support of the optimization method which is capable of effectively evaluating the compressive strength, the split tensile strength, and the flexural strength with the aid of the recognized input values. The quality of the paper is the strength of pumice lightweight concrete. With partial replacement of the sand with the quarry dust the time and cost are minimized [16].

LITERATURE REVIEW

In 2012 Sivakumar and Gomathi [17] got name and fame for projecting the fly ash as one of the godsend gifts which can be used as both supplementary cementitious material and lightweight aggregate. The artificially generated lightweight aggregates were obtained from the industrial byproducts like the fly ash, bottom ash, silica fume, blast furnace slag, rice husk, slag or sludge waste or palm oil shell, shale, slate, and clay. The advent of the modern era witnessed with curious eyes the ever-rocking usage of the cost-conscious building materials triggered by the amazing augmentation in the requirement for the lightweight concrete for a host of applications. Of late, the inclusion of the artificial aggregates paved the way for a realistic cutback in the building outlays and has become the cynosure of attraction in view of its shining quality equivalent to those of the traditional aggregates.

In 2013 Kabir et al. [18] amazingly advocated the potential utilization of the early-day compressive strength outcomes to forecast the characteristic strength of the standard weight concrete which was the subject matter of research. An easy mathematical model equipped with the acumen of forecasting the compressive strength of concrete at any age was elegantly launched for both the stone and the local aggregate concrete. The data deployed in this exploration were gathered from certain earlier researches and modern investigational endeavors. The evaluations performed by means of the innovative technique employing diverse data effectively illustrated convincing forecast of the concrete strength at various ages such as 7, 14, and 28 days with incredible excellence.

In 2013 Afify and Soliman [19] had their glorious days when they valiantly launched the lightweight aggregates and the chemical admixtures which cast a very significant part in the manufacture of the lightweight concrete. It was heartening that the innovative artificial coarse aggregate was in the pipeline, riveting the eager eyes of the enthusiastic experimenter, and it was well-geared to be deployed in the production of the lightweight concrete. The captioned investigation was spearheaded with an eye on ascertaining the viability of the lightweight aggregate type commercially offered in the domain of the concrete industry in Egypt. The configuration performance of the tested beams and slabs was subjected to deep experimentation with special focus on their deflections, longitudinal strain and cracking against the backdrop of various stages of loading in addition to the final loads and modes of breakdown.

In 2013 Devi [20] vividly explained the corrosion as the fundamental durability issue triggering the corrosion of the concrete structures. Thus, the underlying aim behind the investigation is to evaluate the strength and corrosion resisting the qualities of concrete having the quarry dust as the fine aggregate together with calcium nitrite as the corrosion inhibiting admixture at the dosage of 1%, 2%, 3%, and 4% in terms of weight of the cement. Experimentations were performed to ascertain the strength, the water absorption capacity, and the durability and the outcomes were analyzed and contrasted with the natural sand concrete. The corrosion resistance efficiency was evaluated by means of the impressed voltage method, the rapid chloride permeability test, and the gravimetric weight loss technique. The cheering outcomes proclaimed the efficacy of the quarry dust as the most appropriate alternative for the river sand in the concrete which is competent to perk up the strength of the concrete, and especially with the addition of the inhibitor it is found to offer very effective resistance against the corrosion.

In 2013 Nagpal et al. [21] were wise enough to advocate the appropriateness of the crushed stone dust waste as the fine aggregate for the concrete which was analyzed and its fundamental traits contrasted with those of the traditional concrete. Two vital mixes were chosen for the natural sand to attain the M25 and M30 grade concrete. The equivalent mixes were attained by substituting the natural sand with the stone dust in part and also completely. The fascinating outcomes proved without any iota of doubt that the crushed stone dust waste had the quality of being employed extensively and effectually as a substitute for the natural sand in the concrete. In the test investigations regarding the strength qualities of the concrete using the crushed stone dust as the fine aggregate it was unequivocally established that the enhancement in the compressive strength, the flexural strength, and the tensile strength of the concrete had been awesome and amazing.

In 2014 Subbulakshmi and Vidivelli [22] convincingly spelt out the fact that the High Performance Concrete was capable of accomplishing incredibly higher performance from the concrete as against the one achieved by the routing concrete. The ambit of the current investigation was elongated to cover the research regarding the impact of the quarry dust towards the efficiency in accomplishment of the High Performance Concrete. An earnest endeavor was made to keep a keen eye on the mechanical qualities of High Performance Concrete created with quarry dust material. The strength qualities like the compressive strength and flexural strength were subjected to experimentation to ascertain the optimum substitution of the quarry dust. The feat of the concrete ratio and quarry dust substitution level on the compressive strength of the quarry dust concrete also became the subject matter of intensive investigation.

PROPOSED METHODOLOGY

The technique effectively employed forecast the objective function by means of the mighty mathematical modeling. The training and testing technique is deployed to ascertain the objective function of the mathematical model. The coarse aggregate (CA), fine aggregate (FA), cement, water, silica fume (SF), superplasticizer (SP), quarry dust (QD), load (KN), and ultimate load (KN) are the various significant parameters that are made use of to train the mathematical model by exploiting 80% test database and the residual 20% finds itself grossly engaged in authenticating the mathematical model. Once the authentication is over, the mathematical model is well set to pursue its paramount purpose of ushering in the objective function of the concrete mixing task. Thereafter, the optimization algorithm is entrusted with the indubitable task of predicting various parameters such as optimal compressive strength (Mpa), split tensile strength (Mpa), flexural strength (Mpa), and deflection (in mm). With this end in view, a feast of diverse optimization algorithms are used to evaluate the optimal solutions and the most prominent among them are kingpin genetic algorithm (GA), particle swarm optimization (PSO), ant colony optimization (ACO), and artificial bee colony (ABC) optimization algorithm. The related optimization algorithms invariably resort to their own unique optimization procedures for bringing to limelight the optimal value. In the optimization procedure, at the out the constraints such as the minimum cost and time reduction are assigned. The objective functions are effectively evaluated by means of the optimal solutions like coarse aggregate (CA), fine aggregate (FA), cement, water, silica fume (SF), superplasticizer (SP), quarry dust (QD), load (KN), and ultimate load (KN). In fact the characteristic of the ant colony optimization (ACO) algorithm procedure is mainly assigned the paramount task of ascertaining the optimal solutions like compressive strength (Mpa), split tensile strength (Mpa), flexural strength (Mpa), and deflection (in mm) which pave the way for the incredible scaling down of the financial outlays together with fabulous decrease in the time-frame.

Materials Used for Experimental Work

In the experimental work, the Ordinary Portland Cement is used. Normal river sand is used as fine aggregate with specific gravity of 2.82 and the course aggregate used is the pumice lightweight aggregate of size 16 mm. The mix design is based on the ACI method. The chemical properties of the pumice aggregate are presented in Table 1.

Table 1: Properties of aggregate.

Sl. number	Chemical contents	Percentage (%)
1	CaO	4.6
2	SiO_2	59.2
3	Al_2O_3	16.6
4	Fe_2O_3	4.80
5	MgO	1.8
6	SO_3	0.40
7	K_2O	5.40
8	Na_2O	5.60
9	Loss on ignition	1.60

Mix Proportion and Laboratory Testing Program

Eight types of concrete mixtures with different proportions of lightweight aggregate and quarry dust are offered. In addition to the above mixture concrete is mixed with silica fume in replacement of cement up to 10% and chemical admixture of SP430 is added at 2% to cement weight (Table 2).

Table 2: Mixing proportions.

Sl. number	Mix type	Pumice aggregate (%)	Quarry dust (%)	Admixtures
1	M1	100	0	10% of cement is replaced with silica fume & 2% of Conplast SP430 admixture is added to cement weight
2	M2	90	10	
3	M3	80	20	
4	M4	70	30	
5	M5	60	40	
6	M6	50	50	
7	M7	25	75	
8	M8	0	100	

Cube Compressive Strength

The different mixes were moulded in cube of size 150 × 150 × 150 mm for testing the compressive strength. Before moulding, the mould is thoroughly cleaned using a waste cloth and properly oiled along its faces. Then the concrete was poured into the mould, compacted, and cured for 7 days and 28 days. After curing the cube is dried and tested in CTM and the compressive strength is found and listed in Table 3.

Table 3: Compressive strength split tensile strength and flexural strength for mixing concrete.

Sl. number	Mix type	Compressive strength (Mpa)		Split tensile strength (Mpa)		Flexural strength (Mpa)
		7 days	28 days	7 days	28 days	28 days
1	M1	21.6	26.4	1.78	1.880	22.04
2	M2	22.41	27.8	1.828	2.107	23.81
3	M3	23.32	28.35	1.88	2.162	25.17
4	M4	23.97	29.31	1.91	2.230	26.90
5	M5	25.21	30.64	1.97	2.279	28.55
6	M6	24.32	29.67	1.85	2.229	26.65
7	M7	20.65	25.19	1.68	1.883	24.83
8	M8	18.49	23.24	1.62	1.810	22.80

Split Tensile Strength

In order to find the split tensile test the different concrete mixes are casted in a standard cylinder of size diameters 150 mm and 300 mm in length. The casted concrete is demoulded after one day of casting and cured for 7 days and 28 days. The cured specimens were tested in CTM and the split tensile strength is obtained.

Flexural Strength Beams and Reinforcement Details

The size of the beam for finding the flexural strength is 1000 × 150 × 150 mm. Totally 24 beams were casted and tested for their flexural strength. From Figure 1 the load is at 1/3rd of the span and the distance of applying load is 300 mm, the length is 150 mm, and the breadth is 150 mm in the beam.

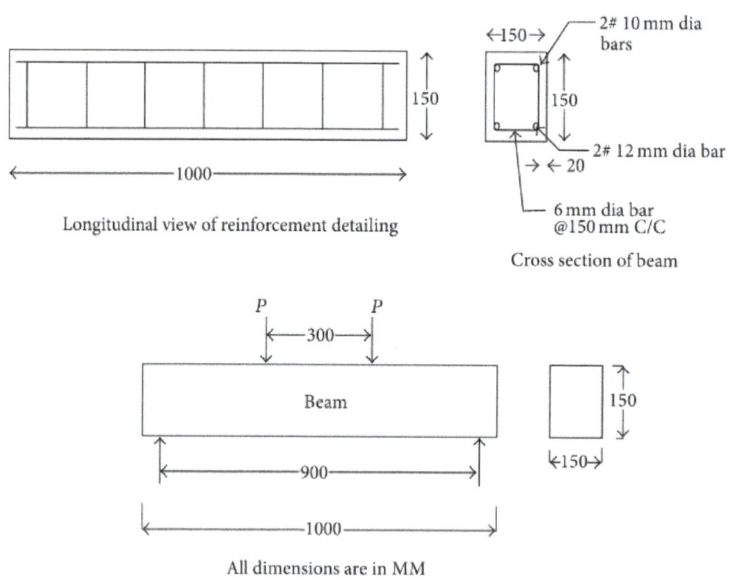

Figure 1: Beam in deflection.

ACO ALGORITHM

The ACO algorithm represents one of the effective methods and possesses the significant features of the state transition regulation and the pheromone update technologies. A colony of ants is sent to locate the solutions in each and every iteration. Each ant performs each and every step strictly in accordance with the state transition regulation. When an ant completes a task, the local pheromone update is initiated on the chosen strength, which significantly scales down the probability for the other ants to select the identical strength continually and thus paves the way for diversity in the search. After every ant has built a perfect solution, the global pheromone update mechanism is initiated on the edges of the best-so-far solution with an eye on amplifying the appeal of the localities. The concrete materials like coarse aggregate (CA), fine aggregate (FA), cement, water, silica fume (SF), superplasticizer (SP), and quarry dust (QD) are assured in the mixing ratio and compressive strength, split tensile strength, flexural strength, and deflection are analyzed after applying the load in the ACO technique.

Flowchart

Figure 2 shows the flowchart of the ant colony optimization (ACO) which is explained below. The solutions are initialized in the flow and the fitness functions are computed. Then the probability transition is formed based on the function of pheromone updating and then the evaporation is performed. Finally the optimum value is predicted under the minimum error value.

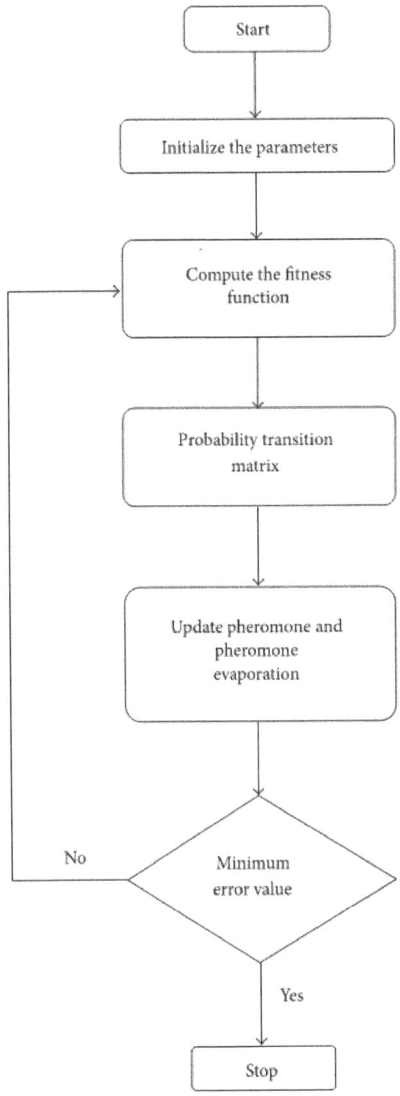

Figure 2: Flowchart for mathematical modeling with ant colony optimization.

Initialize the Parameters

Initialize the controlling parameters α_{ij} and β_{ij} in the solution and transmit them to the fitness calculation:

$$R_i = \{R_{0j}, R_{1j}, \ldots, R_{nj}\}, \tag{1}$$

where R_i defines an initial solution, $i \in [1, 2, \ldots, 10]$, and $j \in [1, 2, \ldots, 160]$. Since ith value is considered as the number of solution and jth value is considered as the length of solution, then

$$R_j = [(\text{No of hidden neuron} * \text{No of input data}) + \text{No of hidden neuron}], \tag{2}$$

where total input = 7 and hidden neuron (h) = 20.

Based on (2), the attained solution length is 160 and the solution range lies within $-10 \le R_{ij} \le 10$. According to the initial solution four outputs such as compressive strength (Mpa), split tensile strength (Mpa), flexural strength (Mpa), and deflection (in mm) are evaluated.

Fitness Function

The Sigmoid function is derived as a function which is incorporated with the artificial neural network process to optimize the mathematical modeling. By optimizing α_{ij} and β_{ij} in the function we can obtain the optimized mathematical modeling. The error value is attained as [original value − obtained value] in the mathematical modeling:

$$f_i = \sum_{j=1}^{h} \alpha_j * \left[\frac{1}{1 + \exp\left(-\sum_{i=1}^{N} R_i \beta_{ij}\right)} \right], \tag{3}$$

$$F_i = \frac{\sqrt{(\text{original} - \text{obtained})^2}}{\text{No of input data}}. \tag{4}$$

Equation (4) is considered for the fitness computation in the optimization. The error value is minimized in the fitness which is computed to predict the optimal solution by using the controlling parameters such as α_{ij} and β_{ij}. N indicates number of data. The outputs considered as the optimal solutions constitute compressive strength (Mpa), split tensile strength (Mpa), flexural strength (Mpa), and deflection (in mm).

Probability Transition Matrix

At each and every configuration phase, ant c applies a probabilistic action choice rule, known as the random proportional rule, to determine which location has to be visited next. Particularly, the probability with which ant c, currently at location i, prefers to go to location j is detailed below:

$$P_{ij}^{c} = \frac{(\tau_{ij})^{\alpha} (\eta_{ij})^{\beta}}{\sum (\tau_{ij})^{\alpha} (\eta_{ij})^{\beta}}, \tag{5}$$

where $\eta_{ij} = 1/d_{ij}$ represents a heuristic value and edges (i, j) where j is a location that the ant c has not visited till now. By this probability regulation, the possibility of choosing a specific arc (i, j) enhances the value of the linked pheromone trail τ_{ij} and of the heuristic data value η_{ij} and the parameters α and β organize the comparative significance of the pheromone versus the heuristic data, which is furnished by η_{ij} (i.e., if α_{ij} = 0, the contiguous cities are further likely to be chosen, while if β_{ij} = 0, only pheromone amplification is functioning).

Update Pheromone and Evaporation Pheromone

The pheromone values are updated when all the s ants have formed solutions after their iteration. The pheromone intensity linked with the edge joining locations i and j is updated below:

$$\tau_{ij} = (1 - \rho) * \tau_{ij} + \sum_{c=1}^{s} \Delta\tau_{ij}^{c}, \tag{6}$$

where ρ is pheromone evaporation rate, s is the number of ants, and $\Delta\tau_{ij}^{c}$ is the quantity of pheromone laid on edge (i, j) by cth ant. Consider

$$\Delta\tau_{ij}^{c} = \begin{cases} \frac{Q}{Lc}; & \text{if ant } c \text{ used connection } (i, j) \text{ in its tour} \\ 0; & \text{otherwise.} \end{cases} \tag{7}$$

Equation (6) comprises two distinct components. The left segment enables the pheromone on all edges decay.The speed of this decay is represented by ρ, which is the evaporation constraint. The right segment enhances the pheromone on all the edges visited by ants. The amount of pheromone an ant c deposits on an edge is indicated by Lc which represents the length of the tour generated by the related ant. In this manner, the enhancement of pheromone for an edge is clearly dependent on the number of ants that utilize this edge and on the excellence of the solutions located by those ants. The updating procedure of ACO technique is shown in above process. Finally the error is forecast until the minimization.

The fitness function is performed in the equations below:

$$F_i = F_1, F_2, F_3, F_4, \tag{8}$$

where F_1 is compressive strength, F_2 is split tensile strength, F_3 is flexural strength, and F_4 is deflection.

The outputs considered as the optimal solutions like compressive strength (Mpa) as (F_1), split tensile strength (Mpa) as (F_2), flexural strength (Mpa) as (F_3), and deflection (in mm) as (F_4) are applied in the fitness function for predicting the optimal solution:

$$f_i = \sum_{j=1}^{h} \alpha_{j(\text{opt})} * \left[\frac{1}{1 + \exp\left(-\sum_{i=1}^{N} R_i \beta_{ij(\text{opt})}\right)} \right]. \tag{9}$$

From (9) the outputs $F2$, $F3$, $F4$ are calculated and the optimal value is obtained.

RESULT AND DISCUSSION

The different inputs represent coarse aggregate (CA), fine aggregate (FA), cement, water, silica fume (SF), superplasticizer (SP), quarry dust (QD), load (KN), and ultimate load (KN). The proposed mathematical modeling with the ant colony optimization is used to find the optimal solutions like compressive strength (Mpa), split tensile strength (Mpa), flexural strength (Mpa), and deflection (in mm). During the optimization procedure, at the outset, the constraints such as the minimum cost and time reduction are assigned. The ant colony optimization is used with the mathematical modeling in the concrete mixing ratios and the strength is analyzed by applying the load in the concrete. In the real time experiment the cost required is high and the time is extended and to solve the problem the mathematical modeling is mentioned to optimize compressive strength (Mpa), split tensile strength (Mpa), flexural strength (Mpa), and deflection (in mm).

Mathematical Modeling with Optimization Techniques

The mathematical modeling with PSO, GA, and ACO yields the minimum error value for finding the optimal equation with the optimal controlling parameters (α_{ij} and β_{ij}). The error is minimum when the ACO algorithm is used compared with the other techniques such as the GA and PSO. Figure 3 shows the error graph for the mathematical modeling with genetic algorithm (GA), particle swarm optimization (PSO), and ant colony optimization (ACO) which are the optimization techniques used to find the error value of the mathematical modeling.

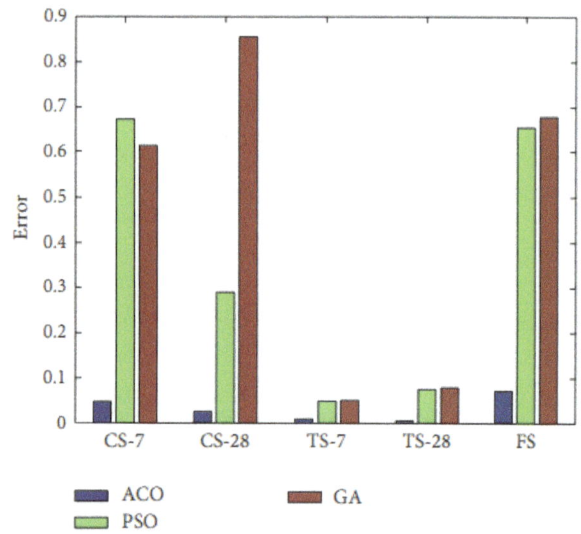

Figure 3: Error value graph.

The error value is determined by using the difference between the predicted values and desired values and it gives the fitness of the mathematical modeling. In Figure 3 the error graph shows that the ant colony optimization algorithm minimizes the error value in the compressive strength (CS-7) (CS-28), the split tensile strength (TS-7) (TS-28), and the flexural strength (FS) to 0.02967 when compared to other techniques like GA and PSO.

Figure 4 shows the deflection error when applying the load in a certain range and the minimization of error value is predicted by the ant colony optimization (ACO) compared to the GA and PSO. The load is taken as 10 to 90 KN and the error value is zero. The prediction performance of the deflection is 94% in the mathematical modeling.

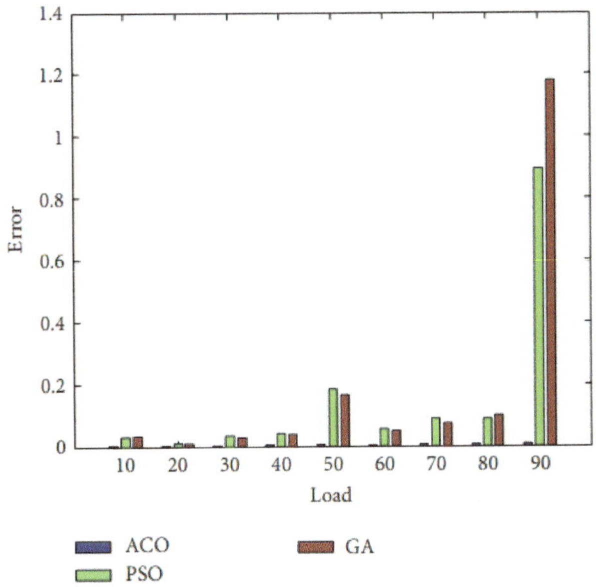

Figure 4: Error graph of deflection.

Experimental Results

The ACO algorithm offers the optimal solutions such as compressive strength (Mpa), split tensile strength (Mpa), flexural strength (Mpa), and deflection (in mm).

Table 3 shows the compressive strength, the split tensile strength, and the flexural strength of concrete while mixing coarse aggregate (CA), fine aggregate (FA), cement, water, silica fume (SF) and 10% of each mixing ratio except M1, superplasticizer (SP) and 2% of each mixing ratio except M1, and quarry dust (QD) in percentage like 10%, 20% up to 100% are added except M1 mixing ratio for 7 days and 28 days. In this process the ACO is used in the mathematical modeling to predict the optimal solutions.

Figure 5 shows the deflection in the center of the concrete beam as M1 to M8 mixing ratio by using the ACI method of mix design with the lightweight concrete and replacement of the quarry dust. The load is applied at the 1/3rd of the span like 10 KN, 20 KN up to 100 KN and the deflection is calculated by the measuring instruments.

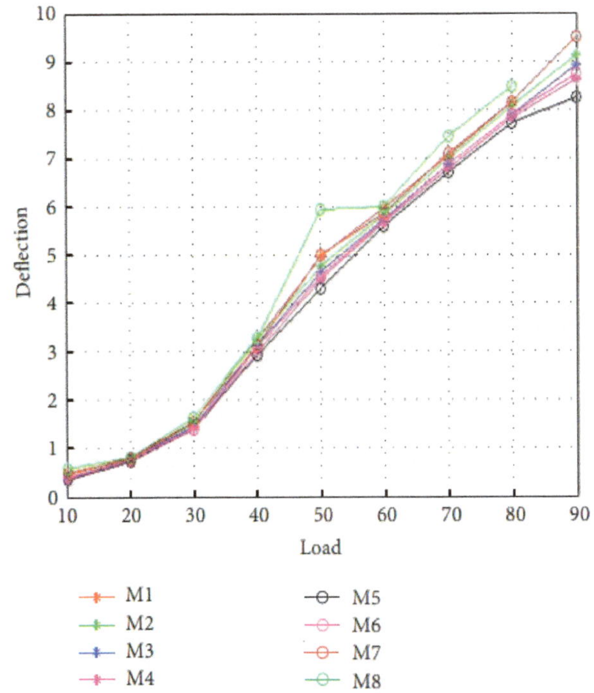

Figure 5: Load versus deflection.

Optimal Solutions of the ACO

In Table 4 the optimal solutions for the different concrete mixings are shown. From the mathematical modeling of the compressive strength, the flexural strength, and the split tensile strength the optimal values are obtained from the ACO algorithm. Based on the mixing coarse aggregate (CA), fine aggregate (FA), cement, water, silica fume (SF) and 10% of each mixing ratio except M1, and superplasticizer (SP) and 2% of each mixing ratio except the optimal solutions are obtained. In 7-day testing of M1 mixing the compressive strength is 22.363 and compared to the 28 days the variation is 4.23%. Similarly in the other days also the optimal strength is attained.

Table 4: Compressive strength split tensile strength and flexural strength for mixing concrete.

Sl. number	Mix type	Compressive strength (Mpa)		Split tensile strength (Mpa)		Flexural strength (Mpa)
		7 days	28 days	7 days	28 days	28 days
1	M1	22.3634	26.4	1.780148	1.983401	21.98486
2	M2	22.3534	27.79988	1.831565	2.149852	23.81681
3	M3	23.39843	29.0571	1.83313	2.149852	25.16909
4	M4	23.39843	29.31015	1.833131	2.224513	26.93635
5	M5	23.38843	30.64	1.833131	2.279675	28.55261
6	M6	26.06786	29.67506	1.833131	2.224513	26.63621
7	M7	17.43668	25.1898	1.709717	1.87364	24.75354
8	M8	17.29327	23.24	1.709268	1.810025	22.79477

Convergence Graphs

In Figure 6, the convergence graph is plotted between the iteration and fitness estimations of the various strategies, such as GA, PSO, and ACO. The graph shows that the ACO procedure gives the greatest fitness using the least possible iteration. Through the graph, the ACO strategy takes the minimum iteration for providing the ideal result and it receives the primary greatest estimation of the fitness. Subsequently, in the compressive strength, the value is minimized in the ACO technique and the ultimate load is utilized to predict the optimal solution and in the same process the split tensile strength and flexural strength are presented. In the convergence the iteration of compressive strength is for 7 days and 28 days and the fitness variation is 0 to 4 and then in split tensile strength for 7 days from 0 to 100 iterations shows from 0 to 1 fitness variation and for 28 days it is varied from 0 to 4; then at the end the flexural strength for 28 days shows from 0 to 4 fitness variations in convergence.

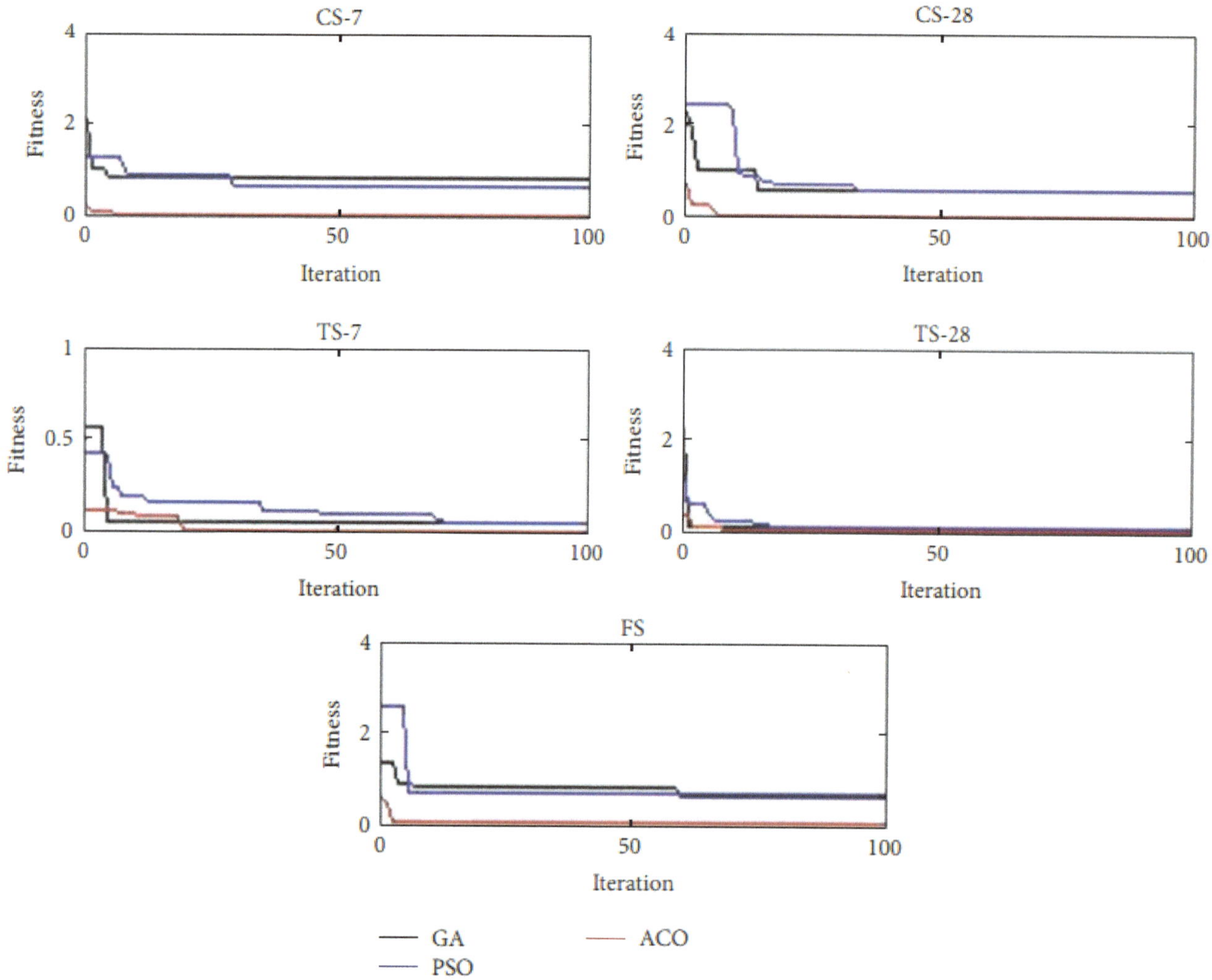

Figure 6: Convergence graph.

Figure 7 shows that in the deflection convergence the fitness and iteration are estimated in the graph by applying load at the 1/3rd of the span. The deflection is formed and the performance is calculated under 10 KN to 90 KN load in each iteration. The ACO is the least in the iteration compared to the other techniques such as the genetic algorithm (GA) and particle swarm optimization (PSO). In 10 KN load the fitness is varied from 0 to 1, in 20 KN load the fitness is varied from 0 to 1.5, in 30 KN load the fitness is varied from 0 to 1.5, in 40 KN load the fitness is varied from 0 to 2, in 50 KN load the fitness is varied from 0 to 1, in 60 KN load the fitness is varied from 0 to 3, in 70 KN load the fitness is varied from 0 to 1, in 80 KN load the fitness is varied from 0 to 1.5, and in 90 KN load the fitness is varied from 0 to 2.

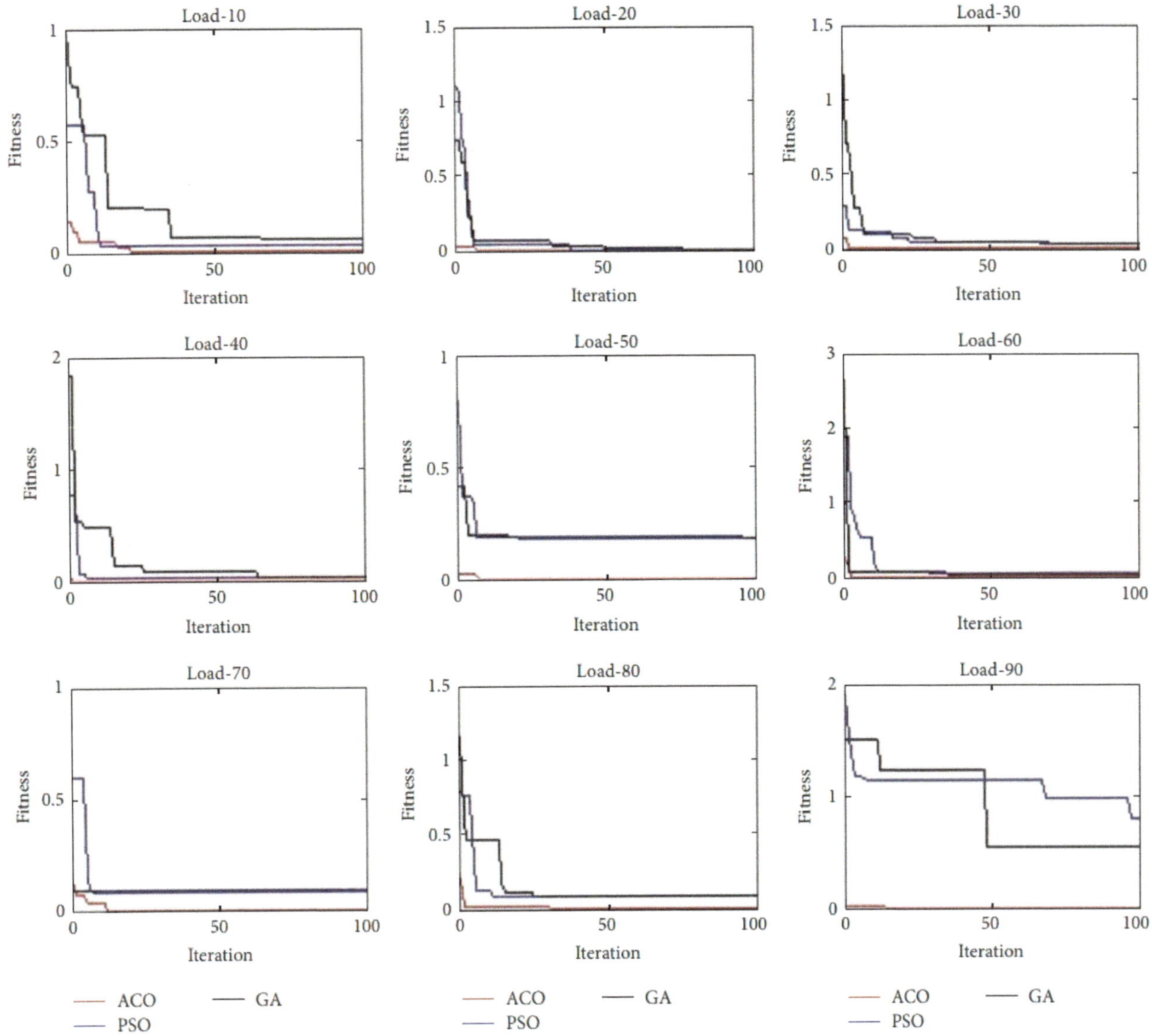

Figure 7: Convergence graph of deflection.

Prediction of Outputs

Figures 8, 9, and 10 and Table 5 show the variation in the actual and predicted results of 7 days and 28 days of the mixes M1 to M8 using the ACO technique for various parameters such as the compressive strength, the split tensile strength, and the flexural strength. The training and testing values are deployed to ascertain the objective functions of the mathematical modeling. The same procedure is performed for 7 days and 28 days, respectively. In the case of the flexural strength, the strength is varied up to 29, and here the predicted value is high whereas the actual value is low. The experimental results are compared to the predicted results and the error has a minimum value. By using the mathematical modeling with the ACO technique the prediction is achieved.

Table 5: Comparison of actual and predicted results for various parameters.

Number of data	Parameter/test	Curing period	Variation	High-level variation mix
Ml to M8	Compressive strength	7 days	2	M5
		28 days	2.5	
Ml to M8	Split tensile strength	7 days	1.7	M5
		28 days	1.8	
Ml to M8	Flexural strength	28 days	2.2	M5

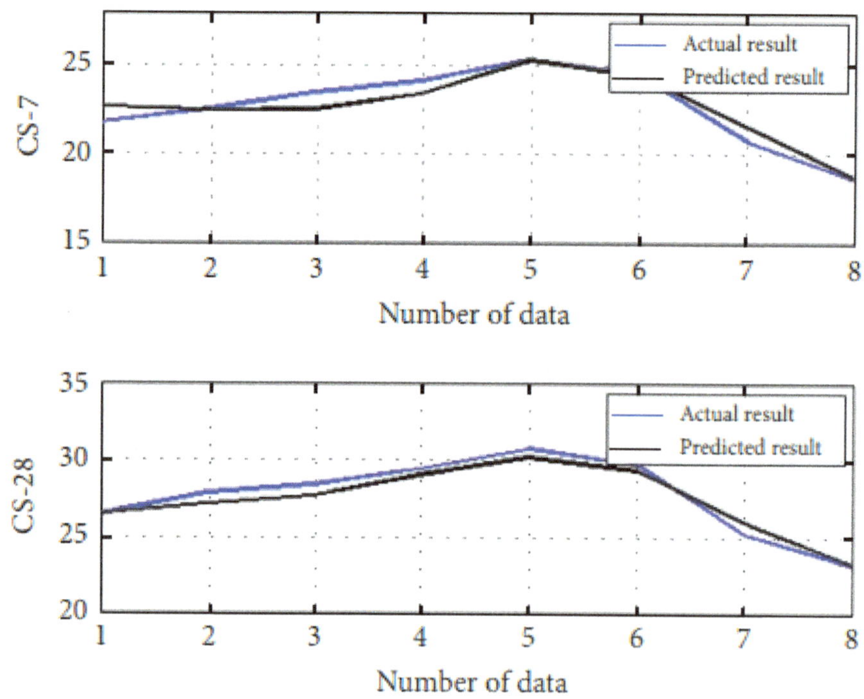

Figure 8: Prediction of compressive strength.

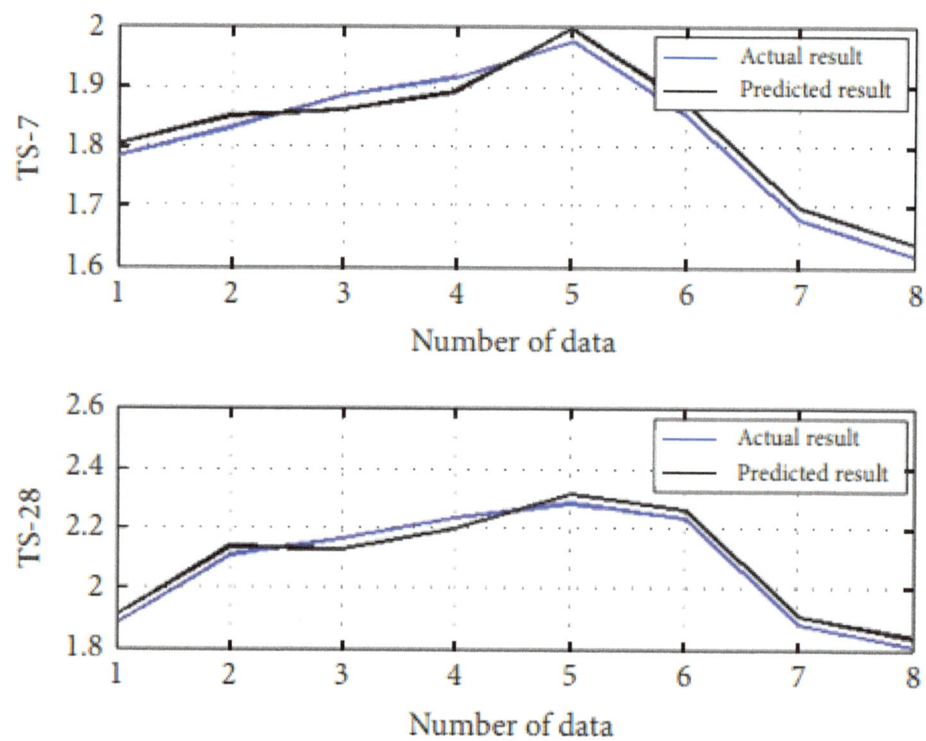

Figure 9: Prediction of split tensile strength.

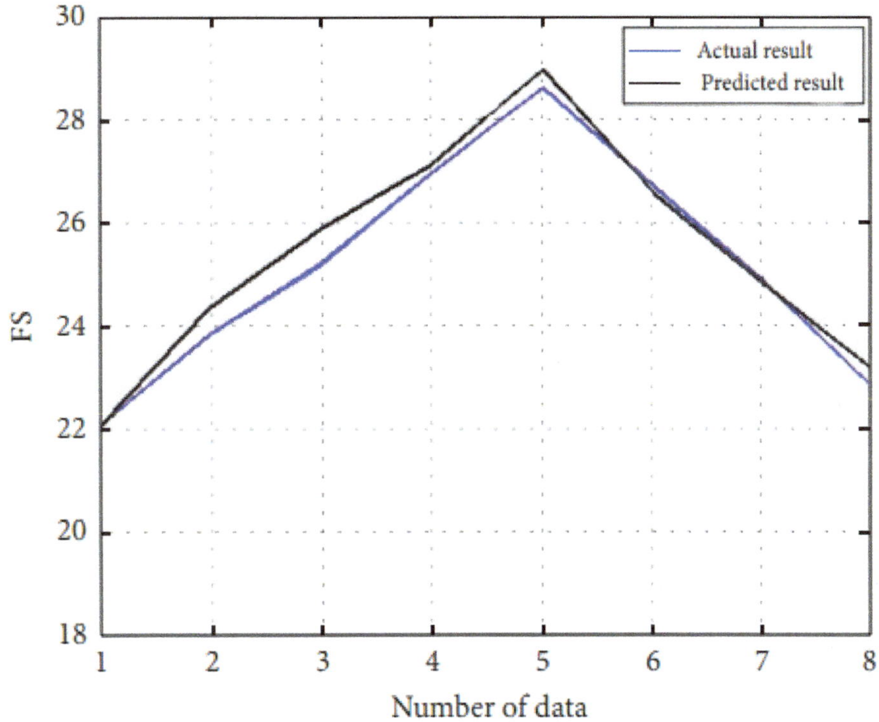

Figure 10: Prediction of flexural strength.

Figure 11 and Table 6 show the deflections by applying the 1/3rd span of the beam. The lightweight concrete with the silica fume (SF), the superplasticizer (SP), and the quarry dust (QD) is added to the mix design, and then the beam is cured for 28 days, respectively, and the specimens are used for finding various parameters such as the compressive strength, the split tensile strength, and the flexural strength. Finally the load is applied at the 1/3rd of the span in the beam and the deflection is specified, and then with the help of the mathematical modeling in the ACO technique, the prediction is analyzed and performed.

Table 6: Variation on actual and predicted values of deflection in flexural strength test.

Load	Number of data	Variation: actual versus predicted
10	M1 to M8	0 to 1
20	M1 to M8	0.7 to 0.9
30	M1 to M8	0 to 4
40	M1 to M8	0 to 10
50	M1 to M8	0 to 10
60	M1 to M8	0 to 10
70	M1 to M8	0 to 10
80	M1 to M8	0 to 10
90	M1 to M8	−20 to 20

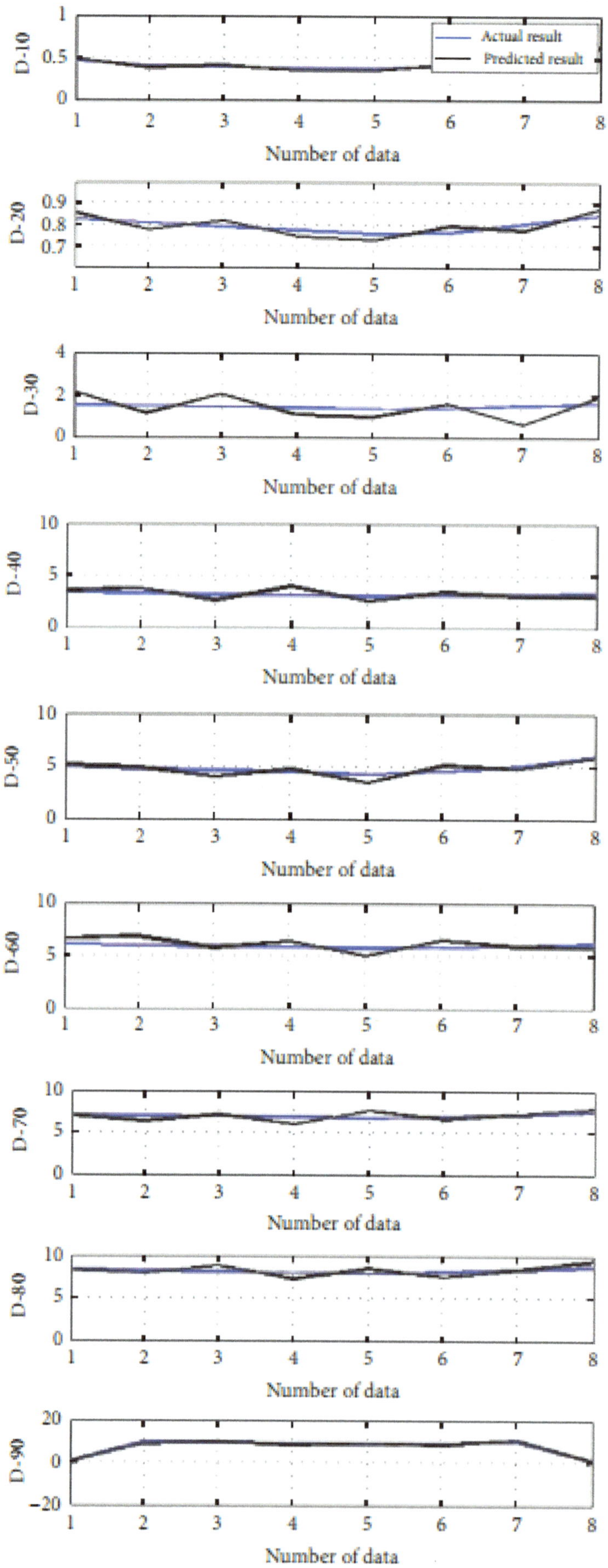

Figure 11: Prediction of load versus deflection.

CONCLUSION

The offered numerical modeling with the ant colony optimization strategy is executed for various parameters such as compressive strength (Mpa), split tensile strength (Mpa), flexural strength (Mpa), and deflection (in mm). The inputs like coarse aggregate (CA), fine aggregate (FA), cement, water, silica fume (SF), superplasticizer (SP), quarry dust (QD), load (KN), and ultimate load (KN) are involved to predict the minimum values. In each output the mathematical modeling with the ant colony optimization is performed and the results are shown. According to the mathematical modeling the fitness is analyzed for various parameters such as compressive strength, split tensile strength, flexural strength, and deflection. The mathematical modeling with the optimization is utilized to minimize the cost and to save the time in the future experiments. The performance obtained in the ACO for various parameters such as compressive strength, split tensile strength, flexural strength, and deflection is 91%, 98%, 87%, and 94% of predicted values, respectively, in the mathematical modeling. It is worth mentioning that the entire procedure gets well implemented in the working platform of the MATLAB software. It is also revealed from the experimental data that the mix proportion of M5 gives good results in partial replacement of the quarry dust in the fine aggregate in the analysis of the compressive strength, the split tensile test, and the flexural properties.

CONFLICT OF INTERESTS

The authors declare that there is no conflict of interests regarding the publication of this paper.

REFERENCES

1. V. Patel and N. Shah, "A survey of high performance concrete developments in civil engineering field,"Journal of Civil Engineering, vol. 3, no. 2, pp. 69–79, 2013.
2. K. W. Hoe and M. Ramli, "Rational mix design approach for high strength concrete using sand with very high fineness modulus," American Journal of Applied Sciences, vol. 7, no. 12, pp. 1562–1568, 2010.
3. M. M. Al Bakri, H. Mohammed, H. Kamarudin, K. Niza, and Y. Zarina, "Review on fly ash-based geopolymer concrete without Portland Cement," Journal of Engineering and Technology Research, vol. 3, no. 1, pp. 1–4, 2011.
4. H. K. Kim, J. H. Jeon, and H. K. Lee, "Workability, and mechanical, acoustic and thermal properties of lightweight aggregate concrete with a high volume of entrained air," Construction and Building Materials, vol. 29, pp. 193–200, 2012.
5. Ben Fraj, M. Kismi, and P. Mounanga, "Valorization of coarse rigid polyurethane foam waste in lightweight aggregate concrete," Construction and Building Materials, vol. 24, no. 6, pp. 1069–1077, 2010.
6. S. O. Franklin, "Modelling of the compressive and tensile strength relationship of concrete in studies on the punching phenomenon in prestressed flat slabs," Journal of Applied Sciences Research, vol. 6, no. 3, pp. 205–211, 2010.
7. P. Ramadoss, "Modeling for the evaluation of strength and toughness of high-performance fiber reinforced concrete," Journal of Engineering Science and Technology, vol. 7, no. 3, pp. 280–291, 2012.
8. P. Appukutty and R. Murugesan, "Substitution of quarry dust to sand for mortar in brick masonry works," International Journal on Design and Manufacturing Technologies, vol. 3, no. 1, pp. 59–63, 2009.
9. P. P. Nanda, A. K. Das, and N. C. Moharana, "Stone crusher dust as a fine aggregate in concrete for paving blocks," Journal of Civil and Structural Engineering, vol. 1, no. 3, pp. 613–620, 2010.
10. C. Sukesh, K. B. Krishna, P. Sri Lakshmi Sai Teja, and S. Kanakambara Rao, "Partial replacement of sand with quarry dust in concrete," International Journal of Innovative Technology and Exploring Engineering, vol. 2, no. 6, pp. 254–258, 2013.
11. J. O. Ukpata, M. E. Ephraim, and G. A. Akeke, "Compressive strength of concrete using lateritic sand and quarry dust as fine aggregate," Journal of Engineering and Applied Sciences, vol. 7, no. 1, pp. 81–92, 2012.
12. J. O. Ukpata and M. E. Ephraim, "Flexural and tensile strength properties of concrete using lateritic sand and quarry dust as fine aggregate," Journal of Engineering and Applied Sciences, vol. 7, no. 3, pp. 324–331, 2012.
13. Jayaraman, V. Senthilkumar, and M. Saravanan, "Compressive and tensile strength of concrete using lateritic sand and lime stone filler as fine aggregate," Journal of Research in Engineering and Technology, vol. 3, no. 1, pp. 79–84, 2014.
14. R. Ilangovana, N. Mahendrana, and K. Nagamanib, "Strength and durability properties of concrete containing quarry rock dust as fine aggregate," Journal of Engineering and Applied Sciences, vol. 3, no. 5, pp. 20–26, 2008.
15. M. S. Hameed and A. S. S. Sekar, "Properties of green concrete containing quarry rock dust and marble sludge powder as fine aggregate," Journal of Engineering and Applied Sciences, vol. 4, no. 4, pp. 83–89, 2009.
16. U. Abubakar and K. S. Baharudin, "Compressive strength of high volume coal bottom ash utilization as fine aggregate in fly ash-cement blended concrete," Journal of Engineering & Technology Sciences, vol. 1, no. 4, pp. 226–239, 2013.
17. Sivakumar and P. Gomathi, "Pelletized fly ash lightweight aggregate concrete: a promising material,"Journal of Civil Engineering and Construction Technology, vol. 3, no. 2, pp. 42–48, 2012.
18. Kabir, M. Hasan, and Md. K. Miah, "Strength prediction model for concrete," Journal of Civil Engineering, vol. 2, no. 1, pp. 14–19, 2013.
19. M. R. Afify and N. M. Soliman, "Feasibility of using lightweight artificial coarse aggregates in the manufacture of R. C. elements," International Journal of Engineering and Advanced Technology, vol. 3, no. 2, pp. 290–306, 2013.
20. M. Devi, "Performance evaluations of calcium nitrite on steel rebar corrosion in quarry dust concrete,"Journal

of Engineering and Science, vol. 2, no. 10, pp. 23–27, 2013. V

21. L. Nagpal, A. Dewangan, S. Dhiman, and S. Kumar, "Evaluation of strength characteristics of concrete using crushed stone dust as fine aggregate," Journal of Innovative Technology and Exploring Engineering, vol. 2, no. 6, pp. 102–104, 2013.

22. T. Subbulakshmi and B. Vidivelli, "Mechanical properties of high performance concrete in corporating with quarry wastes," Journal of Engineering and Advanced Technology, vol. 3, no. 6, pp. 231–236, 2014.

CHAPTER 11 AN OPTIMIZATION MODEL FOR DESIGN OF ASPHALT PAVEMENTS BASED ON IHAP CODE NUMBER 234

Ali Reza Ghanizadeh

INTRODUCTION

Surface transportation is the most widely used mode of transportation, and pavements are an essential part of roads, streets, and parking lots in all over the world. The development of a country is often judged in terms of its total paved road. Like other engineered structures, designed pavements are expected to be adequately strong and durable for their design life. They are expected to function properly by providing a smooth traveling surface for the traffic under various conditions of the environment. In order to ensure this, pavements must be designed, constructed, maintained, and managed properly [1].

Pavements can be mainly classified into flexible and rigid pavements. The wearing layer, which is in contact with the traffic, is an asphalt mix in case of flexible pavements and Portland cement concrete in case of rigid pavements.

Methods for design of flexible pavements can be classified into five main categories as follows [2]:(1)empirical method with or without a soil strength test,(2)limiting shear failure method,(3)limiting deflection method,(4)regression method based on pavement performance or road test,(5)mechanistic-empirical method.The concept of mechanistic-empirical pavement design was firstly introduced in the 1960s [3, 4]. This method is used nowadays by many countries and institutions for the design of flexible pavements [5–11]. However, due to the need for equipped labs and because extensive research is still needed for the use of mechanistic-empirical method, only empirical methods are used for pavement design in many countries [12–15].

Flexible pavements design in Iran commonly is accomplished by method presented in Iranian Highway Asphalt Paving Code Number 234 (IHAP Code 234). This method is categorized as a regression method and has been adopted with minor changes from AASHTO 1993 pavement design method. Flexible pavement design, using AASHTO method, is based on studies and tests that AASHTO carried out in Ottawa and Illinois between the years 1958 and 1960. The first AASHTO pavement design manual was published in 1961 and revised in 1972 and 1981. Again, during the years 1984 and 1985, a committee, consisting of AASHTO experts and some consultant engineers, revised it under NCHRP 20-7/24 Project and, after some modifications, presented the 1986 AASHTO pavement design manual [2]. Then in 1993 the method was revised again and led to the publication of the 1993 version of the manual. Equations proposed in AASHTO 1993 have some significant limitations, because these equations were developed for the specific conditions of the AASHO Road Test. These limitations are as follows.(i)The equations are only valid in case of the specific pavement materials and roadbed soil of the AASHO Road Test.(ii)The equations are only valid in case of environmental conditions at the AASHO Road Test.(iii)The developed equations are based on a two-year testing period rather than a long period. Therefore, it is difficult to extrapolate environmental factors to a longer period.(iv)The traffic was about one million equivalent single axle load (ESAL) and tires inflated to 70 psi.Despite all the abovementioned limitations,

Citation: Ali Reza Ghanizadeh, "An Optimization Model for Design of Asphalt Pavements Based on IHAP Code Number 234," Advances in Civil Engineering, vol. 2016, Article ID 5942342, 8 pages, 2016. doi:10.1155/2016/5942342.

Iran Highway Asphalt Paving Code Number 234 has been proposed for design of flexible pavements in all conditions [15]. Also, due to the lack of information about life-cycle cost parameters in Iran, life cycle analysis has not been considered in Iran Highway Asphalt Paving Code Number 234 and design of flexible pavements is commonly fulfilled by considering construction costs only.

Researchers proposed different models and approaches for optimum design of flexible pavement structures. Rouphail proposed a mixed-integer programming model to identify the number, type, and thicknesses of paving materials required to meet the structural strength requirements of the pavement system at a minimum initial cost [16]. Nicholls presented an optimization program supplement to the DNPS86 pavement design computer program to produce a minimum-cost combination of pavement layer thicknesses. This method enables the obtaining of the least cost design for flexible and rigid pavements without manual iteration [17]. Mamlouk et al. developed a project-level optimization approach to minimize total pavement cost within an analysis period [18]. This approach enables the designer to select the optimum initial pavement thickness, overlay thickness, and overlay timing. The developed model combined the AASHTO design procedure and the mechanistic multilayer elastic solution. Mu-yu and Shao-yi proposed an optimization model for design of flexible pavements based on fatigue and rutting performance [19]. They used genetic algorithms (GAs) to solve the optimization model. Abaza and Abu-Eisheh represented an optimum approach for the design of flexible pavements based on AASHTO method which utilized the anticipated performance of pavement and its life-cycle cost. They showed that pavements should be designed for higher terminal serviceability index values than currently recommended [20]. Ouyang and Madanat presented a mixed-integer nonlinear programming for optimal highway pavement rehabilitation planning which minimizes the life-cycle cost during design period [21]. Fakhri and Ghanizadeh developed an optimization model to determine the optimum structure and thickness of pavement layers, based on the AASHTO method [22]. The proposed model, in the form of a linear programming model, could determine the optimum configuration of pavement layers as well as optimum thickness of pavement layers. It could only consider the optimum structure of pavements consisting of asphalt, granular base, and granular subbase layers. Proposed model did not consider the treated base layers in pavement structure. Also, by employing this model, thickness of layers was determined as real numbers not integer numbers which should be revised for application in construction stage. Sanchez-Silva et al. present a model for reliability cost-based optimization of asphalt pavement structures based on both economic and operation considerations [23]. The proposed model considered the fatigue damage caused on the asphalt surface and the degradation of granular materials caused by repetitive loading cycles. They showed that the reliability based design optimization combined with a long-term maintenance policy of pavements produces appropriate integral designs. Rajbongshi and Das presented a simple methodology to assist a pavement designer in selecting an optimal pavement design thickness which is cost effective yet does not compromise the reliability of the pavement design [24]. They developed pavement design charts as an illustrative example to explain how the proposed methodology can be considered as an improvement over the deterministic design. Santos and Ferreira proposed a pavement design optimization model, called OPTIPAV, which considers pavement performance, construction costs, maintenance and rehabilitation costs, user costs, the residual value of the pavement at the end of the project analysis period, and preventive maintenance and rehabilitation interventions [25].

In this paper, an optimization model was proposed to determine the optimum structure of pavement as well as the thickness of each layer, based on minimizing the initial construction cost of pavement. Also several charts were developed to determine the optimum thickness of pavement layers with respect to road classification, design traffic, and strength of subgrade soil. The proposed optimization method in this research considers stabilized layers in optimization model. Also optimization problem was modeled as an integer programming model.

PAVEMENT DESIGN USING IHAP CODE 234

Equation (1) shows the basic equation for design of flexible pavements using IHAP Code 234 [15]:

$$\log W_{8.2} = Z_R S_0 + 9.36 \log(\mathrm{SN}+1) - 0.2 + \frac{\log(\Delta\mathrm{PSI}/(4.2-1.5))}{0.4 + 1094/(\mathrm{SN}+1)^{5.19}} + 2.3\log\left(\frac{M_r}{0.07}\right) - 8.07, \quad (1)$$

where $W_{8.2}$ is predicted number of 80 kN single axel load applications, Z_R is standard normal deviate, S_0 is combined standard error of traffic prediction and performance prediction, ΔPSI is difference between initial design serviceability index (P_0) and terminal design serviceability index (P_t), and M_r is resilient modulus (kg/cm^2).

In this equation all the parameters are known except the pavement structural number (SN). So, it is possible to find the value of SN by solving (1) using iteration method. It can also be solved with the help of existing graphs.

Design parameters dictated by IHAP Code 234 for design of flexible pavements for different class of roads are given in Table 1.

Table 1: Design parameters for different class of roads [15].

Class of road	Reliability	Standard normal deviate (Z_R)	S_0	P_i	P_t
Expressways and freeways	90	−1.282	0.35	4.2	3.0
Major rural roads	80	−0.841	0.35	4.2	2.5
Secondary rural roads (2nd class)	70	−0.524	0.35	4.2	2.0

After finding SN, it is possible to find the thickness of each layer by converting SN to real thickness of the constituent layers. The thickness of each layer should be found so that the following equation can be totally satisfied:

$$\mathrm{SN} = \frac{1}{2.5}\left(a_1 d_1 + m_2 a_2 d_2 + m_3 a_3 d_3 + \cdots + m_n a_n d_n\right), \tag{2}$$

where m_i is ith layer drainage coefficient. a_i is ith layer coefficient. di is ith layer thickness (cm).

The layer coefficient for an asphalt concrete layer with resilient modulus of 31500 kg/cm² can be assumed as 0.44. Equations (3) and (4) can be used for determination of the layer coefficients for granular base and granular subbase layers, respectively

$$a_{\mathrm{GB}} = 0.249\log_{10}\left(\frac{M_{r\mathrm{GB}}}{0.07}\right) - 0.977, \tag{3}$$

$$a_{\mathrm{GSB}} = 0.227\log_{10}\left(\frac{M_{r\mathrm{GSB}}}{0.07}\right) - 0.839, \tag{4}$$

where $M_{r\mathrm{GB}}$ is the resilient modulus of granular base layer and $M_{r\mathrm{GSB}}$ is the resilient modulus of granular subbase layer in kg/cm².

In this study, the following regression equations were developed to determine the layer coefficient and resilient modulus of different layers based on charts presented in IHAP Code 234 and AASHTO 1993.

Asphalt concrete is as follows:

$$\begin{aligned} a_{\mathrm{AC}} &= 0.16727\ln\left(M_{r\mathrm{AC}}\right) - 1.29682, \\ M_{r\mathrm{AC}} &= 33.511268\,\mathrm{Marshal}_{\mathrm{AC}} - 326.00433. \end{aligned} \tag{5}$$

Asphalt treated base is as follows:

$$\begin{aligned} a_{\mathrm{ATB}} &= 0.1419\ln\left(M_{r\mathrm{ATB}}\right) - 1.13175, \\ M_{r\mathrm{ATB}} &= 0.014391\left(\mathrm{Marshal}_{\mathrm{ATB}}\right)^2 \\ &\quad + 14.969313\,\mathrm{Marshal}_{\mathrm{ATB}} + 5295.655843. \end{aligned} \tag{6}$$

Cement treated base is as follows:

$$\begin{aligned} a_{\mathrm{CTB}} &= 0.2108\ln\left(M_{r\mathrm{CTB}}\right) - 2.08733, \\ a_{\mathrm{CTB}} &= 0.00224\,\mathrm{CS}_{\mathrm{CTB}} + 0.0935. \end{aligned} \tag{7}$$

Granular base is as follows:

$$M_{r\mathrm{GB}} = 299.73347 \times \mathrm{CBR}_{\mathrm{GB}}^{0.43023}. \tag{8}$$

Granular subbase is as follows:

$$M_{r\mathrm{GSB}} = 391.5423 \times \mathrm{CBR}_{\mathrm{GSB}}^{0.28552}. \tag{9}$$

Subgrade soil is as follows:

$$M_{r\mathrm{SG}} = 134.79525 \times \mathrm{CBR}_{\mathrm{SG}}^{0.6846}, \tag{10}$$

where M_r denotes resilient modulus in kg/cm², Marshal denotes marshal stability of asphalt mix materials in kg, a denotes layer coefficient, and CS denotes 7 days compressive strength (kg/cm²) of cement treated materials. In case of all developed equations, the coefficient of determination (R_2) was more than 0.98.

Not only should (2) be satisfied but also the thickness of each layer should be such that the total compressive stress, applied on lower layers, be reduced to the tolerable stress of these layers. To this end, the following equations should be satisfied

$$\begin{aligned} a_1 \cdot d_1 &\geq 2.5\mathrm{SN}_1, \\ a_1 \cdot d_1 + m_2 \cdot a_2 \cdot d_2 &\geq 2.5\mathrm{SN}_2, \\ a_1 \cdot d_1 + m_2 \cdot a_2 \cdot d_2 + m_3 \cdot a_3 \cdot d_3 &\geq 2.5\mathrm{SN}_3. \end{aligned} \tag{11}$$

In these equations, SN_1, SN_2, and SN_3 are the structural number of granular base, granular subbase, and subgrade layers, respectively. Their values are found from (1) with the only difference that instead of resilient modulus of subgrade soil, resilient modulus of granular base is used to find SN_1 and resilient modulus of granular subbase is used to find SN_2. Also, the thickness of asphalt concrete layer and granular base layer should not be taken less than those given in Table 2, considering the construction thickness. For the granular subbase layer, the minimum thickness should be considered as 15 cm.

Table 2: Minimum thickness for asphalt concrete and granular base [15].

Traffic (ESAL)	Asphalt concrete (cm)	Granular base (cm)
Less than 150000	5	10
150000–500000	6	10
500000–2000000	8	15
2000000–7000000	9	15
Greater than 7000000	10	15

OPTIMAL DESIGN OF FLEXIBLE PAVEMENTS BASED ON IHAP CODE 234

Considering a six-layer pavement system as shown in Figure 1, the integer programming model to find the optimum structural configuration as well as optimum thickness of each layer can be written as follows.

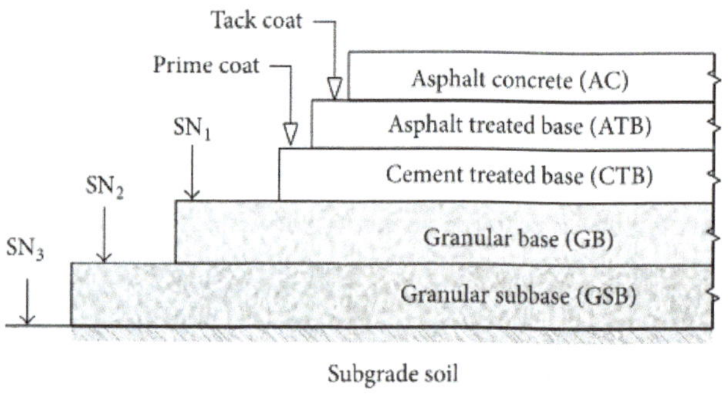

Figure 1: Assumed structure of pavement for optimization.

Objective Function. Consider

$$\begin{aligned}\text{Min } Z &= C_{AC}\cdot d_{AC} + E_{ATB}\cdot C_{ATB}\cdot d_{ATB}\\ &+ \frac{E_{CTB}\cdot C_{CTB}\cdot d_{CTB}}{100} + \frac{C_{GB}\cdot d_{GB}}{100}\\ &+ \frac{C_{GSB}\cdot d_{GSB}}{100} + E_{ATB}\cdot C_{TC}\cdot Q_{TC}.\end{aligned} \tag{12}$$

Constraints

Van Tile Constraints. Consider

$$\begin{aligned}&a_{AC}\cdot d_{AC} + E_{ATB}\cdot a_{ATB}\cdot d_{ATB} + E_{CTB}\cdot a_{CTB}\\ &\quad\cdot d_{CTB} \geq 2.5\text{SN}_1\cdot E_{GB},\end{aligned} \tag{13}$$

$$\begin{aligned}&a_{AC}\cdot d_{AC} + E_{ATB}\cdot a_{ATB}\cdot d_{ATB} + E_{CTB}\cdot a_{CTB}\cdot d_{CTB}\\ &\quad + E_{GB}\cdot m_{GB}\cdot a_{GB}\cdot d_{GB} \geq 2.5\text{SN}_2\cdot E_{GSB},\end{aligned} \tag{14}$$

$$\begin{aligned}&a_{AC}\cdot d_{AC} + E_{ATB}\cdot a_{ATB}\cdot d_{ATB} + E_{CTB}\cdot a_{CTB}\cdot d_{CTB}\\ &\quad + E_{GB}\cdot m_{GB}\cdot a_{GB}\cdot d_{GB} + E_{GSB}\cdot m_{GSB}\cdot a_{GSB}\\ &\quad\cdot d_{GSB} \geq 2.5\text{SN}_3,\end{aligned} \tag{15}$$

$$d_{AC} \geq \min\nolimits_{d_{AC}}, \tag{16}$$

$$d_{ATB} \geq E_{ATB}\cdot \min\nolimits_{d_{ATB}}, \tag{17}$$

$$d_{CTB} \geq E_{CTB}\cdot \min\nolimits_{d_{CTB}}, \tag{18}$$

$$d_{GB} \geq E_{GB}\cdot \min\nolimits_{d_{GB}}, \tag{19}$$

$$d_{GSB} \geq E_{GSB}\cdot \min\nolimits_{d_{GSB}}. \tag{20}$$

Decision Variables. Consider

$$\begin{aligned}&d_{AC}, d_{ATB}, d_{CTB}, d_{GB}, d_{GSB} = \text{integer},\\ &\quad E_{ATB}, E_{CTB}, E_{GB}, E_{GSB} = \text{binery}.\end{aligned} \tag{21}$$

Consider the following:

Z is construction cost of one square meter of pavement;

d_{AC} is thickness of asphalt concrete layer (cm);

d_{ATB} is thickness of asphalt treated base layer (cm);

d_{CTB} is thickness of cement treated base layer (cm);

d_{GB} is thickness of granular base layer (cm);

d_{GSB} is thickness of granular subbase layer (cm);

Q_{TC} is quantity of tack coat for bounding asphalt concrete layer to asphalt treated base layer (kg/m^2);

a_{AC} is layer coefficient of asphalt concrete material;

a_{ATB} is layer coefficient of asphalt treated base material;

a_{CTB} is layer coefficient of cement treated base material;

a_{GB} is layer coefficient of asphalt granular base material;

a_{GSB} is layer coefficient of granular subbase material;

C_{AC} is construction cost of one square meter of asphalt concrete material having a thickness of 1 cm;

C_{ATB} is construction cost of one square meter of asphalt treated base material having a thickness of 1 cm thickness;

C_{CTB} is construction cost of one cubic meter of cement treated base material;

C_{GB} is construction cost of one cubic meter of granular base material;

C_{GSB} is construction cost of one cubic meter of granular subbase material;

C_{TC} is construction cost of one kilogram of tack coat material;

E_{ATB} is variable indicating the presence or absence of the asphalt treated base layer in the optimal structure of the pavement; $E_{ATB} = 1$ means the presence and $E_{ATB} = 0$ means the absence of the asphalt treated base layer;

E_{CTB} is variable indicating the presence or absence of the cement treated base layer in the optimal structure of the pavement; $E_{CTB} = 1$ means the presence and $E_{CTB} = 0$ means the absence of the asphalt treated base layer;

E_{GB} is variable indicating the presence or absence of the granular base layer in the optimal structure of the pavement; $E_{GB} = 1$ means the presence and $E_{GB} = 0$ means the absence of the granular base layer;

E_{GSB} is variable indicating the presence or absence of the granular subbase layer in the optimal structure of the pavement; $E_{GSB} = 1$ means the presence and $E_{GSB} = 0$ means the absence of the granular subbase layer;

$\min_{dATB}$ is minimum thickness of asphalt treated base layer (cm);

$\min_{dCTB}$ is minimum thickness of cement treated base layer (cm);

min $_{dGB}$ is minimum thickness of granular base layer (cm);

min $_{dGSB}$ is minimum thickness of granular subbase layer (cm).

Constraints (13) to (15) are related to maximum allowable compressive stress on granular base, granular subbase, and subgrade layers, respectively. Constraints (16) to (20) represent the constraints for the minimum construction thickness of different layers.

According to Figure 1, if an asphalt treated base layer is used in the pavement structure, application of an additional tack coating becomes necessary; presence of $E_{ATB} \cdot C_{TC} \cdot Q_{TC}$ term in the objective function reflects the additional cost in this case.

SOLVING THE OPTIMIZATION MODEL

A careful study of the model shows that the presence of some terms such as $E{\cdot}d$ obtained from the multiplication of two unknown decision variables converts the optimization model to a nonlinear optimization model. In fact some terms in the objective function and constraints are in the form of product of two unknown decision variables (the thickness of the layer multiplied by the variable indicating the presence or absence of layer).

It is obvious that the value of variables (E) can only be equal to zero or one. So, in order to convert this nonlinear model to a linear model, once the value of the variable (E) is assumed to be equal to zero (the obscene of the desired layer in optimal configuration of pavement) and once the value of the variable (E) is assumed to be equal to one (the presence of the desired layer in optimal configuration of pavement) and in each case, the objective function and constraints are written and the resulting integer programming model is solved. Then it is possible to find the optimum configuration as well as thicknesses of pavement by comparison between the construction cost of different alternatives.

In fact, four layers including asphalt treated base (ATB), cement treated base (CTB), granular base, and granular subbase can be considered or not considered in the pavement structure and by combination of these alternatives, total of 16 alternatives for structural configuration of pavement is achieved. Possible structures for pavement are shown in Table 3. According to this method, the integer programming model is written and solved for each possible case of pavement configuration (total of 16 pavement configurations) and finally optimal configuration of pavement structure as well as the optimal thickness of each layer will be determined based on the minimum cost among 16 alternatives.

Table 3: Different alternatives for pavement structure.

Structure number	Layers in pavement structure				
	HMA	ATB	CTB	Granular base	Granular subbase
1	√	√	√	√	√
2	√	√		√	√
3	√	√		√	
4	√	√			√
5	√	√			
6	√		√	√	√
7	√		√	√	
8	√		√		√
9	√		√		
10	√	√	√		
11	√	√	√	√	
12	√	√	√		√
13	√			√	√
14	√			√	
15	√				√
16	√				

DEVELOPING DESIGN CHARTS

In order to develop pavement design charts for different classes of road, including freeways and expressways, major roads, and secondary roads with respect to the current price of pavement materials in Iran, the proposed optimization model was solved by assuming different levels of design traffic and strengths of subgrade soil. Design parameters and price of pavement materials for developing design charts are given in Tables 4 and 5, respectively. Application of CTB layer is not common in practice in Iran and no specific cost has been provided for construction of this material in "Basic Price List for Road, Runway and Railway." For this reason, CTB layer was excluded for developing design charts ($H_{CTB} = 0$).

Table 4: Design parameters for developing design charts.

Parameter	Value	Parameter	Value	Parameter	Value
a_{AC}	0.44	E_{Base}	1960 kg/cm^2	Q_{TC}	0.5 kg/m^2
a_{ATB}	0.32	$E_{Subbase}$	1050 kg/cm^2	$\min_{d_{AC}}$	Table 2
a_{GB}	0.13	P_i	4.2	$\min_{d_{GB}}$	Table 2
a_{GB}	0.13	P_t	Table 1	$\min_{d_{SGB}}$	15 cm
S_0	0.35	Z_R	Table 1	$\min_{d_{ATB}}$	7.5 cm

Drainage coefficient for all layers is assumed to be equal to 1.

Table 5: Construction cost for different materials [26].

Layer	Unit	Unit cost (Rials)
Asphalt concrete	1 m^2 per cm thickness	18500
Asphalt treated base	1 m^2 per cm thickness	15800
Granular base	m^3	149700
Granular subbase	m^3	84900
Tack coat	kg	15600

At this time 1$ ≅ 30000 Rials.

For developing design charts, the resilient modulus of subgrade soil was assumed between 200 and 1000 kg/cm^2 with increments of 100 kg/cm^2 .This range covers a wide range of possible subgrade soils with different strength.

Also, the number of equivalent single axle loads (ESALs) or $W_{8.2}$ was assumed as 10_5 , $5 \times 10_5$, 10_6 , 3×10^6, 6×10^6 , and 10^7 which can reflect light to heavy traffic.

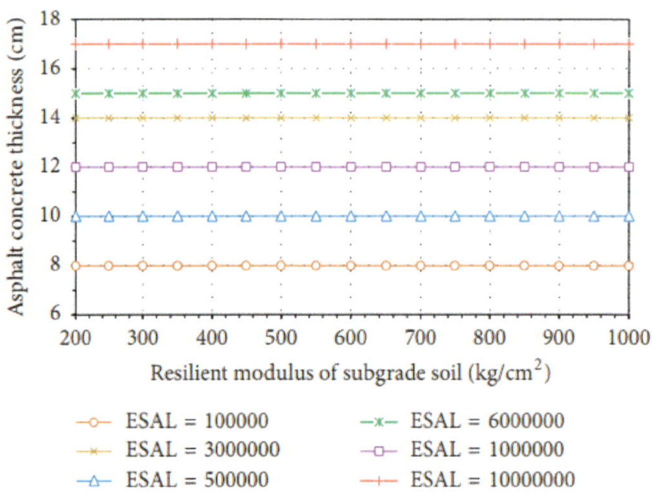

Figure 2: Optimal design thickness of asphalt concrete for secondary rural roads.

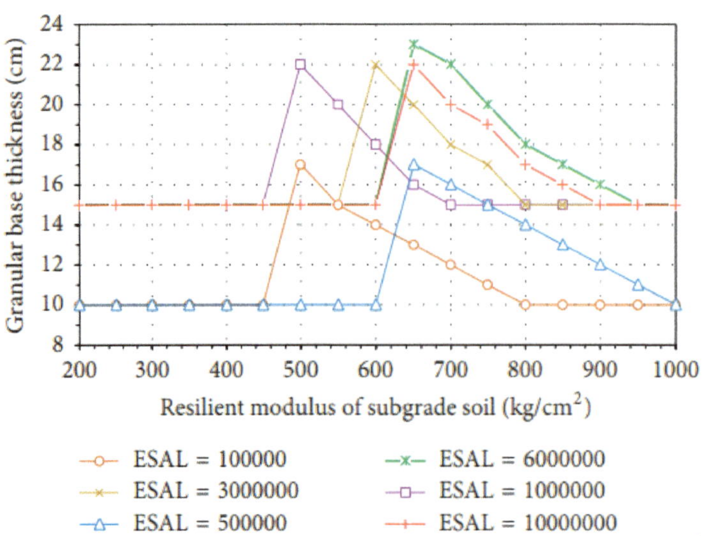

Figure 3: Optimal design thickness of granular base for secondary rural roads.

The results of determining the optimum thickness of each layer of pavement, with respect to the road classification are shown in Figures 2–10. In all cases, the thickness of asphalt treated layer is equal to zero, which indicates the fact that, in the current situation (material prices based on "Basic Price List for Road, Runway and Railway" in 2015), use of asphalt treated base layer in pavement structure is not cost effective. Also, with increasing the strength of subgrade soil, the granular subbase layer may be removed from the optimum structure of pavement. The possibility of removing the granular subbase layer increases by decreasing design traffic and increasing subgrade strength. For example in case of major roads with an ESALs of 1×10^6 and 10^7 , when resilient modulus of subgrade exceeds 500 and 700 kg/cm^2 , respectively, the granular subbase layer is removed from the optimum structure of pavement

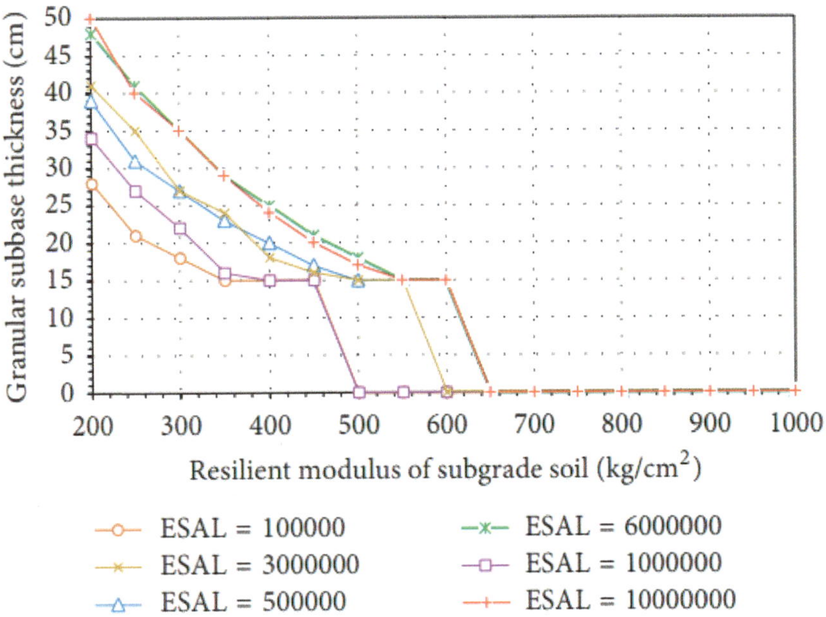

Figure 4: Optimal design thickness of granular subbase for secondary rural roads.

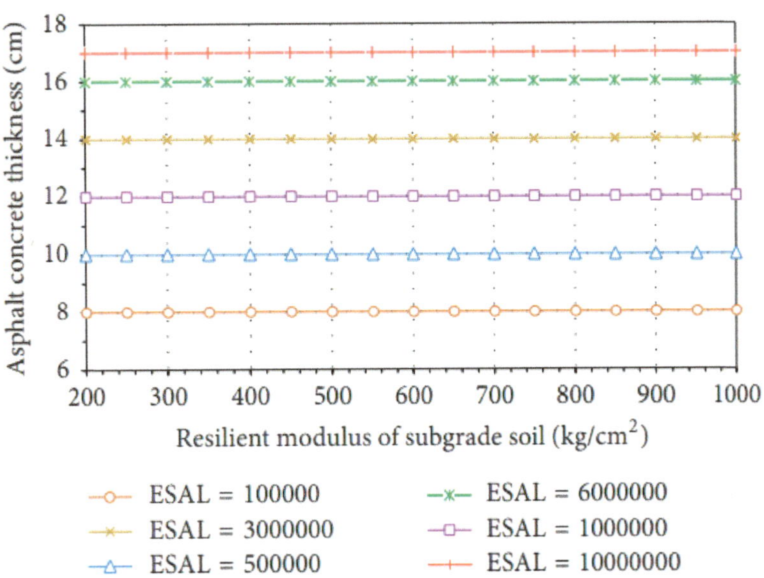

Figure 5: Optimal design thickness of asphalt concrete for major rural roads.

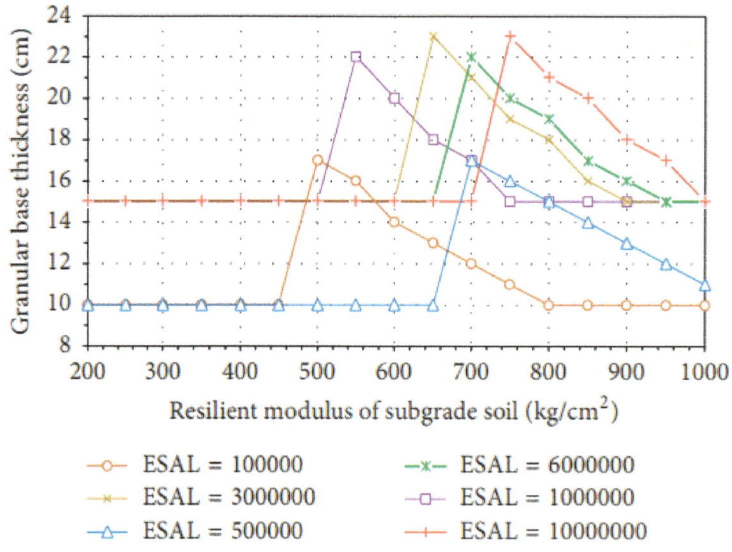

Figure 6: Optimal design thickness of granular base for major rural roads.

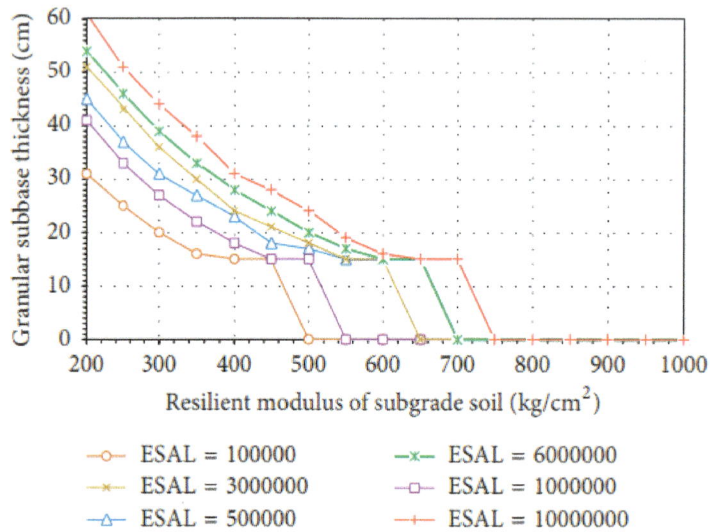

Figure 7: Optimal design thickness of granular subbase for major rural roads.

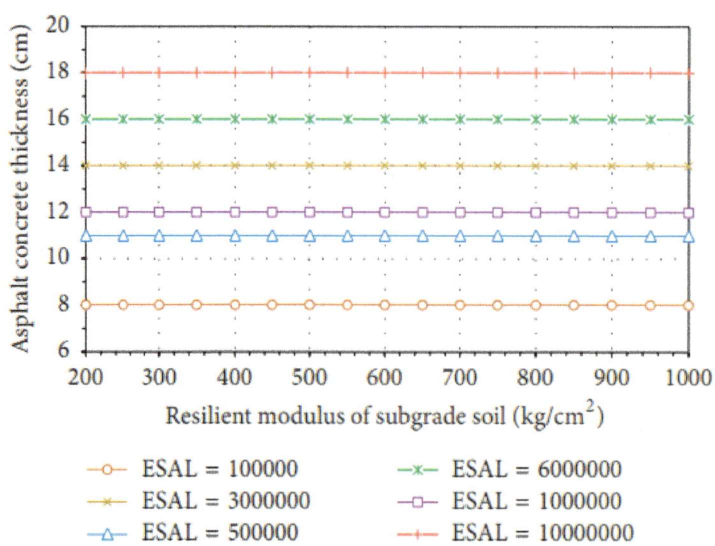

Figure 8: Optimal design thickness of asphalt concrete for expressways and freeways.

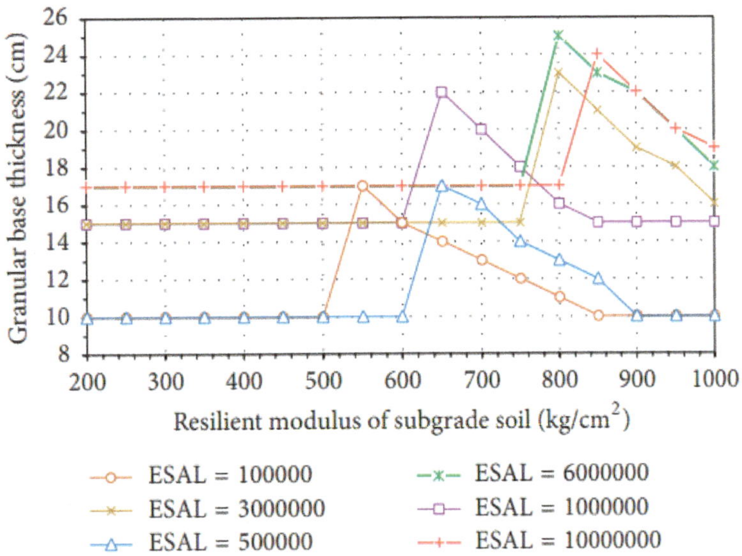

Figure 9: Optimal design thickness of granular base for expressways and freeways.

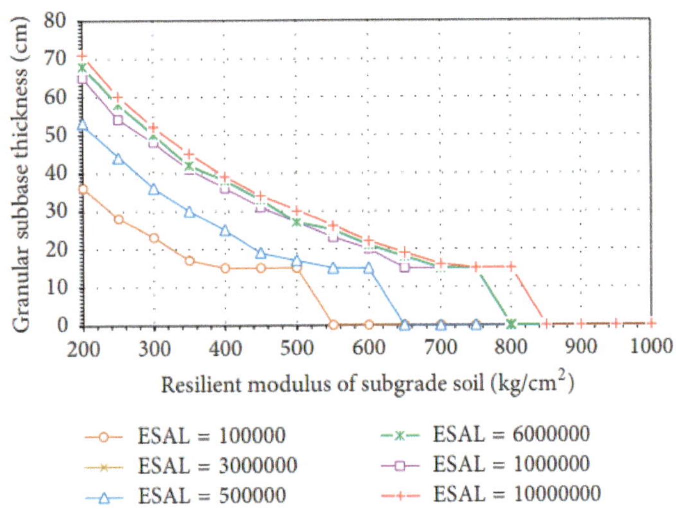

Figure 10: Optimal design thickness of granular subbase for expressways and freeways.

According to road classification, design traffic, and resilient modulus of subgrade, Figures 2–10 can be used to determine the optimum thickness of pavement layers including asphalt concrete, granular base, and granular subbase.

CONCLUSION

In this paper an optimization model was proposed to determine the optimal configuration as well as the optimum thickness of each layer of pavement structure based on the Iran Highway Asphalt Paving Code 234. By utilizing the proposed optimization model and according to the "Basic Price List for Road, Runway and Railway" of Iran in 2015, the optimum thickness of pavement layers for secondary rural roads, major rural roads, and freeways were determined and then design charts were developed. Developed charts help designers to determine the optimum thickness of pavement layers including asphalt concrete, granular base, and granular subbase with respect to road classification, design traffic, and resilient modulus of subgrade. This study showed that in the current situation (material prices in 2015), use of asphalt treated layer in pavement structure is not cost effective. Also it was shown that, with increasing the strength of subgrade soil, the granular subbase layer may be removed from the optimum structure of pavement.

CONFLICT OF INTERESTS

The author declares that there is no conflict of interests regarding the publication of this paper.

REFERENCES

1. R. Mallick and T. El-Korchi, Pavement Engineering: Principles and Practice, CRC Press, New York, NY, USA, 2013.
2. Y. H. Huang, Pavement Analysis and Design, Prentice Hall, New York, NY, USA, 2nd edition, 2004.
3. C. L. Monismith, K. E. Secor, and W. Blackmer, "Asphalt mixture behaviour in repeated flexure," inProceedings

of the Association of Asphalt Paving Technologists, vol. 30, pp. 188–222, 1961.

4. G. M. Dorman, "The extension to practice of a fundamental procedure for design of flexible pavements," in Proceedings of the 1st International Conference of Structural Design of Asphalt Pavements, pp. 785–793, Ann Arbor, Mich, USA, 1962.
5. Austroads, Guide to Pavement Technology—Part 2: Pavement Structural Design, Austroads, Sydney, Australia, 2010.
6. NCHRP, "Mechanistic-empirical design of new & rehabilitated pavement structures," NCHRP 1-37A, 2008.
7. IRC, Guidelines for the Design of Flexible Pavements, 3rd Revision, Indian Road Congress, New Delhi, India, 2012.
8. Shell International Petroleum Company Limited, Shell Pavement Design Manual—Asphalt Pavements and Overlays for Road Traffic, London, UK, 1978.
9. H. Theyse, J. Maina, and L. Kannemeyer, "Revision of the South African flexible pavement design method; mechanistic-empirical components," in Proceedings of the 9th Conference on Asphalt Pavements for Southern Africa (CAPSA '07), Gaborone, Botswana, 2007.
10. The Asphalt Institute, Thickness Design—Asphalt Pavements for Highways and Streets, Manual Series No. 1 (MS-1), The Asphalt Institute, 9th edition, 1999.
11. LCPC and SETRA, French Design Manual for Pavement Structures, Guide Technique, Union des Syndicats de l'Industrie Routière Française, 1997.
12. American Association of State Highway Official, AASHTO Guide for Design of Pavement Structures, American Association of State Highway Official, Washington, DC, USA, 1993.
13. Japan Road Association, Manual for Asphalt Pavement, Japan Asphalt Association, 1989.
14. HMSO, "A guide to the structural design for new roads," Tech. Rep. RN-29, Department of the Environment, London, UK, 1970.
15. Iran Management and Planning Organization, Iran Highway Asphaltic Pavements (IHAP) Code, vol. 234, Tehran, Iran, 2nd edition, 2010.
16. N. M. Rouphail, "Minimum-cost design of flexible pavements," Journal of Transportation Engineering, vol. 111, no. 3, pp. 196–207, 1985.
17. R. Nicholls, "Optimization of AASHTO DNPS86 pavement design program," Journal of Transportation Engineering

CHAPTER 12 SEISMIC ACTIVE EARTH PRESSURE BY VARIATIONAL LIMIT EQUILIBRIUM METHOD

Jiangong Chen, Zejun Yang, Richeng Hu, and Haiquan Zhang

INTRODUCTION

The magnitude and distribution of active earth pressure on the retaining wall under the seismic loading are the theoretical premises of the aseismic design for the retaining wall and play a vital role in evaluating the stability of the retaining wall in the seismic area. The common calculation method of seismic active earth pressure is the Mononobe-Okabe (M-O) theory which is based on the Coulomb theory and believes that the sliding soil wedge is a rigid body under the seismic loading. The pseudo-static approach is adopted to simplify the seismic force into an inertia force acting on the sliding soil wedge and transcribe the dynamic problem as a static problem [1]. The M-O theory assumes that the backfill behind the wall is cohesion-less soil and the slip surface is a plane, and the theory cannot obtain the real action point of resultant force without considering the equation of moment equilibrium; thus, restrictions are formed on this theory. The M-O theory has been improved by a number of scholars to expand its application scope [2–6].

Both the M-O theory and the improvement method based thereon are on the basis of the assumption that the slip surface is a plane, which does not conform to the practical situation. One rigorous method in math is the variational limit equilibrium method, by which the seismic active earth pressure on the retaining wall is attributed as the functional extremum problem of two undetermined functions. One undetermined function is expressed by the shape of the slip surface, while the other is the function of the normal stress distribution on the slip surface. They are numerically solved by the variational method.

The variational limit equilibrium method was first proposed by Kopáscy [7–9]. Then it was introduced to the stability analysis of slope and foundation [10–22]. And some scholars utilized the variational limit equilibrium method to study the lateral earth pressure on the retaining wall [23–26]. According to the variational method, Shaojun studied the shape of the slip surface of the sliding soil wedge behind the retaining wall and obtained the analytical solution of the shape of the slip surface and the magnitude of seismic active earth pressure [27]. However, only the cases where the retaining wall is vertical and the backfill surface is horizontal without surcharge are considered in the calculation models, while the influences of wall-movement modes of the retaining wall on the magnitude and the action point position of seismic active earth pressure fail to be taken into account. In fact, the distribution of earth pressure on the retaining wall is nonlinear. The magnitude and the action point position of seismic active earth pressure depend on the coordinated deformation of soil-wall contact surface and vary with the change of wall-movement modes of the retaining wall [26]. However, in the design process, it is often hard to accurately estimate the wall-movement modes of the retaining wall. For the static and dynamic ultimate load acting on the retaining wall, a reasonable approach is to contain the seismic active earth pressure under different wall-movement modes in a certain range as possible for the engineering designers to select and use. In this paper, the variational limit equilibrium method is used to study the

Citation: Jiangong Chen, Zejun Yang, Richeng Hu, and Haiquan Zhang, "Seismic Active Earth Pressure by Variational Limit Equilibrium Method," Shock and Vibration, vol. 2016, Article ID 4158785, 10 pages, 2016. doi:10.1155/2016/4158785.

seismic active earth pressure on the gravity retaining wall under general conditions (the retaining wall is inclined and coarse; the backfill is cohesive soil; the backfill surface is a curved surface with nonuniform surcharge). The interval of the seismic active earth pressure under different wall-movement modes can be effectively estimated by the proposed approach.

VARIATIONAL ANALYSIS OF SEISMIC ACTIVE EARTH PRESSURE

Basic Assumption

(1) The research problem is a plane strain problem; (2) the soil behind the wall is Coulomb material, which can be represented by the intensity parameter cohesion c and the internal friction angle φ; (3) when the backfill soil is in the critical active state, a sliding soil wedge is formed and its slip surface passes through the wall heel; (4) the retaining wall is rigid and its motion forms are unconstrained, wherein the motion displacement can be ignored compared with the wall height; (5) the seismic action is simplified to static load acting on the sliding wedge, with horizontal seismic coefficient k_H and vertical seismic coefficient k_V.

Limit Equilibrium Equation of Sliding Soil Wedge

The calculation model of active earth pressure under the seismic loading is shown in Figure 1, wherein the height of the retaining wall is H; the retaining wall is inclined and coarse; the slope angle of the wall to vertical is α; the friction angle between soil and wall is δ; the unit weight of soil is γ; the cohesion is c; the internal fraction angle is φ, the expression of the backfill surface is $g(x)$, and the expression of the slip surface is $s(x)$; the expression of the vertical surcharge distribution on the backfill surface is $q(x)$, the expression of the tangential stress distribution on the slip surface is $\tau(x)$, and the expression of the normal stress distribution on the slip surface is $\sigma(x)$; the seismic soil pressure on the wall is E_a.

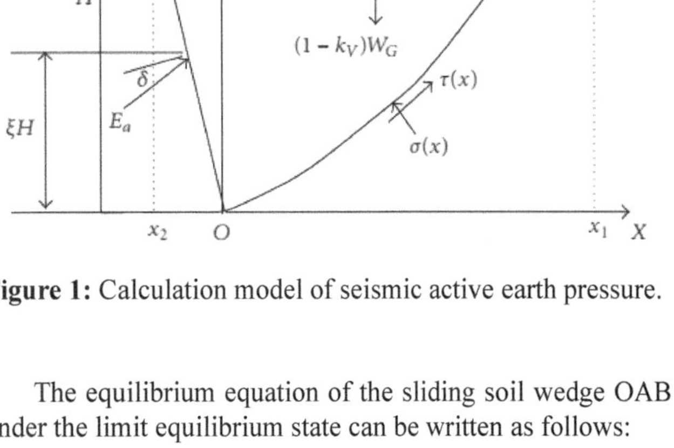

Figure 1: Calculation model of seismic active earth pressure.

The equilibrium equation of the sliding soil wedge OAB under the limit equilibrium state can be written as follows:

$$E_a \cos(\alpha+\delta) + \int_0^{x_1} \tau - \sigma s' - k_H [\gamma(g-s)+q]\,dx - \int_{x_2}^{0} k_H [\gamma(g-kx)+q]\,dx = 0, \tag{1}$$

$$E_a \sin(\alpha+\delta) + \int_0^{x_1} \tau s' + \sigma - (1-k_V)[\gamma(g-s) + q]\,dx - \int_{x_2}^{0} (1-k_V)[\gamma(g-kx)+q]\,dx = 0, \tag{2}$$

$$-\int_{x_2}^{0} k_H q g\,dx - \int_0^{x_1} k_H q g\,dx - \int_{x_2}^{0} \frac{1}{2} k_H \gamma \left(g^2 - k^2x^2\right) dx - \int_0^{x_1} \frac{1}{2} k_H \gamma \left(g^2 - s^2\right) dx - \int_0^{x_1} \left\{\left(xs'-s\right)\tau + \left(x+ss'\right)\sigma - (1-k_V)\left[qx + \gamma(g-s)x\right]\right\} dx = 0, \tag{3}$$

wherein $s' = ds/dx$, $k = -\cot\alpha$, $x1$ is X coordinate of B, $x2$ is X coordinate of A, and $x_2 = -H\tan\alpha$. ξ is the position coefficient of the action point of seismic active earth pressure, and its value is the ratio of the vertical distance (from the action point position to the wall heel) to the height of the retaining wall.

It is assumed that the tangential stress τ and the normal stress σ on the slip surface are subject to the Mohr-Coulomb failure criterion:

$$\tau = c + n_1 \sigma, \tag{4}$$

wherein $n_1 = \tan\varphi$.

Variational Limit Equilibrium Model of Seismic Active Earth Pressure

The problem to solve the seismic active earth pressure E_{ae} can now be stated as follows. The seismic active earth pressure E_{ae} is the maximum value of E_a, corresponding to the most dangerous (critical) slip surface. Realizing the maximum value of E_a is to find the function of the slip surface $s(x)$ and the function of normal stress distribution $\sigma(x)$ on the slip surface, subject to the three equations of LE ((1)~(3)). Among these three equations, (3) is changed as objective function and the other equations are the constraint conditions according to the variational method.

The following objective function J can be obtained after (4) is substituted into (3). Obviously, to solve the maximum of E_a means the same as to solve the maximum of objective function J:

$$J = \int_0^{x_1} F_0\,dx, \tag{5}$$

wherein

$$J = \frac{E_a \xi H \cos\delta}{\cos\alpha} + z_0,$$

$$F_0 = \left(n_1 s' x + x + s s' - n_1 s\right)\sigma + c x s' - c s - \left(1 - k_V\right) \cdot \left[(g - s)\gamma x + q x\right] + k_H q g + \frac{1}{2} k_H \gamma \left(g^2 - s^2\right),$$

$$z_0 = \int_{x_2}^{0} \left\{\left(1 - k_V\right)\left[\gamma (g - kx) + q\right] x - k_H q g - \frac{1}{2} k_H \gamma \left(g^2 - k^2 x^2\right)\right\} dx = \text{const.} \tag{6}$$

Two constraint conditions can be obtained after (4) and (5) are substituted into (1) and (2):

$$\int_0^{x_1} \left(z_1 F_0 + F_1\right) dx = z_2,$$

$$\int_0^{x_1} \left(z_3 F_0 + F_2\right) dx = z_4, \tag{7}$$

wherein

$$F_1 = \left(n_1 - s'\right)\sigma + c - k_H\left[\gamma (g - s) + q\right],$$

$$F_2 = \left(n_1 s' + 1\right)\sigma + c s' - \left(1 - k_V\right)\left[q + \gamma (g - s)\right],$$

$$z_1 = \frac{\cos\alpha \cos(\alpha + \delta)}{\xi H \cos\delta} = \text{const},$$

$$z_2 = z_0 z_1 + \int_{x_2}^{0} k_H \left[\gamma (g - kx) + q\right] dx = \text{const},$$

$$z_3 = \frac{\cos\alpha \cos(\alpha + \delta)}{\xi H \cos\delta} = \text{const},$$

$$z_4 = z_0 z_3 + \int_{x_2}^{0} \left(1 - k_V\right)\left[\gamma (g - kx) + q\right] dx = \text{const.} \tag{8}$$

Equations (5) and (7) show the isoperimetric model of the constrained variational extremum with an undetermined boundary. The starting point of the slip surface O is a fixed point on the coordinates of ($x_0 = 0$, $y_0 = 0$) and the end point B moves on the backfill surface (x) with the undetermined coordinates of (x_1, (x_1)). The undetermined boundary is a variational boundary. The variational boundary refers to an unfixed boundary on which one or two ends, respectively, move on the given function.

According to the variational method of the functional with constraints, the following auxiliary functional J^* is constructed by Lagrange multipliers to convert the functional extremum problem under constraint conditions into a functional extremum problem without constraint conditions:

$$J^* = \int_0^{x_1} F\, dx,$$

$$F = F_0 + \lambda_1 F_1 + \lambda_2 F_2, \tag{9}$$

wherein λ_1 and λ_2 are Lagrange multipliers.

Now, the solution of the maximum of E^a is converted into the solution of the maximum of auxiliary function J^*. According to the necessary conditions for the existence of the extremum of auxiliary functional J^*, the function of the slip surface (x) and the function of the normal stress distribution $\sigma(x)$ on the slip surface must meet the following requirements.

(1) The Euler differential equation of the auxiliary function F is

$$\frac{\partial F}{\partial \sigma} - \frac{d}{dx}\left(\frac{\partial F}{\partial \sigma'}\right) = 0, \tag{10}$$

$$\frac{\partial F}{\partial s} - \frac{d}{dx}\left(\frac{\partial F}{\partial s'}\right) = 0. \tag{11}$$

(2)The integral constraint equations are (7).

(3)Boundary conditions are

$$\text{Fixed boundary condition: } s(0) = 0,$$

$$\text{Variational boundary condition: } s(x_1) = g(x_1). \tag{12}$$

(4)Transversality condition at the variational boundary is

$$\left.\left(F - s'\frac{\partial F}{\partial s'} + g'\frac{\partial F}{\partial s'}\right)\right|_{x=x_1} = 0. \tag{13}$$

VARIATIONAL SOLUTION OF SEISMIC ACTIVE EARTH PRESSURE

Shape Function of the Slip Surface

The following equation can be obtained after (5), (7), and (9) are substituted into (10):

$$\frac{ds}{dx} = \frac{x - n_1 s + \lambda_1 n_1 + \lambda_2}{-n_1 x - s + \lambda_1 - \lambda_2 n_1}. \tag{14}$$

Polar coordinate transformations $r\cos\theta = x + \lambda_2$ and $r\sin\theta = s - \lambda_1$ are brought in. For the calculation model in polar coordinates, see Figure 2. Equation (14) is transcribed as a differential equation:

$$\frac{dr}{d\theta} = -n_1 r. \tag{15}$$

The general solution of the differential equation is

$$r = z_5 e^{-n_1\theta}, \tag{16}$$

wherein z_5 is an arbitrary integration constant

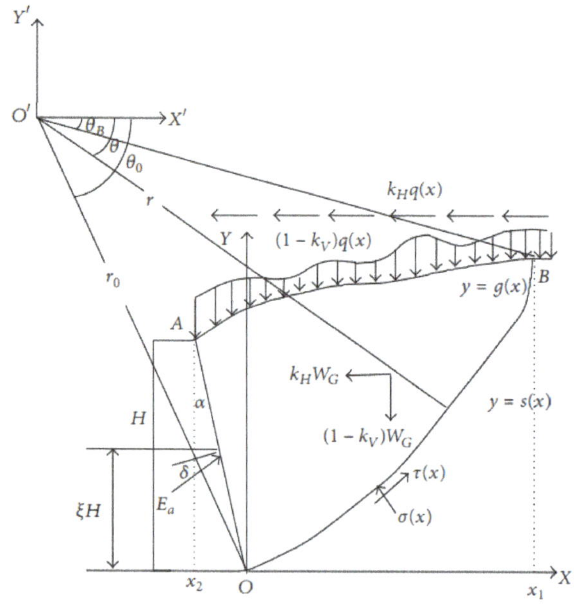

Figure 2: Calculation model of seismic active earth pressure in polar coordinates.

The fixed boundary condition (0) = 0 shows that the pole of the polar coordinate is (r_0, θ_0). After it is substituted into (16), the expression of the logarithmic spiral of the slip surface can be obtained:

$$r = r_0 e^{n_1(\theta_0-\theta)}, \tag{17}$$

wherein

$$r_0 = \sqrt{\lambda_1^2+\lambda_2^2},$$
$$\theta_0 = -\arctan\left(\frac{\lambda_1}{\lambda_2}\right), \quad \lambda_2 > 0,$$
$$\theta_0 = -\pi - \arctan\left(\frac{\lambda_1}{\lambda_2}\right), \quad \lambda_2 \le 0. \tag{18}$$

Normal Stress Distribution on the Slip Surface

After (5), (7), and (9) are substituted into (11), the following equation can be obtained:

$$2n_1\sigma + (n_1x + s - \lambda_1 + n_1\lambda_2)\sigma' - (1-k_V)\gamma x - (1-k_V)\lambda_2\gamma + 2c + k_H\gamma s - k_H\lambda_1\gamma = 0. \tag{19}$$

Introducing the Polar coordinate transformations $r\cos\theta = x+\lambda_2$ and $r\sin\theta = s-\lambda_1$, (19) is transcribed as a differential equation:

$$\frac{d\sigma}{d\theta} - 2n_1\sigma = 2c - (1-k_V)\gamma r_0 e^{n_1(\theta_0-\theta)}\cos\theta + k_H\gamma r_0 e^{n_1(\theta_0-\theta)}\sin\theta. \tag{20}$$

The general solution of the differential equation is (when $n_1 \neq 0$)

$$\sigma = e^{\int_{\theta_1}^{\theta} 2n_1 d\theta}\left(z_6 + \int_{\theta_1}^{\theta}\left(2c - (1-k_V)\gamma r_0 e^{n_1(\theta_0-\theta)}\cos\theta + k_H\gamma r_0 e^{n_1(\theta_0-\theta)}\sin\theta\right) \cdot e^{\int_{\theta_1}^{\theta} -2n_1 d\theta} d\theta\right), \tag{21}$$

wherein z_6 is an integration constant; θ_1 is an arbitrary angle; $\theta_1 = 0$ is available

$$\sigma = z_6 e^{2n_1\theta} - \frac{(1-k_V)\gamma r_0 e^{n_1(\theta_0-\theta)}}{1+9n_1^2}\left(\sin\theta - 3n_1\cos\theta + 3n_1 e^{3n_1\theta}\right) - \frac{k_H\gamma r_0 e^{n_1(\theta_0-\theta)}}{1+9n_1^2}\left(3n_1\sin\theta + \cos\theta - e^{3n_1\theta}\right) - \frac{c}{n_1}. \tag{22}$$

According to the transversality condition at the variational boundary, the normal stress distribution at B of the slip surface can be obtained:

$$\sigma(x_1) = \frac{\sin\theta_1 (c - k_H q(x_1)) + \cos\theta_1 \left[(1-k_V)q(x_1) - cg'(x_1)\right]}{\sin\theta_1 (g'(x_1) - n_1) + \cos\theta_1 (n_1 g'(x_1) + 1)}. \tag{23}$$

After it is substituted into (22), the following equation is obtained:

$$z_6 = e^{-2n_1\theta_1}\left\{\frac{c}{n_1} + \frac{k_H\gamma r_0 e^{n_1(\theta_0-\theta_1)}}{1+9n_1^2}\left(3n_1\sin\theta_1 + \cos\theta_1 - e^{3n_1\theta_1}\right) + \frac{(1-k_V)\gamma r_0 e^{n_1(\theta_0-\theta_1)}}{1+9n_1^2}\left(\sin\theta_1 - 3n_1\cos\theta_1 + 3n_1 e^{3n_1\theta_1}\right) + \frac{\sin\theta_1 (c - k_H q(x_1)) + \cos\theta_1\left[(1-k_V)q(x_1) - cg'(x_1)\right]}{\sin\theta_1 (g'(x_1) - n_1) + \cos\theta_1 (n_1 g'(x_1) + 1)}\right\}. \tag{24}$$

Optimal Solution of Seismic Active Earth Pressure

The analysis carried out so far indicates that the maximum value of E_a can be obtained according to a pair of functions $r(\theta)$ and $\sigma(\theta)$ determined by two undetermined Lagrange multipliers λ_1 and λ_2. The problem is converted into the search of a pair of constants λ_1 and λ_2 through the two constraint equations. The two constraint equations—(7)—must be satisfied:

$$f_1(\lambda_1,\lambda_2) = \int_0^{x_1}(z_1F_0 + F_1)\,dx - z_2 = 0,$$
$$f_2(\lambda_1,\lambda_2) = \int_0^{x_1}(z_3F_0 + F_2)\,dx - z_4 = 0. \tag{25}$$

The following equation is always correct:

$$\begin{matrix} f_1(\lambda_1,\lambda_2) = 0 \\ f_2(\lambda_1,\lambda_2) = 0 \end{matrix} \Longleftrightarrow \Phi(\lambda_1,\lambda_2) = f_1^2(\lambda_1,\lambda_2) + f_2^2(\lambda_1,\lambda_2) = 0. \tag{26}$$

Therefore, two integral equations—(25)—are equivalent to the following equation:

$$\Phi(\lambda_1,\lambda_2) = f_1^2(\lambda_1,\lambda_2) + f_2^2(\lambda_1,\lambda_2)$$
$$= \left[\int_0^{x_1}(z_1F_0 + F_1)\,dx - z_2\right]^2$$
$$+ \left[\int_0^{x_1}(z_3F_0 + F_2)\,dx - z_4\right]^2 = 0. \quad (27)$$

For $\Phi(\lambda_1, \lambda_2) \geq 0$, so

$$\min\left[\Phi(\lambda_1,\lambda_2)\right] = 0. \quad (28)$$

The above analysis indicates that the solution of the two constraint equations—(7)—can be obtained through the solution of the minimum value of the function Φ (the minimum value is 0). When the minimum value of Φ is not 0, it shows that the sliding soil wedge cannot keep balance and is improper

In this paper, the fminsearch function provided by MATLAB is used to find the optimal solution. For this function, it only requires establishing a function module of Φ and giving an initial value to search the minimum value of the function. Fminsearch applies the derivative-free method to find the minimum value of the unconstrained multivariable function, which is generally called unconstrained nonlinear optimization. Fminsearch finds the minimum of a scalar function of several variables starting at an initial estimate. Therefore, one major drawback of this optimization tool is that the search would be very slow if the initial estimate is improper. In order to overcome this drawback and in the consideration of the condition that the center of the logarithmic spiral is usually at the top left corner of the retaining wall, the initial value is $\lambda_1 = \lambda_2 = H$ in the optimization analysis of this paper and the search is quite rapid.

CALCULATED RESULT AND PARAMETER ANALYSIS

Interval Estimation of Seismic Active Earth Pressure.

Under the general conditions that the retaining wall is inclined and coarse, the backfill is cohesive soil, and the backfill surface is inclined with surcharge, the interval of the position coefficient ξ of the action point under seismic loading is numerically solved. Besides, the influence of the action point position on the shape of the slip surface and the magnitude of seismic active earth pressure is studied.

The retaining wall has a height of 10 m, with the slope angle of the wall to vertical $\alpha = 10°$, the friction angle between soil and wall $\delta = 15°$, the unit weight of soil $\gamma = 18$ kN/m³, cohesion $c = 10$ kPa, internal friction angle $\varphi = 20°$, slope angle of the backfill soil β=5°, surcharge $q = 50$ kN/m, and seismic coefficients $k_H = 0.05$ and $k_V = 0.0$.

Figure 3 shows the curve of the minimum value of the function Φ varying with the position coefficient ξ of the action point. As shown in the figure, the position coefficient ξ of the action point of seismic active earth pressure is in an interval (lower limit value $\xi d = 0.449$; upper limit value $\xi_u = 0.612$). In the interval, the minimum value of the function Φ is 0. In other words, when the action point of seismic active earth pressure falls within the range of 4.49~6.12 m from the wall bottom, the counterforce provided by the retaining wall can balance the soil mass behind the wall; otherwise, the soil mass will inevitably lose its balance no matter how big the counterforce provided by the retaining wall is.

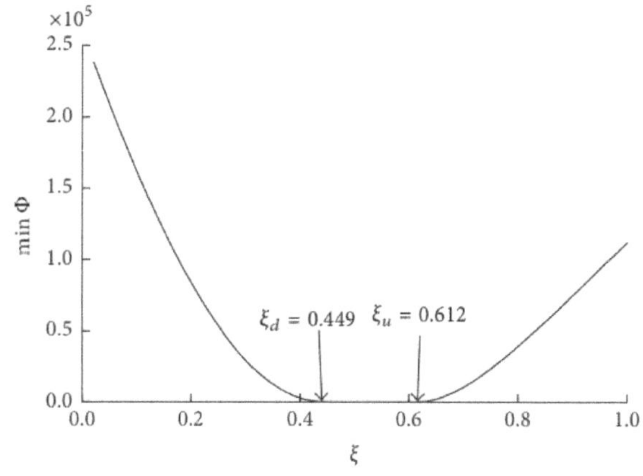

Figure 3: The minimum value of the function Φ.

There is a one-to-one corresponding relation between the magnitude of seismic active earth pressure and the shape of slip surface (Figure 5). For the curves varying with the position coefficient ξ of the action point of seismic active earth pressure, see Figures 3 and 4. At the lower limit of ξ, the slip surface is a plane, with the minimal value of seismic active earth pressure. As the action point moves upward, the slip surface becomes a logarithmic spiral surface and its curvature is gradually increased; that is, in the logarithmic spiral equation, $r0$, is gradually decreased, and the corresponding value of seismic active earth pressure is increased in a nonlinear manner. At the upper limit of ξ, the seismic active earth pressure is maximal. The minimum and maximum values can be as an interval estimation of the seismic active earth pressure under different wall-movement modes of the retaining wall.

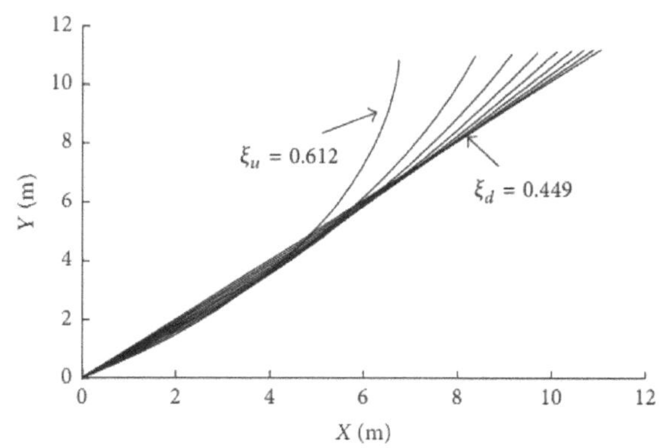

Figure 4: The shape of the slip surface.

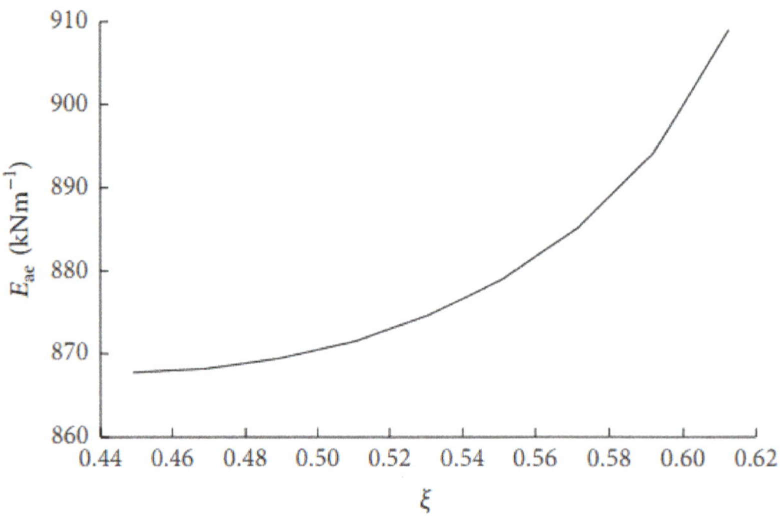

Figure 5: The magnitude of seismic active earth pressure

Comparison with the M-O Theory Solution.

To verify the correctness of the calculation method proposed in this paper, a comparison is made between the calculated result of the proposed method and the solution of M-O theory. According to the assumption of M-O theory, the values of cohesion c and the surcharge q are both 0 in the contrastive analysis. For other parameters and specific comparison results, see Table 1.

Table 1: Comparison of calculated results of active earth pressure with that of M-O.

$\alpha/°$	$\beta/°$	k_H	k_V	The proposed method		M-O method	
				E_{ae}/kN	ξ	E_a/kN	ξ
0	0	0	0	391.0~412.1	0.265~0.408	391.0	1/3
0	5	0.05	0.05	424.6~450.2	0.286~0.449	424.5	1/3
0	10	0.1	0.1	470.8~495.0	0.306~0.490	470.7	1/3
10	0	0.05	0.1	460.5~475.2	0.327~0.408	460.1	1/3
10	5	0.1	0	504.3~525.3	0.327~0.449	504.3	1/3
10	10	0	0.05	559.9~572.8	0.367~0.469	559.6	1/3
20	0	0.1	0.05	550.4~562.8	0.347~0.408	550.3	1/3
20	5	0	0.1	604.9~616.1	0.367~0.429	604.7	1/3
20	10	0.05	0	682.7~696.3	0.388~0.469	682.5	1/3

When the slip surface is a plane, the seismic active earth pressure is the lower limit value of the interval in the table. This value is equivalent to the solution of seismic active earth pressure calculated by the M-O theory. The solution obtained in the proposed method is degraded into M-O theory solution, thereby proving the correctness of the proposed method.

When the slip surface is a logarithmic spiral surface, the seismic active earth pressure calculated by the M-O theory is relatively small, and the assumed action point position is lower than the actual action point position under most possible wall-movement modes. The underestimation of the magnitude and the action point position of seismic active earth pressure cause the potential safety hazard of overturning of the retaining wall designed by M-O theory.

Parameter Analysis

In this section, discussions are made regarding the influence of relevant parameters on the magnitude and the action point position of seismic active earth pressure, as well as the overturning moment on the retaining wall. Such parameters include the slope angle α of the retaining wall, the slope angle β of backfill surface, the surcharge q on the backfill surface, the horizontal seismic coefficient k_H, and the vertical seismic coefficient k_V

Influence of Parameters on the Action Point Position.

The influence of relevant parameters on the position coefficient ξ of action point of seismic active earth pressure is shown in Figures 6(a)–6(c). With the increase of the slope angle β of backfill surface, the action point position moves up; with the increase of α, the action point position slightly moves down; with the increase of the horizontal seismic coefficient k_H and the vertical seismic coefficient k_V, the action point position moves up, wherein the horizontal seismic coefficient k_H has a significant influence on it; with the increase of the surcharge q, the action point position moves up in a nonlinear way.

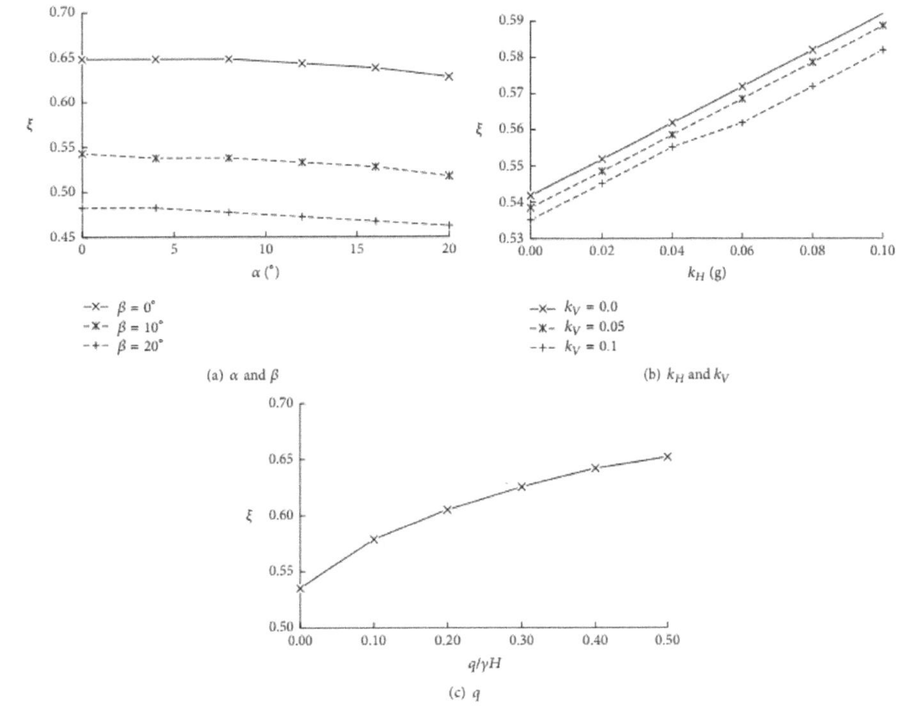

Figure 6: Effects of relevant parameters on ξ.

Influence of Parameters on the Seismic Active Earth Pressure *Eae.*

Figures 7(a)–7(c) show the influence of relevant parameters on the dimensionless seismic active earth pressure $E_{ae}/(\gamma H^2)$. The seismic active earth pressure E_{ae} is increased with the increase of the slope angle α of the retaining wall and the slope angle β of backfill surface; E_{ae} is increased with the increase of the horizontal seismic coefficient k_H and decreased with the increase of the vertical seismic coefficient k_V; E_{ae} is increased linearly with the increase of the dimensionless surcharge $q/(\gamma H)$ of slope surface.

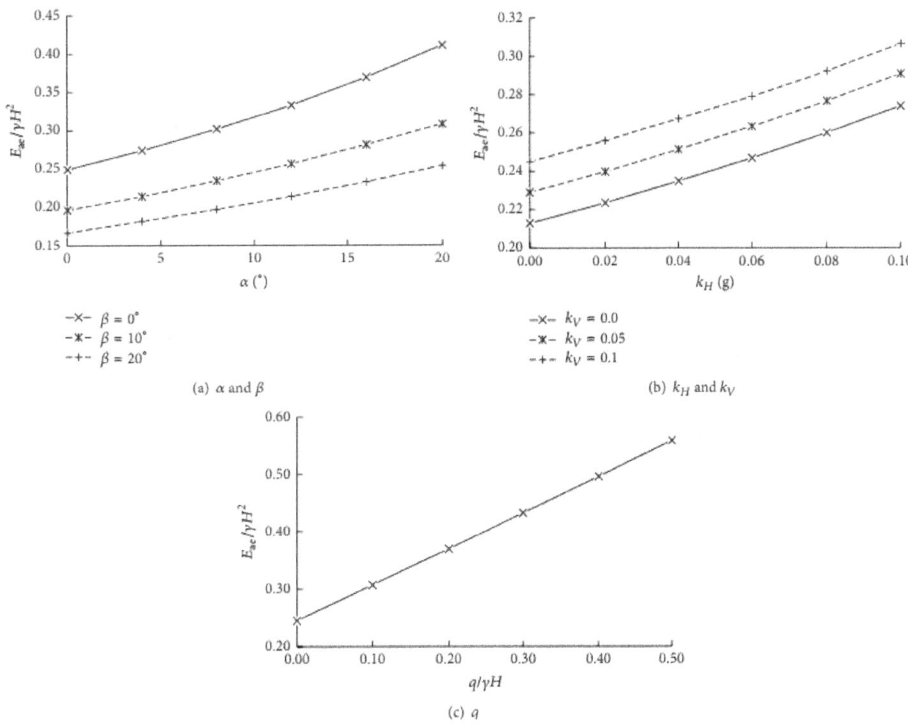

Figure 7: Effects of relevant parameters on E_{ae}.

Influence of Parameters on the Overturning Moment M_a.

Figures 8(a)–8(c) show the influence of relevant parameters on the dimensionless overturning moment $M_a/\gamma H^3$. The overturning moment M_a is increased with the increase of the slope angle α of the retaining wall and the slope angle β of backfill surface; M_a is increased with the increase of the horizontal seismic coefficient k_H and decreased with the increase of the vertical seismic coefficient k_V; M_a is increased linearly with the increase of the dimensionless surcharge load $q/(\gamma H)$ of slope surface.

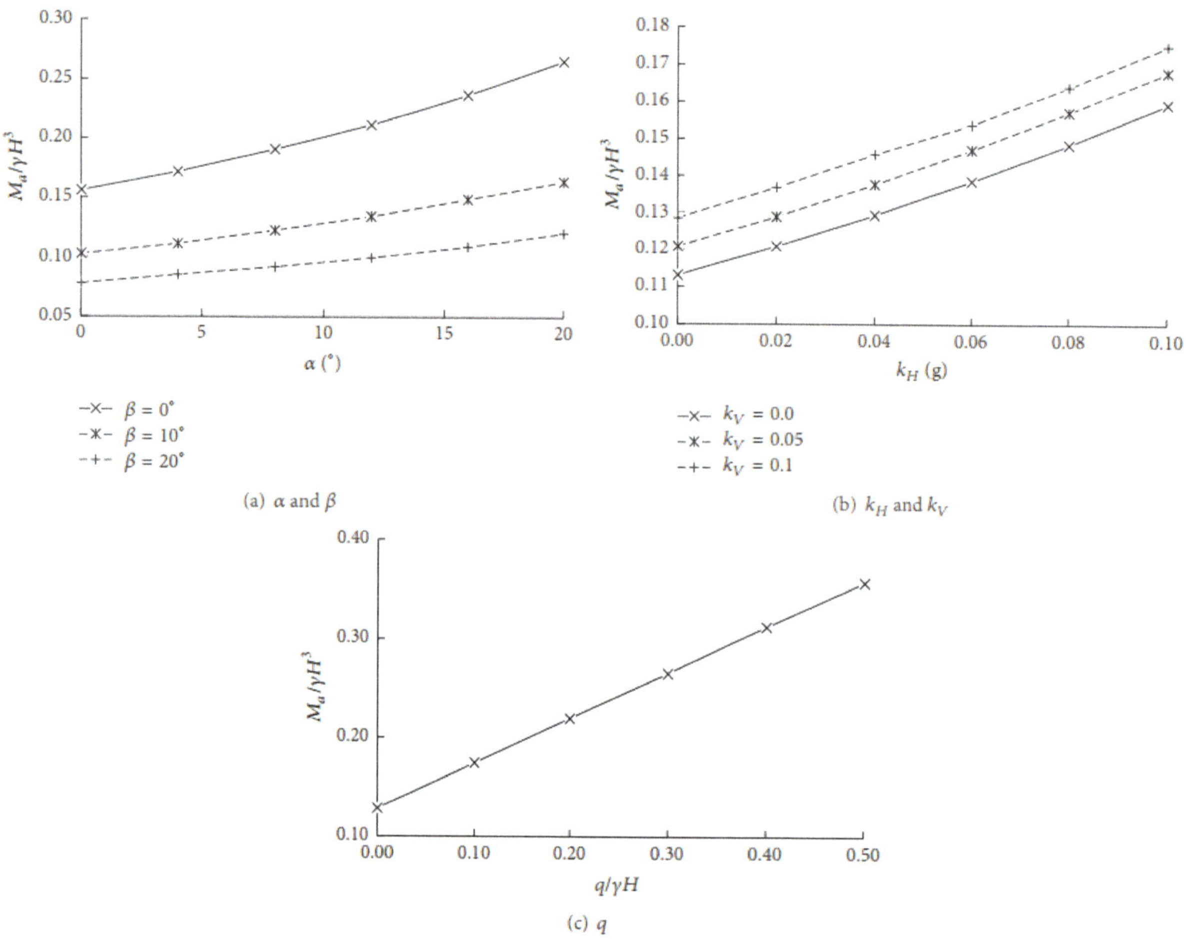

Figure 8: Effects of relevant parameters on M_a.

CONCLUSION

The seismic active earth pressure is studied in the paper based on the variational limit equilibrium method, and the following conclusions are obtained.

(1) In the framework of limit equilibrium theory, the isoperimetric model of functional extremum regarding the seismic active earth pressure under general conditions (the retaining wall is inclined and coarse; the backfill is cohesive soil; the backfill surface is a curved surface with nonuniform surcharge) is deduced.

(2) With the introduction of Lagrange multipliers and in combination with the necessary conditions required for the existence of functional extremum, the solution of seismic active earth pressure is transcribed as an optimization problem of two undetermined Lagrange multipliers; with the fminsearch function provided by MATLAB, an optimization solution of two Lagrange multipliers is founded to obtain the magnitude and the action point position of seismic active earth pressure, as well as the shape of slip surface.

(3) With the continuous changes of wall-movement modes of the retaining wall, the position coefficient ξ of the action point of seismic active earth pressure is an interval with lower and upper limit values. When ξ is the lower limit value, the slip surface is a plane and the seismic active earth pressure is minimal. As the action point moves upward, the slip surface becomes a logarithmic spiral surface and its curvature is gradually increased, with the corresponding value of seismic active earth pressure increased in a nonlinear manner. At the upper limit of ξ, the seismic active earth pressure is maximal; the minimum and maximum values can be estimated as an interval of the seismic active earth pressure under different wall-movement modes of the retaining wall.

(4) The contrastive analysis shows that when the slip surface is a plane, the solution of seismic active earth pressure obtained by the proposed method is in line with the M-O theory solution, verifying the correctness of the proposed method; when the slip surface is a logarithmic spiral surface, the underestimation of the

magnitude and the action point position of seismic active earth pressure cause the potential safety hazard of overturning of the retaining wall designed by M-O theory

(5) The parameter analysis shows that the action point position moves up with the increase of the slope angle β of backfill surface, the horizontal seismic coefficient k_H, and the surcharge q on the backfill surface. However, the vertical seismic coefficient k_V and the slope angle α of the retaining wall have little influence on it, so they can be neglected. The seismic active earth pressure E_{ae} and the overturning moment M_a are increased with the increase of the slope angle α of the retaining wall, the slope angle β of backfill surface, the horizontal seismic coefficient k_H, and the surcharge q on the backfill surface and decreased with the increase of the vertical seismic coefficient k_V.

COMPETING INTERESTS

The authors declare that they have no competing interests.

REFERENCES

1. S. Caltabiano, E. Cascone, and M. Maugeri, "Static and seismic limit equilibrium analysis of sliding retaining walls under different surcharge conditions," Soil Dynamics and Earthquake Engineering, vol. 37, pp. 38–55, 2012.
2. S. Saran and R. P. Gupta, "Seismic earth pressures behind retaining walls," Indian Geotechnical Journal, vol. 33, no. 3, pp. 195–213, 2003.
3. G. Mylonakis, P. Kloukinas, and C. Papantonopoulos, "An alternative to the Mononobe–Okabe equations for seismic earth pressures," Soil Dynamics and Earthquake Engineering, vol. 27, no. 10, pp. 957–969, 2007.
4. S. Ghosh and S. Sengupta, "Extension of Mononobe-Okabe theory to evaluate seismic active earth pressure supporting c-φ backfill," Electronic Journal of Geotechnical Engineering, vol. 17, pp. 495–504, 2012.
5. M. Yazdani, A. Azad, A. H. Farshi, and S. Talatahari, "Extended 'mononobe-okabe' method for seismic design of retaining walls," Journal of Applied Mathematics, vol. 2013, Article ID 136132, 10 pages, 2013.
6. Y.-L. Lin, W.-M. Leng, G.-L. Yang, L.-H. Zhao, L. Li, and J.-S. Yang, "Seismic active earth pressure of cohesive-frictional soil on retaining wall based on a slice analysis method," Soil Dynamics and Earthquake Engineering, vol. 70, pp. 133–147, 2015.
7. J. Kopáscy, "Über die Bruchflächen und Bruchspannungen in den Erdbauten," in Gedenkbuch Für Prof. Dr. J. Jáky, K. Széchy, Ed., pp. 81–99, Akadémiaikiadó, Budapest, Hungary, 1955.
8. J. Kopáscy, "Three-dimensional stress distribution and slip surfaces in earth works at rupture," inProceedings of the 4th International Conference on Soil Mechanics and Foundations Engineering, vol. 1, pp. 339–342, London, UK, 1957.
9. J. Kopáscy, "Distribution des contraintesàla rupture, forme de la surface de glissement et hauteur théorique des talus," in Proceedings of the 5th International Conference on Soil Mechanics and Foundations Engineering, vol. 2, pp. 641–650, Paris, France, 1961.
10. W. F. Chen and N. Snitbhan, "On slip surface and slope stability analysis," Soils and Foundations, vol. 15, no. 3, pp. 41–49, 1975.
11. M. Garber and R. Baker, "Bearing capacity by variational method," Journal of Geotechnical Engineering Division, vol. 103, no. 11, pp. 1209–1225, 1977.
12. M. Garber and R. Baker, "Extreme-value problems of limiting equilibrium," Journal of the Geotechnical Engineering Division, vol. 105, no. 10, pp. 1155–1171, 1979.
13. J. Spencer and T. C. O'Mahony, "An application of the calculus of variations to rectilinear flow of granular materials," International Journal for Numerical and Analytical Methods in Geomechanics, vol. 9, no. 3, pp. 225–235, 1985.
14. D. Leshchinsky, "Slope stability analysis: generalized approach," Journal of Geotechnical Engineering, vol. 116, no. 5, pp. 851–867, 1990.
15. D. Leshchinsky and C.-C. Huang, "Generalized three-dimensional slope-stability analysis," Journal of Geotechnical Engineering, vol. 118, no. 11, pp. 1748–1764, 1992.
16. R. Baker, "Sufficient conditions for existence of physically significant solutions in limiting equilibrium slope stability analysis," International Journal of Solids and Structures, vol. 40, no. 13-14, pp. 3717–3735, 2003.
17. R. Baker, "Stability chart for zero tensile strength Hoek-Brown materials—the variational solution and its engineering implications," Soils and Foundations, vol. 44, no. 3, pp. 125–132, 2004.
18. R. Baker, "Variational slope stability analysis of materials with nonlinear failure criterion," Electronic Journal of Geotechnical Engineering, vol. 10, pp. 1–22, 2005.
19. R. Baker, "A relation between safety factors with respect to strength and height of slopes," Computers and Geotechnics, vol. 33, no. 4-5, pp. 275–277, 2006.
20. L.-Y. Wu and Y.-F. Tsai, "Variational stability analysis of cohesive slope by applying boundary integral equation method," Journal of Mechanics, vol. 21, no. 3, pp. 187–198, 2005.
21. Y. M. Cheng, D. Z. Li, N. Li, Y. Y. Lee, and S. K. Au, "Solution of some engineering partial differential equations governed by the minimal of a functional by global optimization method," Journal of Mechanics, vol. 29, no. 3, pp. 507–516, 2012.
22. X. Li, "Bearing capacity factors for eccentrically loaded strip footings using variational analysis,"Mathematical Problems in Engineering, vol. 2013, Article ID 640273, 17 pages, 2013.
23. H. Soubra and R. Kastner, "Influence of seepage flow on the passive earth pressures," in Proceeding of the International Conference on Retaining Structures, pp. 67–76, ICE, Amsterdam, The Netherlands, 1992.
24. A.-H. Soubra, R. Kastner, and A. Benmansour, "Passive earth pressures in the presence of hydraulic gradients,"

Géotechnique, vol. 49, no. 3, pp. 319–330, 1999.

25. O. Puła, W. Puła, and A. Wolny, "On the variational solution of a limiting equilibrium problem involving an anchored wall," Computers and Geotechnics, vol. 32, no. 2, pp. 107–121, 2005.

26. L. Xinggao and L. Weining, "Study on the action of the active earth pressure by variational limit equilibrium method," International Journal for Numerical and Analytical Methods in Geomechanics, vol. 34, no. 10, pp. 991–1008, 2010.

27. M. Shaojun, Study on Calculation of Earth Pressure behind Retaining Wall under Static and Dynamic Loads and Its Related Issues, Zhejiang University, Zhejiang, China, 2012 (Chinese).

CHAPTER 13

APPLYING THE TAGUCHI METHOD TO RIVER WATER POLLUTION REMEDIATION STRATEGY OPTIMIZATION

Tsung-Ming Yang, Nien-Sheng Hsu, Chih-Chiang Chiu and Hsin-Ju Wang

Keywords: Water pollution remediation, optimization, Taguchi method, orthogonal arrays

INTRODUCTION

River basin pollution can be classified into four types according to the sources, which are domestic sewage, livestock wastewater, industrial wastewater, and non-point source pollution. If the river flows through a large metropolitan area, the main pollution source will be the domestic sewage. To remedy the domestic sewage pollution, the most effective strategy is the construction of sewerage systems. However, the sewerage system construction period is long, and domestic sewage continues to be discharged without treatment until the entire system is completely constructed. Therefore, besides building the sewerage system, the remediation strategy should also concern the water pollution treatment problem during the construction period.

This study aims to use the verified water quality model, from tens of water pollution remediation strategies or facilities, and integrate a method that can quickly determine the optimal remediation measure combination, in order to reduce the number of operations, and obtain the optimal combination of remediation strategy operations. Previous studies have used the optimization model to obtain the optimal solution. However, in practice, tens of combinations of water pollution remediation measures can result in an enormous solution space. Moreover, the calculation of water quality models is time consuming, thus leading to poor efficiency and infeasibility for field application due to the long solving process. Therefore, this study applies the Taguchi method, which has high efficiency in solving problems, in the optimization model, in order to obtain the influence coefficient of decision variable combinations on the reached targets. Then, the optimal strategy combination is determined according to the influence coefficient. This study takes the Danshui River basin in Taiwan for case analysis. The research process includes four parts: collection of basic data and river remediation strategies, construction of optimization model, input of the Taguchi method, and result analysis.

Citation: Tsung-Ming Yang, Nien-Sheng Hsu, Chih-Chiang Chiu and Hsin-Ju Wang, Applying the Taguchi Method to River Water Pollution Remediation Strategy Optimization, doi:10.3390/ijerph110404108.

RESEARCH AREA

The Danshui River is the third longest river in Taiwan, running for 158 km and draining a watershed of approximately 2,726 km^2 (Figure 1). Because the Danshui River has had a long history of water pollution from municipal and industrial wastewaters, the government's cleanup effort focused chiefly on rendering the polluted river odorless and ensuring that the ambient water has sufficient dissolved oxygen (DO) to sustain aquatic life, while also breaking down biological oxygen demand (BOD). The government uses a River Pollution Index based on the DO, BOD, ammonia, and suspended solids levels. Prior to 2002, however, Taiwan's previous effort to curb water pollution progressed slowly. In 2012, the government switched its focus from building sewerage systems with central collection and treatment processes to treating wastewaters on site along the river's upstream and midstream sections. Compared with conventional systems, which require significant time and money to build, onsite treatment systems, through the construction of wetlands and gravel beds in the river, can yield efficient cleanup at a much faster pace [1,2,3,4,5,6,7]. As shown in Figure 2, the overall water quality trend of Danshui River has changed significantly since 2003, and its condition has shifted towards light pollution.

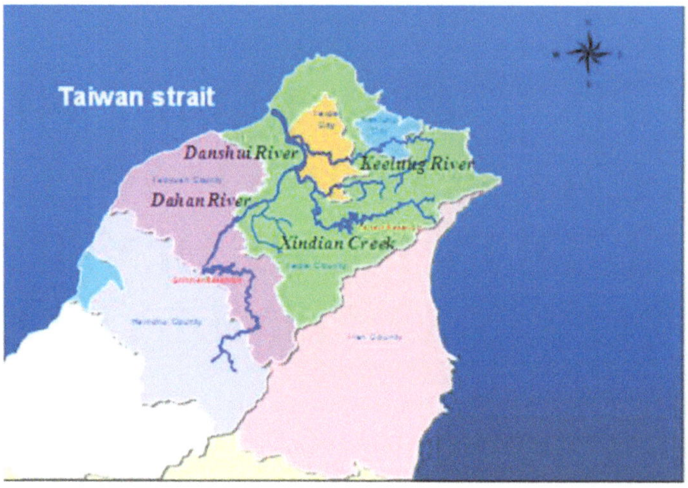

Figure 1: Danshui River basin range.

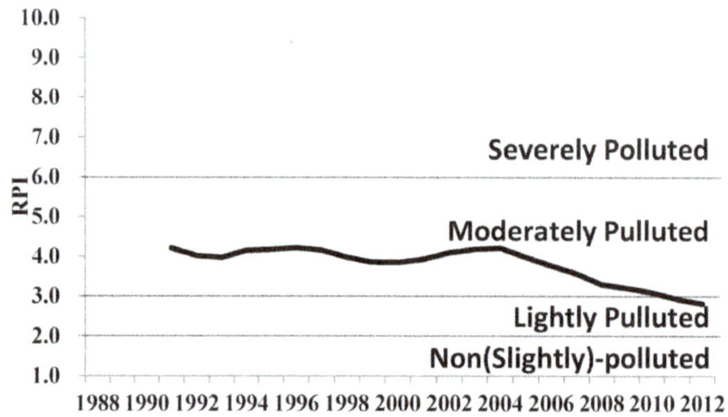

Figure 2: RPI trend map of Danshui River basin (RPI: River Pollution Index).

According to the key remediation measures, the rate of household connection in the cities around the Danshui River basin, by the end of 2012, is 70.4% in Taipei City and 49.5% in New Taipei City. There are three main sewage treatment plants, 35 interception stations and 21 onsite treatment facilities in the whole basin. The locations are shown in Figure 3.

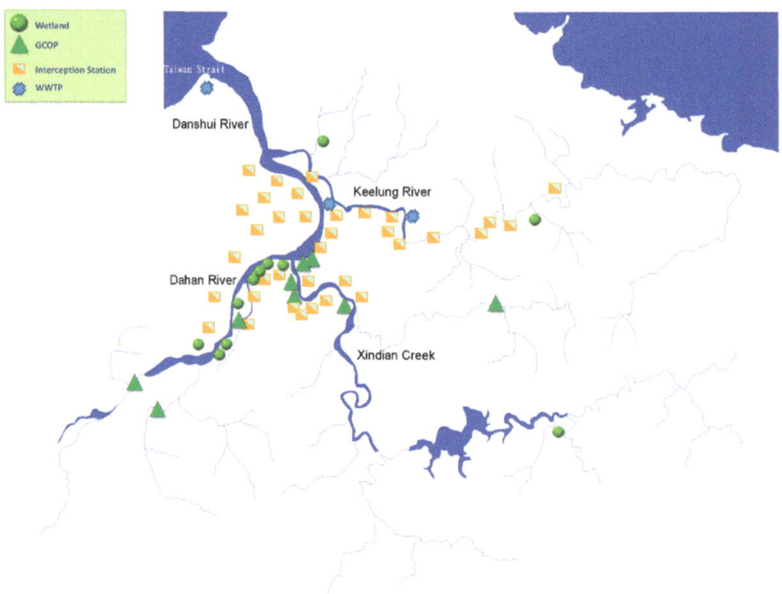

Figure 3: Danshui River basin sewage treatment installation distribution diagram.

The total designed capacity and onsite treatment capacity of interception stations in the Danshui River basin, the total household connection capacity, and the total treatment capacity of sewage treatment plants in 2012 are shown inTable 1. As seen, the total designed interception capacity and total household connection capacity of interception stations in the Danshui River basin have exceeded the treatment capacity of sewage treatment plants. Therefore, it is infeasible to operate all the interception stations. Secondly, in the drainage channels of the basin, a part of sewage network fails to meet or requires emergency treatment, and onsite treatment facilities are implemented. The total capacity and the total interception capacity have already exceeded the load capacity of sewage plants. Therefore, how to determine the start/stop of interception and onsite treatment facilities is a challenge to the present water pollution remediation in Danshui River basin.

Table 1: Comparison of treatment facility and sewage plant capacities.

Item	Flow Rate (m^3/day)
Total Interception Capacity	2,320,184
Total Onsite Treatment Capacity	247,054
Household connection capacity in 2012	1,174,621
Waste water treatment capacity	1,970,000

METHODOLOGY

In practice, the remediation of river water pollution first sets the objectives and period, and then analyzes the effect of remediation strategy, in order to select the optimal strategy combination for attaining the objectives.

In the optimization model solving process, the solution space formed by the decision variables is too large, and the water quality model computation is time consuming, resulting in poor model solving efficiency. Among studies on connecting optimization to simulation models, the speed of optimal solution search is the key factor influencing the overall computing efficiency. In order to improve the computing efficiency, this study uses the Taguchi method and regards the optimization model as a system. The decision variables are experimental parameters, and the objective function is system function. The experimental parameters are changed to obtain the variable parameters for overall function variation, thus accelerating the computational speed for the optimal combination of decision variables, and facilitating the decision-making on the optimal remediation strategy.

This research process includes: (1) determining the key features among domestic sewage—sewage treatment strategy—rivers; (2) developing a river water quality model to simulate river water quality under different strategies; (3) applying the Taguchi method to build an optimization model for selecting optimal strategy. The steps are described below:

Sewage Treatment Correlation and Water Quality Modeling

The sewer policy of Taiwan is based on a rain and sewage diversion system. Before the household connection is completed, the domestic sewage is discharged to the rivers by the rainwater collection system. On sunny days, the high concentration domestic sewage discharged into rivers in this way deteriorates the river water quality. In order to avoid the direct discharge of high concentration sewage, the current improvement measures, besides promoting household connection, include: (1) special pipes at the drainage end intercepts sewage for onsite treatment before being discharged; (2) the interception facilities intercept the sewage into the main sewer with remaining capacity, and the sewage is delivered to the downstream sewage treatment plant to drain after treatment. The overall regional sewage treatment strategy, including the above two remediation strategies, is illustrated in Figure 4.

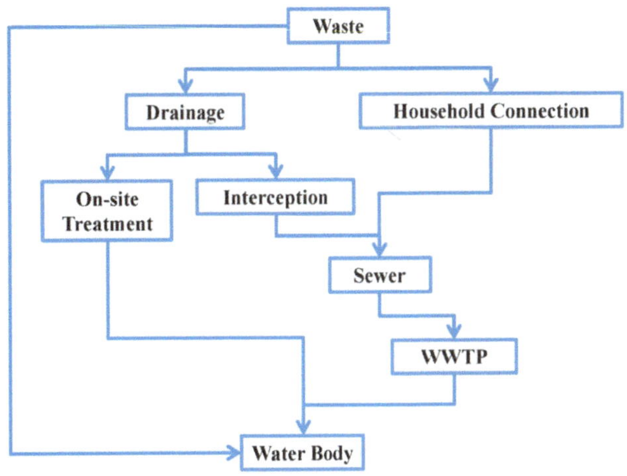

Figure 4: Schematic diagram of sewage treatment strategy.

In Taiwan, there is no domestically developed river water quality model, and all the above models are adopted from foreign experiences. The commonly used water quality models include WASP, QUAL2K, and SWMM. In terms of the Danshui River basin, in order to unify the water quality model tool, the Environmental Protection Administration (EPA) adopted WASP as the main model for simulating the Danshui River water quality. Since then, numerous studies have used WASP as the tool for Danshui River basin. This study uses the Danshui River model built by Chen *et al.* [8,9] as the main simulation tool, and includes the completed sewage treatment related facilities, domestic sewage connection, interception into main sewer, onsite treatment facilities, and direct discharge into river. The situation simulation and effect evaluation under various strategies are discussed [1,2,3,4,5,6,7].

Taguchi Method

The Taguchi method is a quality engineering proposed by Genichi Taguchi. The method includes off-line and on-line quality engineering and MTS (Mahalanobis-Taguchi System). This study uses off-line quality engineering method, which aims to improve R&D, design and production technology for solving optimization problem. The purposes are to obtain the optimal improvement measures, including functional evaluation of quality, determination of specifications (tolerance), parameter design and tolerance design [10].

Orthogonal arrays and S/N ratios are two main components of the Taguchi method. An orthogonal array is used to reduce testing time/cost. If an experiment has 15 control factors with two levels, all possible $n = 215 = 32,768$ mixtures are required to test for assessing the optimal mixture by using a full factorial design of experiment. After using the orthogonal array L16215, only 16 mixtures are required to estimate the optimal mixture *vs.* the optimal determination via the full factorial design of experiment. To significantly reduce the number of tests, while still gaining significant insight into the important factors and optimal settings, Taguchi recommended the use of eighteen basic orthogonal fractional factorial arrays, known as the standard orthogonal arrays. Taguchi treated the S/N ratio as a single indicator that considers the statistical characteristics of results to determine the importance of the factors under study. The S/N ratio can be categorized into three types as follows. Selection of the appropriate S/N ratio depends on the features of responses.

The smaller-the-better (STB) type:

$$SN_{STB} = -10\log_{10}\left(\frac{1}{t}\sum_{i=1}^{m} y^2(i)\right) \tag{1}$$

The larger-the-better (LTB) type:

$$SN_{LTB} = -10\log_{10}\left(\frac{1}{t}\sum_{i=1}^{m} \frac{1}{y^2(i)}\right) \tag{2}$$

The nominal-the-better (NTB) type:

$$SN_{NTB} = -10\log_{10}\left(\frac{1}{t}\sum_{i=1}^{m} (y^2(i) - v)^2\right) \tag{3}$$

where *m*, *y*(*i*), and m are the number of mixtures, the testing result of the i-th test, and the target value of the response, respectively [11].

Construction and Solving of an Optimization Model

The solving process proposed by this study is to build the optimization model first, and then determine the number of system parameters and level. The level of parameters is changed according to the orthogonal arrays of the Taguchi method, and the system is simulated repeatedly. Finally, the sensitivity of parameters to the system and the optimal solution of system are obtained. The solving process is shown in Figure 5.

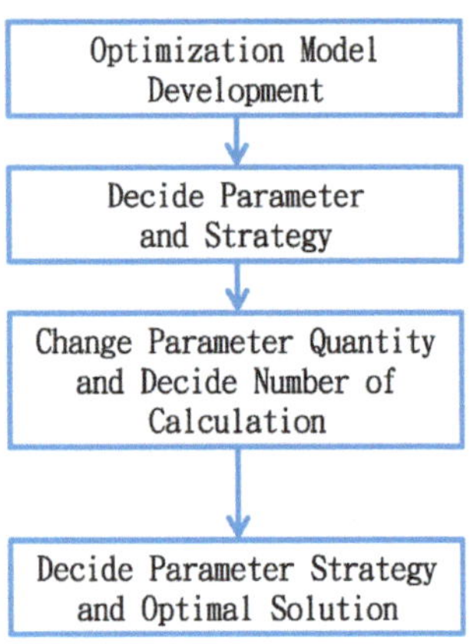

Figure 5: Solving process of this study.

Optimization Model Setting

The objective function of this optimization model is set as the minimum average RPI of the entire Danshui River basin; the purpose of which is to optimize the remediation strategy for the lower river pollution level. The decision variables are the strategies of various treatment facilities, and the constraints are the limitations of treatment facility strategies (the total interception capacity of interception stations along main sewer does not exceed the remaining capacity of the main sewer and the onsite treatment capacity does not exceed the designed capacity). The amount of sewage received by various treatment facilities is the amount of domestic sewage in the corresponding sewage collection area. The total flow of main sewer does not exceed the treatment capacity of sewage treatment plant. The basin pollution concentration is the computing result of WASP water quality model. The optimization model is expressed as follows:

$$Z = \min\sum_{i=1}^{N} F(q_i, \boldsymbol{C}_i) \tag{4}$$

$$s.t.\, q_i = \sum_{j=1}^{m} q_{ijk}\cdot X_{ij} \tag{5}$$

$$\boldsymbol{C}_i = H_{j=1}^{m}(\boldsymbol{C}_{ij}\cdot X_{ij}) \tag{6}$$

$$q_{ijk} = L_{ij}\cdot S_{ij} \tag{7}$$

$$\sum q_{ijk}\cdot(1 - X_{ij}) \le Q_{kT} - L_{kT} \tag{8}$$

$$\sum_{k=1}^{P_T} Q_{kT} \le W_T \tag{9}$$

Equation (4) is the objective function Z, representing basin-wide minimum average PRI, *i* is the water quality model computation node, N is the number of nodes of water quality model, $F(q_i,C_i)$ is the water quality model simulation result, displayed as basin-wide average RPI. The q_i and C_i are the flow rate and water quality at the i-th computation node.

Equations (5) and (6) show the flow rate and water quality at the *i*-th computation node and the effect of the facility strategy in the computation node on the node, where q_{ijk} is the water quantity discharged to the river by the *j*-th facility of the *k*-th main pipe system in the *i*-th node, X_{ij} is the facility operating strategy, as well as the decision variable of this study; C_{ij} is the quality of water discharged to the river by the *j*-th facility in the *i*-th node (given constant); if $k = 0$, it is onsite treatment facility.

Equation (7) is the amount of domestic sewage of the *j*-th facility in the *i*-th node, where L_{ij} is the sewage collection area corresponding to the *j*-th facility in the *i*-th computation node; and S_{ij} is the sewage delivery rate of the sewage collection area (given constant).

In Equation (8), $\Sigma q_{ijk} \cdot (1 - X_{ij})$ is the flow rate into the *k*-th sewer pipe through the *j*-th interception station of the *i*-th computation node; $Q_{kT} - L_{kT}$ is the remaining capacity of the k-th sewer pipe of the T-th sewage treatment plant (given constant).

In Equation (9), $\sum_{k=1}^{P_T} Q_{kT}$ is the total flow into the k-th sewer pipe of the T-th sewage treatment plant; W_T is the treatment capacity of the T-th sewage treatment plant (given constant).

Determine System Parameters and Strategies

The decision variable of this study is the facility operating strategy X_{ij}, *i.e.*, the start-stop condition of the facility. To determine the number of system parameters, this study considers the interception stations and onsite treatment facilities with high operating flexibility in the Danshui River basin as system parameters. There are 31 treatment facilities determined as system parameters after comparing the treatment facility operating data. In addition, in order to reduce the solution space of the optimization model, each system parameter has two operating strategies X_{ij} (*i.e.*, 0% and 100%).

Change Parameters, Repeat Simulating System

When the number of system parameters and operating strategies are determined, the number of repeated simulations and the values of parameters in each simulation can be determined according to the orthogonal array of the Taguchi method. This study selects 31 system parameters and each parameter has two operating strategies. The optimum parameter combination can be obtained according to the simulation result of only 32 repeated system simulations. In comparison to general optimization, which needs to find the optimal solution among $2^{31} = 2{,}147{,}483{,}648$ permutations, the Taguchi method can shorten the computation time greatly. The 31 system parameters selected by this study are indicated as codes A,B~AE, while 1 means that the operating strategy is 100%, 2 means that the operating strategy is 0%. For example, the value of parameter A in the first strategy combination is 1, meaning that the operating strategy of parameter A in the strategy combination is 100%, and so forth. The parameters in various strategy permutations are shown in Table 2.

Determine Parameter Impact Factor and Optimal Solution of System

The basin-wide average water quality of each simulation is determined after 32 simulations. The basin-wide average water quality under the same operating strategy value of the same facility is calculated. Finally, the values under different strategies are subtracted to obtain the influence coefficient of the facility (system parameters), as shown in Figure 6 and Figure 7, and Table 3. The influence coefficient in this study represents the influence of treatment facility changing strategy on basin-wide water quality. Therefore, a larger influence coefficient indicates greater improvement on the basin-wide water quality. This study reviews the sequence of influence coefficients of facilities on river water quality and the remaining capacity in their locations. If the remaining capacity is sufficient, the facility is switched on (100%), otherwise it is switched off (0%).

Table 2: System simulation parameter permutations.

Strategy / Parameter	No. 1	No. 2	No. 3	No. 4	No. 5	No. 6	No. 7	No. 8	No. 9	No. 10	No. 11	No. 12	No. 13	No. 14	No. 15	No. 16	No. 17	No. 18	No. 19	No. 20	No. 21	No. 22	No. 23	No. 24	No. 25	No. 26	No. 27	No. 28	No. 29	No. 30	No. 31	No. 32
A	1	1	1	1	1	1	1	1	1	1	1	1	1	1	1	1	2	2	2	2	2	2	2	2	2	2	2	2	2	2	2	2
B	1	1	1	1	1	1	1	1	2	2	2	2	2	2	2	2	1	1	1	1	1	1	1	1	2	2	2	2	2	2	2	2
C	1	1	1	1	1	1	1	1	2	2	2	2	2	2	2	2	2	2	2	2	2	2	2	2	1	1	1	1	1	1	1	1
D	1	1	1	1	2	2	2	2	1	1	1	1	2	2	2	2	1	1	1	1	2	2	2	2	1	1	1	1	2	2	2	2
E	1	1	1	1	2	2	2	2	1	1	1	1	2	2	2	2	2	2	2	2	1	1	1	1	2	2	2	2	1	1	1	1
F	1	1	1	1	2	2	2	2	2	2	2	2	1	1	1	1	1	1	1	1	2	2	2	2	2	2	2	2	1	1	1	1
G	1	1	1	1	2	2	2	2	2	2	2	2	1	1	1	1	2	2	2	2	1	1	1	1	1	1	1	1	2	2	2	2
H	1	1	2	2	1	1	2	2	1	1	2	2	1	1	2	2	1	1	2	2	1	1	2	2	1	1	2	2	1	1	2	2
I	1	1	2	2	1	1	2	2	1	1	2	2	1	1	2	2	2	2	1	1	2	2	1	1	2	2	1	1	2	2	1	1
J	1	1	2	2	1	1	2	2	2	2	1	1	2	2	1	1	1	1	2	2	1	1	2	2	2	2	1	1	2	2	1	1
K	1	1	2	2	1	1	2	2	2	2	1	1	2	2	1	1	2	2	1	1	2	2	1	1	1	1	2	2	1	1	2	2
L	1	1	2	2	2	2	1	1	1	1	2	2	2	2	1	1	1	1	2	2	2	2	1	1	1	1	2	2	2	2	1	1
M	1	1	2	2	2	2	1	1	1	1	2	2	2	2	1	1	2	2	1	1	1	1	2	2	2	2	1	1	1	1	2	2
N	1	1	2	2	2	2	1	1	2	2	1	1	1	1	2	2	1	1	2	2	2	2	1	1	2	2	1	1	1	1	2	2
O	1	1	2	2	2	2	1	1	2	2	1	1	1	1	2	2	2	2	1	1	1	1	2	2	1	1	2	2	2	2	1	1
P	1	2	1	2	1	2	1	2	1	2	1	2	1	2	1	2	1	2	1	2	1	2	1	2	1	2	1	2	1	2	1	2
Q	1	2	1	2	1	2	1	2	1	2	1	2	1	2	1	2	2	1	2	1	2	1	2	1	2	1	2	1	2	1	2	1
R	1	2	1	2	1	2	1	2	2	1	2	1	2	1	2	1	1	2	1	2	1	2	1	2	2	1	2	1	2	1	2	1
S	1	2	1	2	1	2	1	2	2	1	2	1	2	1	2	1	2	1	2	1	2	1	2	1	1	2	1	2	1	2	1	2
T	1	2	1	2	2	1	2	1	1	2	1	2	2	1	2	1	1	2	1	2	2	1	2	1	1	2	1	2	2	1	2	1
U	1	2	1	2	2	1	2	1	1	2	1	2	2	1	2	1	2	1	2	1	1	2	1	2	2	1	2	1	1	2	1	2
V	1	2	1	2	2	1	2	1	2	1	2	1	1	2	1	2	1	2	1	2	2	1	2	1	2	1	2	1	1	2	1	2
W	1	2	1	2	2	1	2	1	2	1	2	1	1	2	1	2	2	1	2	1	1	2	1	2	1	2	1	2	2	1	2	1
X	1	2	2	1	1	2	2	1	1	2	2	1	1	2	2	1	1	2	2	1	1	2	2	1	1	2	2	1	1	2	2	1
Y	1	2	2	1	1	2	2	1	1	2	2	1	1	2	2	1	2	1	1	2	2	1	1	2	2	1	1	2	2	1	1	2
Z	1	2	2	1	1	2	2	1	2	1	1	2	2	1	1	2	1	2	2	1	1	2	2	1	2	1	1	2	2	1	1	2
AA	1	2	2	1	1	2	2	1	2	1	1	2	2	1	1	2	2	1	1	2	2	1	1	2	1	2	2	1	1	2	2	1
AB	1	2	2	1	2	1	1	2	1	2	2	1	2	1	1	2	1	2	2	1	2	1	1	2	1	2	2	1	2	1	1	2
AC	1	2	2	1	2	1	1	2	1	2	2	1	2	1	1	2	2	1	1	2	1	2	2	1	2	1	1	2	1	2	2	1
AD	1	2	2	1	2	1	1	2	2	1	1	2	1	2	2	1	1	2	2	1	2	1	1	2	2	1	1	2	1	2	2	1
AE	1	2	2	1	2	1	1	2	2	1	1	2	1	2	2	1	2	1	1	2	1	2	2	1	1	2	2	1	2	1	1	2

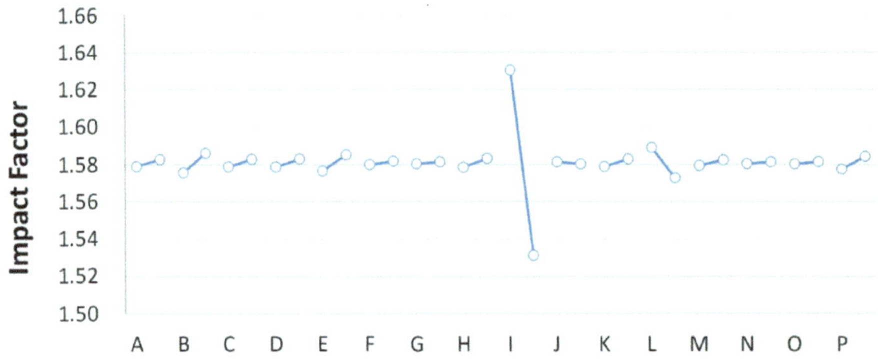

Figure 6: Danshui River water pollution remediation strategy simulation results- Dahan River BOD_5 concentration.

Figure 7: Danshui River water pollution remediation strategy simulation results- Dahan River BOD_5 concentration.

Table 3: System impact factors.

Rank	Impact Factor	Facility	Rank	Impact Factor	Facility
1	0.1248	X	17	0.0037	A
2	0.0993	I	18	0.0030	M
3	0.0343	Q	19	0.0023	Y
4	0.0258	AC	20	0.0020	U
5	0.0238	T	21	0.0018	F
6	0.0198	AA	22	0.0017	V
7	0.0164	L	23	0.0016	Z
8	0.0104	B	24	0.0012	S
9	0.0086	E	25	0.0012	O
10	0.0082	R	26	0.0011	J
11	0.0069	P	27	0.0011	G
12	0.0047	H	28	0.0010	N
13	0.0043	D	29	0.0009	AE
14	0.0041	C	30	0.0008	AD
15	0.0040	K	31	0.0002	W
16	0.0038	AB			

RESULTS

The analyzed optimal solution can be determined according to the impact factor and the remaining capacity of system; the result is shown in Table 4. The "Current Condition" in Table 4 means actual operation percentage compared to design flow rate. This study assumes that there are 2 strategies in each facility (*i.e.*, 0% and 100%), so the priority of each facility can be influenced not only by its capacity but also other facilities'. Although some facilities rank high, they are limited by the remaining capacity; in other words, despite of the high influence coefficient, they cannot be switched on (100%). Some facilities rank rather low, but due to sufficient remaining capacity, they are switched on (100%). The model simulation result is displayed as BOD_5 concentration and RPI, as shown in Figure 8,Figure 9, Figure 10, Figure 11, Figure 12 and Figure 13. The "existing circumstance" refers to the average operation amount of various facilities in 2012. The figure shows the RPI under the existing operating strategy, given that the severe pollution length in the middle reach of the Dahan River is 18 km, while other reaches are below moderate pollution. The model simulation result of this study shows that the length of severe pollution in the middle reach of the Dahan River can be shortened to about 5 km. It is the basin with the most significant effectiveness; whereas there is slight water quality improvement effectiveness on the Xindian Creek and the Keelung River.

Table 4: Optimal solution of optimization model.

Rank	Faculity	Creek	Current Condition	Simulation Result	Rank	Faculity	Creek	Current Condition	Simulation Result
1	X	Dahan	10%	100%	17	A	Xindian	31%	0%
2	I	Dahan	24%	100%	18	M	Dahan	26%	0%
3	Q	Dahan	100%	100%	19	Y	Dahan	68%	100%
4	AC	Dahan	56%	100%	20	U	Dahan	0%	0%
5	T	Keelung	42%	100%	21	F	Dahan	5%	0%
6	AA	Dahan	41%	100%	22	V	Keelung	47%	100%
7	L	Dahan	12%	100%	23	Z	Dahan	100%	100%
8	B	Xindian	35%	0%	24	S	Xindian	26%	0%
9	E	Dahan	16%	100%	25	O	Dahan	100%	100%
10	R	Xindian	77%	0%	26	J	Keelung	47%	0%
11	P	Dahan	100%	100%	27	G	Dahan	19%	0%
12	H	Dahan	64%	0%	28	N	Dahan	22%	0%
13	D	Xindian	83%	0%	29	AE	Xindian	5%	0%
14	C	Xindian	19%	0%	30	AD	Keelung	49%	0%
15	K	Dahan	0%	0%	31	W	Keelung	23%	100%
16	AB	Keelung	38%	100%					

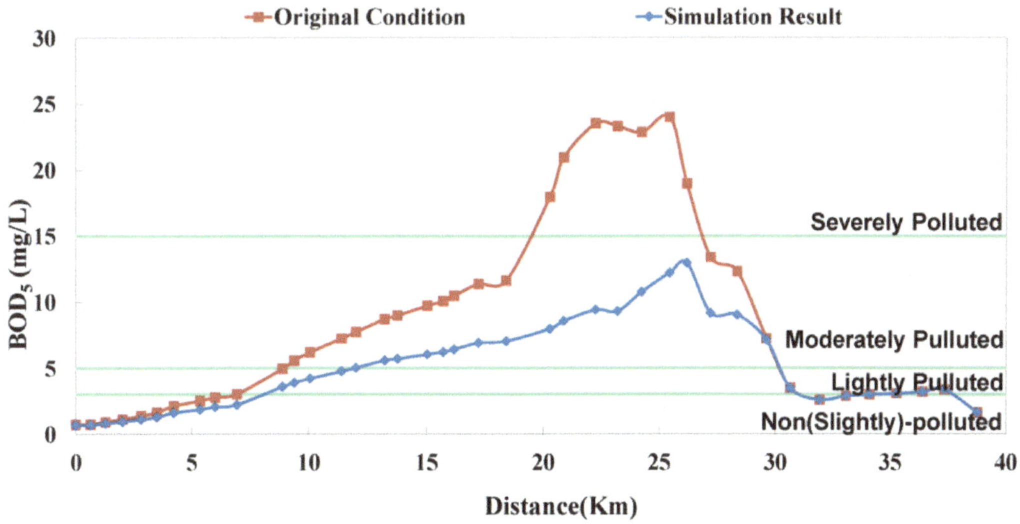

Figure 8: Danshui River water pollution remediation strategy simulation results- Dahan River BOD_5 concentration.

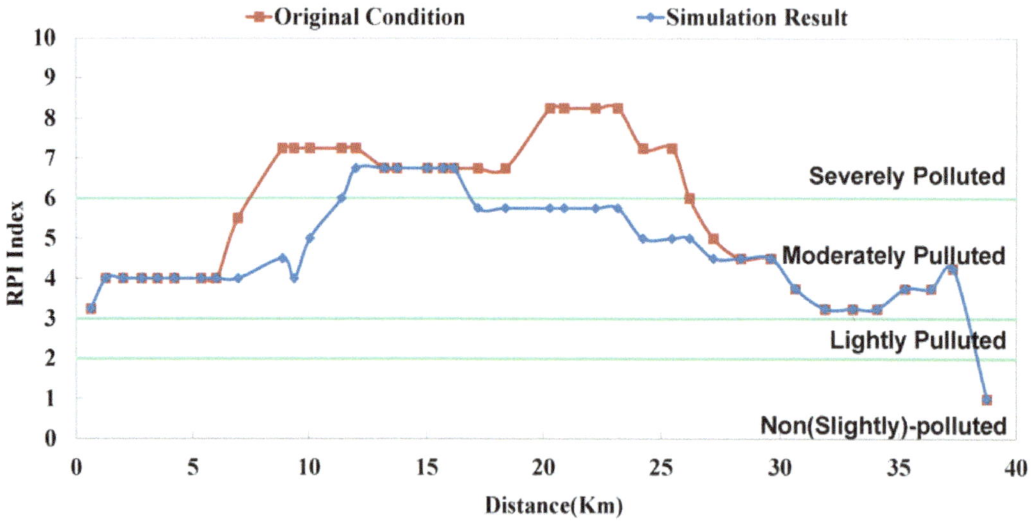

Figure 9: Danshui River water pollution remediation strategy simulation results- Dahan River RPI.

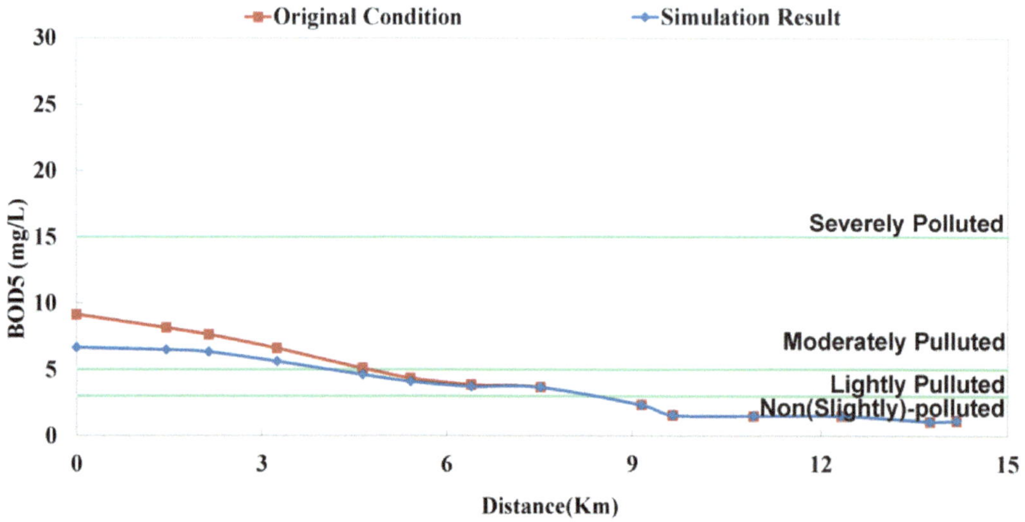

Figure 10: Danshui River water pollution remediation strategy simulation results—Xindian Creek BOD_5 concentration.

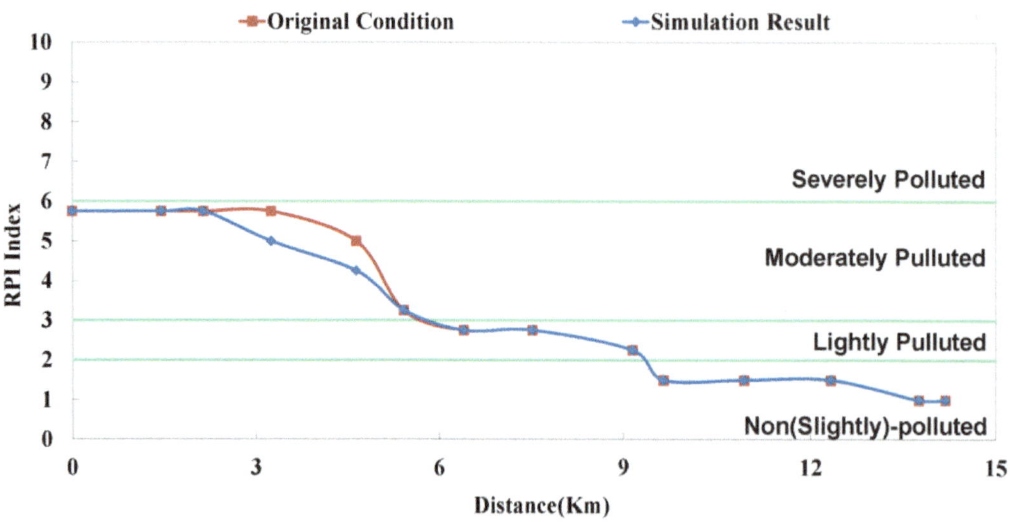

Figure 11: Danshui River water pollution remediation strategy simulation results—Xindian Creek RPI.

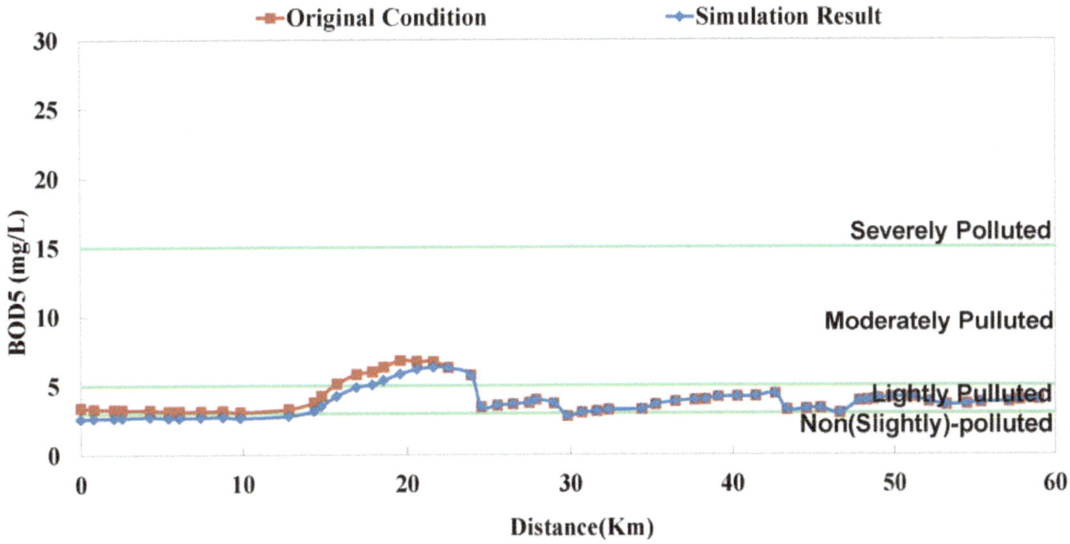

Figure 12: Danshui River water pollution remediation strategy simulation results—Keelung River BOD_5 concentration.

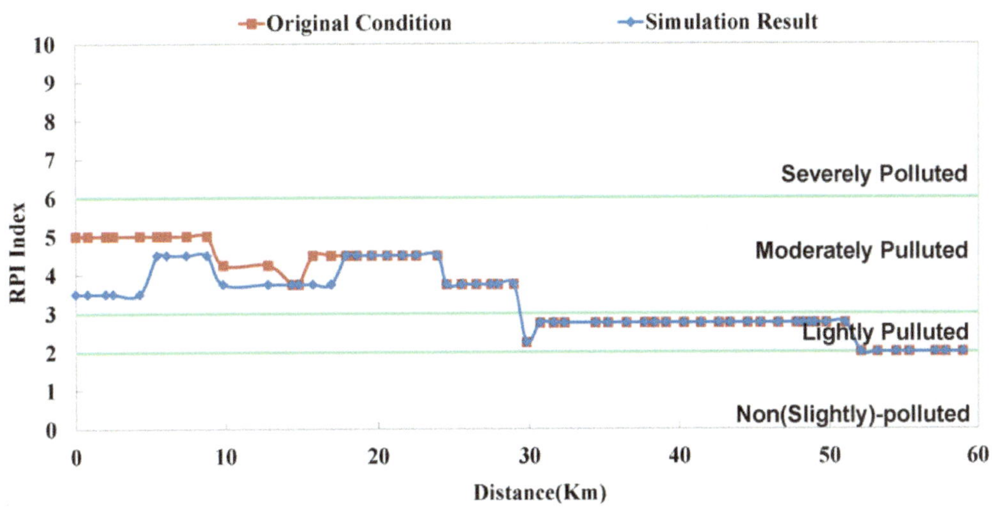

Figure 13: Danshui River water pollution remediation strategy simulation results—Keelung River RPI.

Table 5: Comparison between optimal strategy combination and current condition.

Tributary	Q(CMD)		Number of Operating Facilities		Rate of Household Connection
	Current Condition	Simulation Result	Current Condition	Simulation Result	
Dahan River	474,842	796,700	28	23	25.1%
Hsindian Creek	298,251	39,700	11	4	36.7%
Keelung River	237,378	203,080	14	12	69.1%
Total	1,010,471	1,039,480	53	39	--

The optimal strategy combination of this study and current condition are compared in Table 5. It is observed that after optimization, the facility operation amount is less than the current condition, but the total sewage treatment capacity is slightly higher than the current condition. In other words, the treatment capacity tends to concentrate, and the basin average water quality is apparently improved. Considering the changes in treatment capacity distribution, the optimized treatment capacity of the Dahan River is much higher than the current condition, that of the Xindian Creek is much lower than the current condition, and that of the Keelung River is slightly lower than the current condition. The simulation result is compared with the rate of household connection of various basins. It is found that the rate of household connection of the Dahan River basin is the lowest, suggesting that the domestic sewage in the reach has a high impact on the river water quality. Therefore, the highest effectiveness can be obtained by implementing remediation strategy on the reach. As the Keelung River basin has a higher rate of household connection, there are fewer optimized facilities in the reach. Moreover, the domestic sewage influences the river slightly, the reach pollution condition is slighter than the Dahan River, and the water quality improvement effectiveness is low. The Xindian Creek is a special case. Figure 6, Figure 7, Figure 8, Figure 9, Figure 10 and Figure 11indicate that the simulation result is similar to the current condition. However, Table 5 shows that the sewage operation capacity and the quantity of facilities switched on in the simulation results are reduced greatly. In other words, although the rate of household connection of the Xindian Creek basin is relatively low, the river base discharge is large, and the start-stop of facilities influences the river water quality slightly.

CONCLUSIONS

Combination of Taguchi Method into the Optimization Model

This study incorporated the Taguchi method into the optimization model for solution. According to the sequence of influence of parameters on the system and the limitation of optimization model system capacity, the optimal solution of the optimization problem was determined. The model computing time was shortened greatly.

Model Result

The facility influence weights were obtained using the Taguchi method, and the optimum combination was screened according to the remaining capacity. The result suggested that the simulation result of this study is better than the current condition, and the Dahan River is the basin with the highest water quality improvement effectiveness. Under the optimal strategy of this study, the severe pollution length was shortened from about 18 km to about 5 km.

Relationship between Model Result and Household Connection

According to the comparison between simulation result and the rate of household connection of basins, the basin water quality improvement effectiveness was proportional to the river pollution condition, and inversely proportional to the rate of household connection of basins.

Suggestions for Future Research

This study used two operating strategies for strategy configuration. Future studies can increase the operating strategies, from the On (100%) and Off (0%) strategies to operating altitudes of different openings, allowing the facilities to be used more flexibly and effectively.

AUTHOR CONTRIBUTIONS

Study conception and design: Tsung-Ming Yang

Acquisition of data: Chih-Chiang Chiu, Hsin-Ju Wang

Analysis and interpretation of data: Tsung-Ming Yang, Nien-Sheng Hsu

Drafting of manuscript: Chih-Chiang Chiu, Hsin-Ju Wang

Critical revision: Nien-Sheng Hsu

CONFLICTS OF INTEREST

The authors declare no conflict of interest.

REFERENCES

1. New Taipei City Water Resources Department, Montgomery Watson Harza (MWH). *Water Quality Monitoring, Hydraulic Geometry Survey, Performance Evaluation and Management in Danshui River in New Taipei City*; New Taipei City Water Resources Department: Taipei, Taiwan, 2011.
2. Taipei County Environmental Protection Bureau, Montgomery Watson Harza (MWH). *Dahan River Ecological Corridor and On-Site Treatment Facilities Development Project in Taipei County*; Taipei County Environmental Protection Bureau: Taipei, Taiwan, 2007.
3. Taipei County Water Resources Bureau, Montgomery Watson Harza (MWH). *Water Purification and On-Site Treatment Project in Xindian Creek Basin*; Taipei County Water Resources Bureau: Taipei, Taiwan, 2008.
4. Taipei City Government, Montgomery Watson Harza (MWH). *Technical Services of On-Site Treatment Installation in the Zhong Xiao, Zhengzhou, Guiyang, Chengdu Drainage (Planning and Design Stage)*; Taipei City Government: Taipei, Taiwan, 2009.
5. Taipei County Water Resources Department, Montgomery Watson Harza (MWH). Water Quality Monitoring, Hydraulic Geometry Survey and Model Evaluation in Danshui River in Taipei County, Taipei County Water Resources Department, Taipei, Taiwan, 2010.
6. Taiwan Environmental Protection Administration, Montgomery Watson Harza (MWH). *Program Management of Danshui River Pollution Control Plan*; Taiwan Environmental Protection Administration: Taipei, Taiwan, 2008.
7. Taiwan Environmental Protection Administration, Montgomery Watson Harza (MWH). *Improvement and Management of Water Environment and Master Plan*; Taiwan Environmental Protection Administration: Taipei, Taiwan, 2012.
8. Chen, C.H.; Lung, W.S.; Li, S.W.; Lin, C.F. Technical challenges with BOD/DO modeling of rivers in Taiwan. *J. Hydro. Environ. Res.* 2012, *6*, 3–8.
9. Chen, C.H.; Lung, W.S.; Yang, C.H.; Lin, C.F. Spatially variable deoxygenation in the Danshui River: Improvement in model calibration. *J. Water Environ. Res.* 2013, *85*, 2243–2253.
10. Bureau of Standards, Metrology & Inspection, M.O.E.A. Guidelines for TAGUCHI Methods and Their Application. Available online: http://www.bsmi.gov.tw/wSite/index.jsp (accessed on 11 April 2014).
11. Chang, C.Y.; Huang, R.; Lee, P.C.; Weng, T.L. Application of a weighted Grey-Taguchi method for optimizing recycled aggregate concrete mixtures. *J. Cem. Concr. Compos.* 2011, *33*, 1038–1049.

CHAPTER 14

MULTIOBJECTIVE OPTIMIZATION OF WATER DISTRIBUTION NETWORKS USING FUZZY THEORY AND HARMONY SEARCH

Zong Woo Geem

Keywords: Multiobjective optimization, water distribution network, fuzzy theory, harmony search, reliability

INTRODUCTION

Up to date, diverse phenomenon-mimicking algorithms (PMAs) have been proposed for the design optimization of water distribution networks (WDNs). These algorithms include genetic algorithm [1], simulated annealing [2], tabu search [3], shuffled frog-leaping algorithm [4], ant colony optimization algorithm [5], harmony search [6], cross entropy [7], scatter search [8], hybrid algorithm [9], honey-bee mating optimization [10], differential evolution [11], adaptive cluster covering with local search [12], and Non-Dominated Sorting Genetic Algorithm-II [13].

Although various PMAs have been proposed, these were mostly applied to basic bench-mark WDNs, such as two-loop and Hanoi networks, which are gravity-fed systems with the following optimization formulation:

$$C = \sum_{i=1}^{P} f(D_i, L_i) \tag{1}$$

where f(·) is cost function which has two arguments of pipe diameter Di and pipe length Li; and P is total number of pipe i. This cost function is to be minimized while satisfying the following constraints:

$$\sum Q_{in} - \sum Q_{out} = Q_e \tag{2}$$

where Q_{in} is inflow amount to a specific node; Q_{out} is outflow amount from the specific node; and Q_e is external demand amount at the node. This continuity constraint must be satisfied for every node in a network.

$$\sum h_f = 0 \tag{3}$$

where hf is head loss due to pipe friction, which is calculated using Hazen-Williams or Darcy-Weisbach formulae. This energy conservation constraint must be satisfied for every loop in the network.

$$H_j \geq H_j^{min} \tag{4}$$

where Hj is pressure head at node j; H_j^{min} is minimal pressure head required for node j. This minimum pressure constraint must be satisfied for every node in the network.

In addition to the above optimization formulation, one more reliability constraint was recently considered for the water network optimization. Geem *et al.* [14] added flow velocity constraint for every pipe as follows:

$$v_i^{min} \leq v_i \leq v_i^{max} \tag{5}$$

where vi is flow velocity through pipe i; and v_i^{min} (0.1 m/s in this study) and v_i^{min} (3.0 m/s in this study) are respectively lower and upper bounds of flow velocities in pipe i. Here, in order to analyze the water network and to obtain hydraulic

Citation: Zong Woo Geem, Multiobjective Optimization of Water Distribution Networks Using Fuzzy Theory and Harmony Search, doi:10.3390/w7073613.

values such as flowrate, pressure head, and flow velocity, a popular simulator EPANET [15] is used.

Although this flow velocity constraint has been seldom adopted in previous optimization researches [16], it is critical with respect to pipe reliability. Excessive high flow velocity causes pipe erosion and structural vulnerableness when water hammer occurs. Additionally, it causes significant head loss [17] because of the following relationships:

$$h_f \propto v^{1.85} \text{ (Hazen-Williams formula)} \quad (6)$$

$$h_f \propto v^{m} \text{ (Darcy-Weisbach formula)} \quad (7)$$

Here, it should be noted that the friction coefficient in the Darcy-Weisbach formula is a function of the velocity. Thus, the relationship between head loss and velocity depends on the hydraulic regime (the hydraulic exponent m is 2 only in "fully developed turbulent flow"; it ranges from 1.75 to 2 in "transition flow"; it is 1.75 in "hydraulically smooth flow"; and it becomes 1 in "laminar flow").

Meanwhile, minimum velocity condition is required to prevent fine material sedimentation, which sometimes blocks the pipe.

This study intends to move forward one more step in terms of the reliability of flow velocity. Instead of being considered as the constraint, the flow velocity can be more precisely considered using fuzzy theory.

Thus far, researchers have proposed some reliability indexes: Todini [18] enumerated various failures due to mechanical reason (such as pipe break, pump stop, and power outage) and hydraulic reason (such as demand change and pipe aging), and then proposed a resilience index which is a measure to quantify how much a network can overcome the failures; Prasad and Park [19] also adopted the resilience index and simultaneously optimized cost and reliability using genetic algorithm; Prasad and Tanyimboh [20] proposed flow entropy as a surrogate reliability measure, which alleviates the drawbacks of resilience index; Ghajarnia *et al.* [21] proposed a reliability index based on nodal pressure; and Liu *et al.* [22] suggested a flow entropy-based reliability index. Atkinson *et al.* [23] categorized various reliability indicators into four groups such as Resilience Index [18,19], Entropy [24,25], Minimum Surplus Head [26], and Performance [27,28].

However, no reliability index has been proposed yet with respect to flow velocity. Thus, the key point of this work is to present a new reliability index with respect to flow velocity using fuzzy theory.

FUZZY THEORY AND MULTI-OBJECTIVE FUNCTION

If we just use the velocity constraint (Equation (5)) in the optimization formulation, there is no difference between the velocity adjacent to minimal allowed velocity (v_i^{max}) or maximal allowed velocity (v_i^{max}) and the in-between one. However, if we use fuzzy theory, we can consider how good a velocity is in terms of reliability. For example, the velocity in the center of minimum and maximum bounds should have much higher preference than that located slightly after minimum bound because the velocity near minimum bound may partially have the problems of low velocity. Fuzzy theory can quantify this preference.

Fuzzy theory can be represented as a membership function in Figure 1 if we construct it in symmetric triangular shape. Actually the membership function can have more complicated shape. However, devising more complex shape of membership function requires another extensive task by surveying various hydraulic and practical aspects. Thus, this study adopts the most popular triangular shape to start with. As seen in the figure, velocity reliability (VR) or fuzzy membership degree is zero at $v_i = v_i^{min}$ or $v_i = v_i^{max}$ while it becomes 100% at $v_i = v_i^{mean} = \left(v_i^{min} + v_i^{max}\right)/2$. Any in-between velocity has the reliability calculated as follows:

$$VR = \begin{cases} 0 & \text{if } v_i < v_i^{min} \\ \dfrac{v_i - v_i^{min}}{v_i^{mean} - v_i^{min}} & \text{if } v_i^{min} \le v_i \le v_i^{mean} \\ \dfrac{v_i - v_i^{max}}{v_i^{mean} - v_i^{max}} & \text{if } v_i^{mean} < v_i \le v_i^{max} \\ 0 & \text{if } v_i^{max} < v_i \end{cases} \quad \text{where } v_i^{mean} = \frac{v_i^{min} + v_i^{max}}{2} \quad (8)$$

Then, total velocity reliability for a network, called velocity reliability index (VRI) in this study, can be calculated by summing all velocity reliabilities in each pipe, as follows:

$$VRI = \sum_{i=1}^{P} VR_i \quad (9)$$

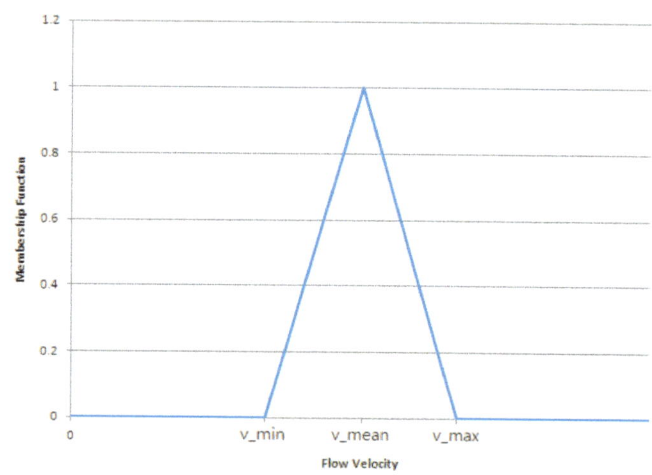

Figure 1: Fuzzy membership function for flow velocity Reliability.

And this optimization problem becomes to have multiobjective functions as follows:

$$\text{Minimize } C = \sum_{i=1}^{P} f(D_i, L_i) \text{ and Maximize } VRI = \sum_{i=1}^{P} VR_i \quad (10)$$

Equation (10) substitutes both Equation (1) and Equation (5) for this new bi-objective optimization problem.

HARMONY SEARCH ALGORITHM

Harmony search algorithm, first proposed by Geem *et al.* [29], has been continuously developed in terms of both theory and application. HS applications include structural design [30], geotechnical analysis [31], road property analysis [32], groundwater management [33], and project scheduling [34].

HS is similar to other evolutionary or phenomenon-inspired algorithms as a meta-heuristic algorithm. It possesses two basic operators (exploration operator for global search and exploitation operator for local search) as other algorithms also have. However, each algorithm has its own uniqueness in term of how it balances the two operators.

Furthermore, HS has its own theoretical background of stochastic derivative [35] to search for a global optimum, which is a new paradigm from the traditional gradient-based derivative in differential calculus. This novel derivative, which is applicable to a discrete variable rather than a continuous one, gives different selection rate to each candidate value for the discrete decision variable, as follows:

$$\left.\frac{\partial f}{\partial x_i}\right|_{x_i=x_i(k)} = \frac{1}{K_i}\cdot P_{Random} + \frac{n(x_i(k))}{HMS}\cdot P_{Memory} + \frac{n(x_i(k\mp m))}{HMS}\cdot P_{Pitch} \tag{11}$$

where f is objective function; x_i is discrete decision variable; $x_i(k)$ is a specific discrete value for the variable xi, which is the element of value set $\{x_i(1), x_i(2), \ldots, x_i(k), \ldots, x_i(K_i)\}$; K_i is total number of discrete values for the variable xi; P_{Random}; is probability to consider random selection operation; P_{Memory} is probability to consider memory consideration operation; P_{Pitch} is probability to consider pitch adjustment operation; m is neighboring index which has normally one; n(·) is count function which counts the number of the specific discrete value stored in harmony memory (**HM**); and HMS is size of **HM** which is expressed as follows:

$$\mathbf{HM} = \left[\begin{array}{cccc|c} x_1^1 & x_2^1 & \cdots & x_n^1 & f(\mathbf{x}^1) \\ x_1^2 & x_2^2 & \cdots & x_n^2 & f(\mathbf{x}^2) \\ \vdots & \cdots & \cdots & \cdots & \vdots \\ x_1^{HMS} & x_2^{HMS} & \cdots & x_n^{HMS} & f(\mathbf{x}^{HMS}) \end{array}\right] \tag{12}$$

Here, the summation of three probabilities P_{Random}, P_{Memory}, and P_{Pitch} becomes unity (or 100%) as follows:

$$P_{Random} + P_{Memory} + P_{Pitch} = 1 \tag{13}$$

Additionally, the cumulative value of the derivatives of all candidate discrete values for each variable xi should be equal to unity:

$$\sum_{k=1}^{K_i}\left.\frac{\partial f}{\partial x_i}\right|_{x_i=x_i(k)} = \frac{\sum_{k=1}^{K_i} 1}{K_i} P_{Random} + \frac{\sum_{k=1}^{K_i} n(x_i(k))}{HMS} P_{Memory} + \left[\frac{\sum_{k=1+m}^{K_i+1} n(x_i(l-m))}{HMS} + \frac{\sum_{k=0}^{K_i-m} n(x_i(l+m))}{HMS}\right]\frac{P_{Pitch}}{2} = 1 \tag{14}$$

In addition to this stochastic derivative, HS has a parameter-setting-free (PSF) process, which gives algorithm users user-friendliness without requiring tedious and time-consuming task of assigning proper values for algorithm parameters [36,37,38].

The PSF-HS simultaneously manages a new matrix, named OTM (operation type memory), to memorize the type of the operation:

$$\begin{bmatrix} ot_1^1 & ot_2^1 & \cdots & ot_n^1 \\ ot_1^2 & ot_2^2 & \cdots & ot_n^2 \\ \vdots & \cdots & \cdots & \cdots \\ ot_1^{HMS} & ot_2^{HMS} & \cdots & ot_n^{HMS} \end{bmatrix} \tag{15}$$

where ot_k^l denotes operation type, which has one of three values (random selection, memory consideration, or pitch adjustment). Based on OTM, algorithm parameters P_{Random}, P_{Memory}, and P_{Pitch} are automatically determined.

HS also has a multiobjective optimization strategy [34]. While the original HS algorithm updates HM based on single objective function as follows:

$$\mathbf{x}^{Worst} \notin \mathbf{HM} \quad \wedge \quad \mathbf{x}^{New} \in \mathbf{HM} \tag{16}$$

where x^{Wors}_t is worst solution vector stored in **HM** and x^{New} is newly generated vector using stochastic derivative of Equation (11), the multiobjective HS algorithm updates HM based on two (or more) objective functions as follows:

$$\mathbf{x}^{Worse} \notin \mathbf{HM} \quad \wedge \quad \mathbf{x}^{New} \in \mathbf{HM} \quad if\ \mathbf{x}^{New} \succ \mathbf{x}^{Worse} \tag{17}$$

where the symbol ≻ represents strict domination. That is, only if the newly generated vector x^{New} strictly dominates another vector x^{Worse} stored in **HM**, these two vectors are swapped. Here, strict domination can be expressed as in Equation (18) or (19):

$$C(\mathbf{x}^{New}) < C(\mathbf{x}^{Worse}) \text{ And } VRI(\mathbf{x}^{New}) \geq VRI(\mathbf{x}^{Worse}) \tag{18}$$

$$C(\mathbf{x}^{New}) \leq C(\mathbf{x}^{Worse}) \text{ And } VRI(\mathbf{x}^{New}) > VRI(\mathbf{x}^{Worse}) \tag{19}$$

Figure 2 shows a flowchart for the aforementioned multiobjective optimization procedure using fuzzy theory and harmony search.

APPLICATIONS AND RESULTS

The proposed multiobjective HS model was applied to a popular bench-mark problem, two-loop WDN [39]. As shown in Figure 3, it has one reservoir, two loops, six demand nodes, and eight pipes. Because each pipe can choose any diameter from 14 candidates, total solution space becomes 148=1.48×109. With harmony memory size (HMS) = 30, harmony memory consideration rate (HMCR) = 0.9, pitch adjustment rate (PAR) = 0.2, and maximum iterations = 200,000, HS obtained the results, as shown in Figure 4, after 10 s. The structure of Pareto curve was not changed any more when the maximum iterations were extended up to 1,000,000.

As seen in Figure 4, the best solution in terms of cost is A (C = \$419,000, VRI = 5.58) and that in terms of reliability is D (C = \$510,000, VRI = 7.17). Figure 3 also provides individual velocity (unit = m/s) and corresponding VR for the least cost solution (C = \$419,000, VRI = 5.58), which is the identical best solution from the conventional cost-wise single objective optimization.

Figure 5 and Table 1 compares four different Pareto solutions (A to D) chosen from Figure 4. As predicted, the flow velocities of solution A have higher variances from the mean velocity (1.55 m/s) than those of solutions B, C, and D in Figure 5. As observed in Table 1, different Pareto solutions have different pipe diameters and flow velocities. For example, pipe 8 can have the diameter of one inch (corresponding velocity is 0.32 m/s) from solution A, that of 2 inches (0.51 m/s) from solution B, that of 4 inches (0.94 m/s) from solution C, and that of 8 inches (1.41 m/s) from solution D. The less the design cost is, the narrower the diameter is and the slower the velocity is in this case.

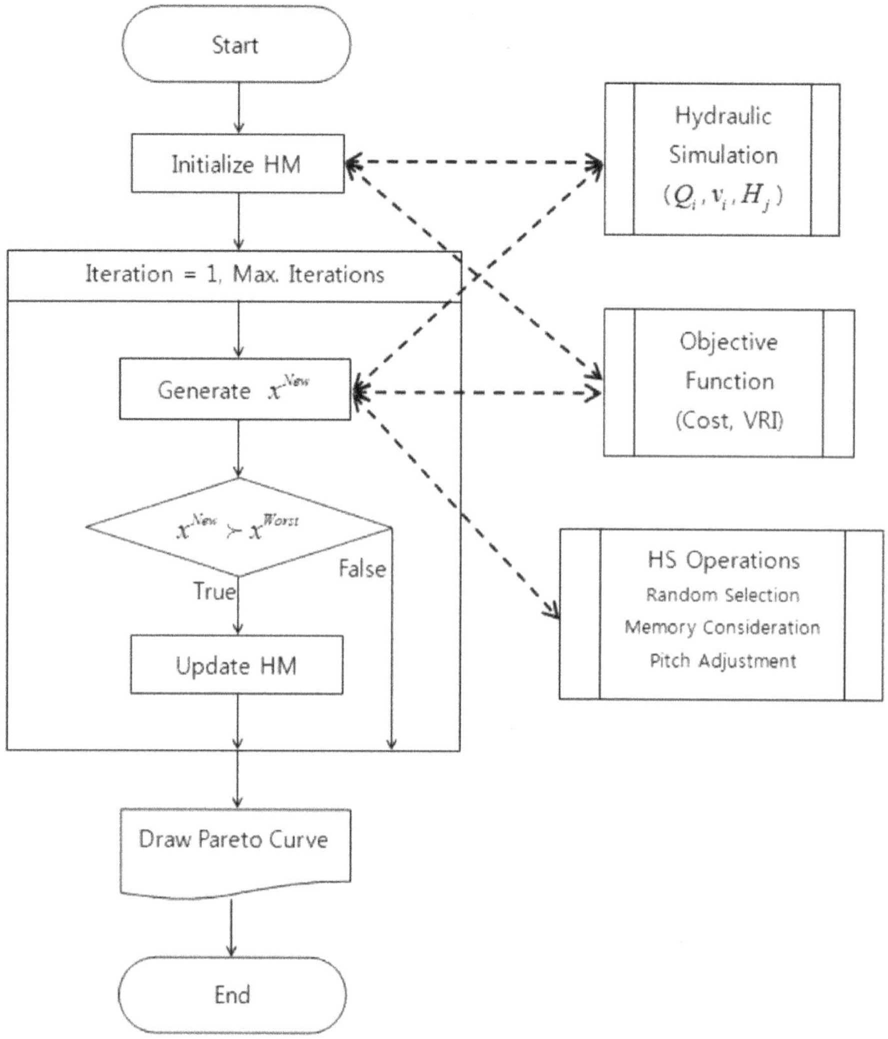

Figure 2: Flowchart of multiobjective optimization.

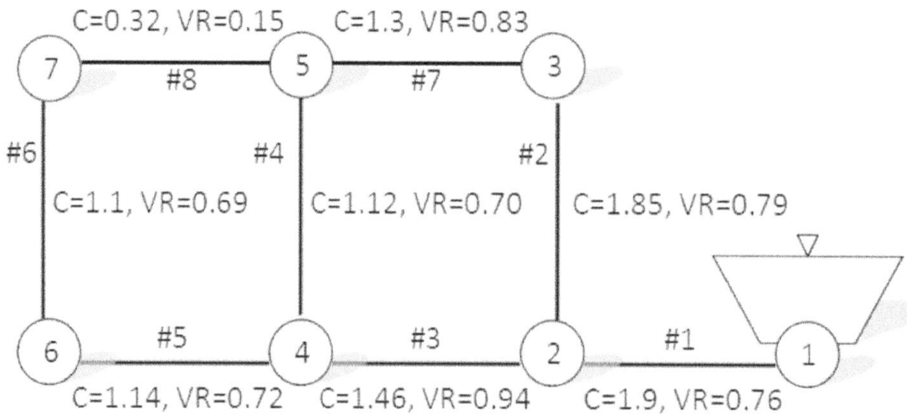

Figure 3: Schematic of two-loop network.

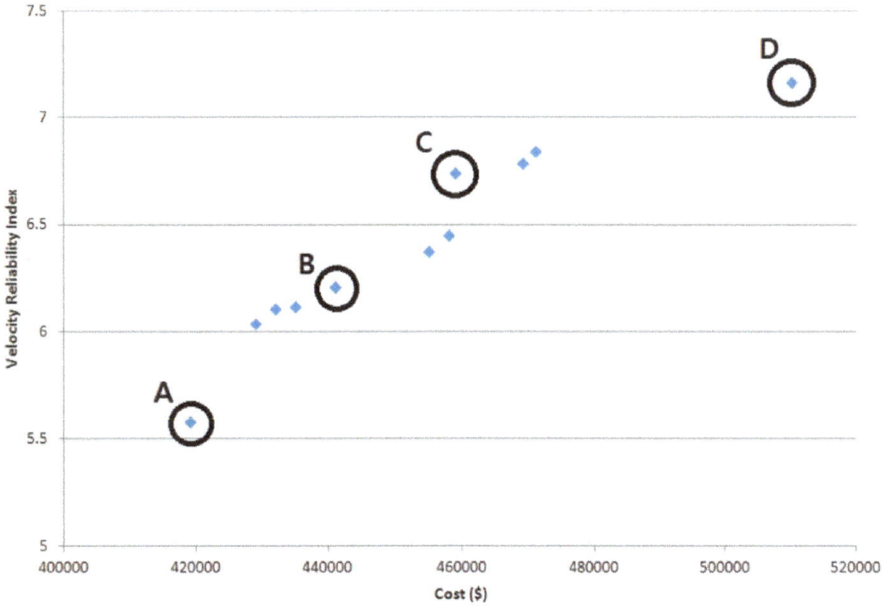

Figure 4: Pareto solutions of two-loop network.

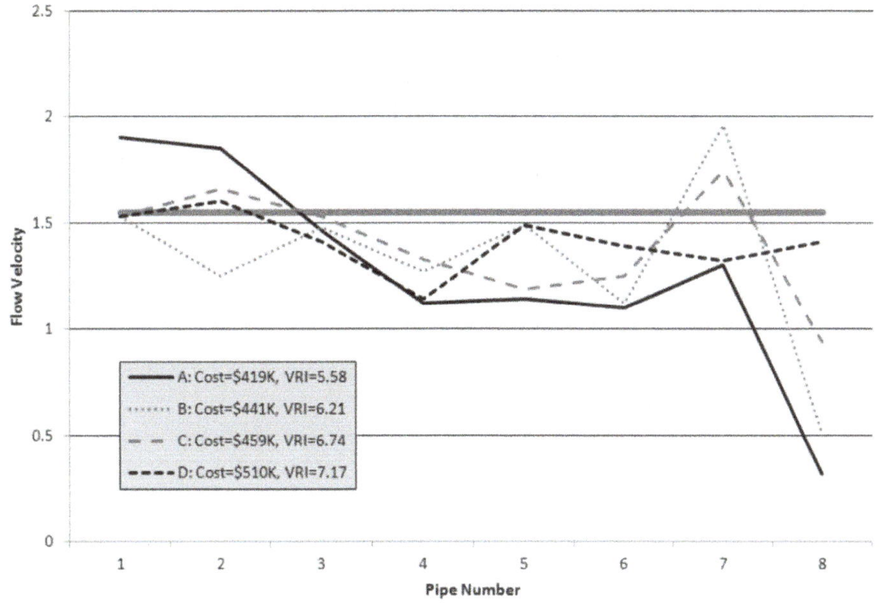

Figure 5: Pareto solutions of two-loop network.

Table 1: Information of four solutions of two-loop network.

Pareto Solution	Pipe Diameters (inch) and Velocities (m/s)							
(Cost, VRI)	1	2	3	4	5	6	7	8
A ($419 K, 5.58)	18 (1.9)	10 (1.85)	16 (1.46)	4 (1.12)	16 (1.14)	10 (1.1)	10 (1.3)	1 (0.32)
B ($441 K, 6.21)	20 (1.53)	12 (1.25)	16 (1.48)	4 (1.27)	14 (1.49)	10 (1.12)	8 (1.96)	2 (0.51)
C ($459 K, 6.74)	20 (1.53)	10 (1.66)	16 (1.53)	4 (1.33)	16 (1.25)	10 (1.25)	8 (1.74)	4 (0.94)
D ($510 K, 7.17)	20 (1.53)	8 (1.6)	18 (1.41)	3 (1.14)	16 (1.49)	12 (1.39)	6 (1.32)	8 (1.41)

When another loading condition (for example, every nodal demand is reduced by 30%) was tested for the least cost solution, VRI was reduced into 4.36 from 5.58 and flow velocity was also reduced in every pipe.

The model was further applied to a real-world problem, Yeosu WDN. As shown in Figure 6, it has one reservoir, 11 loops, 18 demand nodes, and 29 pipes [14]. Because each pipe can choose any diameter from six candidates, total solution space becomes 629=3.68×1022.

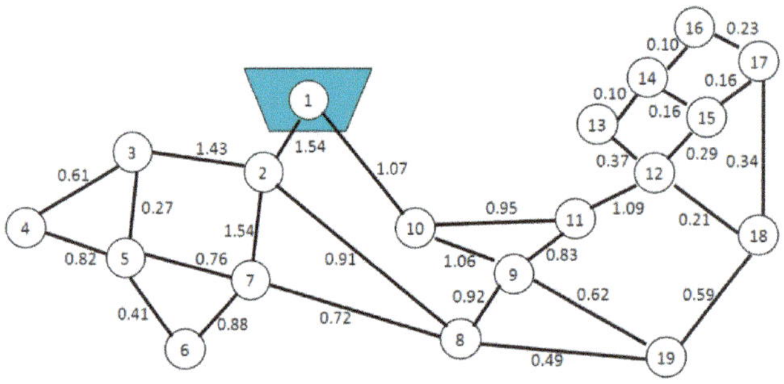

Figure 6: Schematic of Yeosu water distribution network.

With HMS = 30, HMCR = 0.95, PAR = 0.05, and maximum iterations = 106, HS obtained the results after 2'54". Figure 7 and Table 2 show Pareto solutions of the problem and Figure 6 also provides individual velocity (m/s) for the least cost solution (C = 211.6 million Korean Won, VRI = 12.28). Decision makers can choose any one design out of five Pareto solutions by considering its budget availability and reliability level. Additionally, if it has special constraint such that it should construct narrow diameter for pipe 12, it may choose the solution 3 (diameter of pipe 12 = 250 mm) rather than other solutions, which have higher diameter (300 mm) for pipe 12.

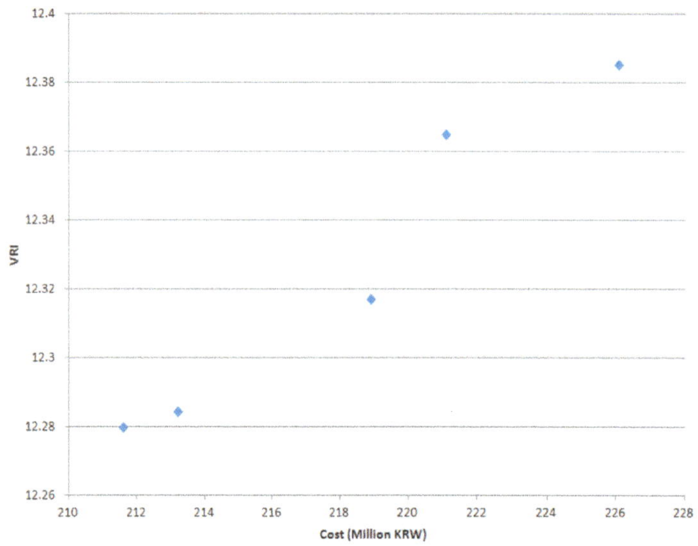

Figure 7: Pareto solutions of Yeosu network.

Table 2: Information of pareto solutions of Yeosu network.

Pipe Number	Pipe Diameters (mm)				
	Solution 1	Solution 2	Solution 3	Solution 4	Solution 5
1	400	400	400	400	400
2	200	200	200	200	200
3	200	200	200	200	200
4	200	200	200	200	200
5	200	200	200	200	200
6	300	300	300	300	300
7	200	200	200	200	200
8	200	200	200	200	200
9	200	200	200	200	200
10	300	300	300	300	300
11	200	250	250	200	200
12	300	300	250	300	300
13	200	200	200	200	200
14	200	200	200	200	200
15	250	200	200	200	200
16	200	200	200	200	200
17	200	200	200	200	200
18	300	300	300	300	300
19	300	300	350	350	400
20	200	200	200	200	200
21	200	200	200	200	200
22	300	300	250	300	250
23	200	200	200	200	200
24	250	250	250	200	250
25	200	200	200	200	200
26	200	200	200	200	200
27	200	200	200	250	200
28	200	200	200	200	200
29	200	200	250	250	250
Design Cost (Korean Won)	211.6×10^6	213.2×10^6	218.9×10^6	221.1×10^6	226.1×10^6
VRI	12.280	12.284	12.317	12.365	12.385

When this model was applied to another popular benchmark Hanoi network, it could not find proper VRI because velocities of first couple pipes are extremely higher ($v1 = 6.83$ m/s and $v2 = 6.53$ m/s) than maximum allowable velocity (3.0 m/s) although maximum candidate diameters (40 inch) were selected for the pipes [14]. Thus, pipe velocities could not be wedged between the lower and upper bounds.

CONCLUSIONS

This study performed WDN design optimization in terms of velocity reliability and cost minimization using fuzzy theory and harmony search. Using fuzzy theory, different degree of reliability was given to flow velocity of each pipe, and this expanded multiobjective formulation was optimized using HS. As a result, a Pareto set of WDN design solutions were successfully obtained.

The novelty of this study can be the fact that it proposes "preventive" reliability measure while traditional popular resilience indexes are "post-accident" reliability measure. Traditional resilience indexes increase when we add extra energy (for example, more energy to a pump for faster flowrate) to cope with failure situations. However, this surplus energy can be a burden on the system, causing quicker system deterioration. On the contrary, velocity reliability index in this study tries to avoid any excessive velocity, causing fewer burdens on the system.

Based on this pioneering approach in velocity-wise reliability index, more practical researches are expected to enhance system reliability, as well as minimize design cost of real-world WDNs in the future. For example, the less-cost design, with less VRI, may cause more maintenance costs from sediment cleaning or structure break. Thus, if we also consider the maintenance cost, as well as construction cost with respect to life-cycle cost, it will become a more practical design optimization.

Also, other related hybrid techniques such as controlled genetic algorithm [40] and genetic doping algorithm [41] can be considered in the future for devising another velocity-wise reliability index.

CONFLICTS OF INTEREST

The author declares that there is no conflict of interests regarding the publication of this paper.

REFERENCES

1. Savic, D.A.; Walters, G.A. Genetic algorithms for least-cost design of water distribution networks. *J. Water Resour. Plan. Manag.* 1997, *123*, 67–77.
2. Cunha, M.C.; Sousa, J. Water distribution network design optimization: Simulated annealing approach. *J. Water Res. Plan. Manag.* 1999, *125*, 215–221.
3. Lippai, I.; Heaney, J.P.; Laguna, M. Robust water system design with commercial intelligent search optimizers.*J. Comput. Civil.* 1999, *13*, 135–143.
4. Eusuff, M.; Lansey, K.E. Optimization of water distribution network design using the shuffled frog leaping algorithm. *J. Water Res. Plan. Manag.* 2003, *129*, 210–225.
5. Maier, H.R.; Simpson, A.R.; Zecchin, A.C.; Foong, W.K.; Phang, K.Y.; Seah, H.Y.; Tan, C.L. Ant colony optimization for design of water distribution systems. *J. Water Resour. Plan. Manag.* 2003, *129*, 200–209.
6. Geem, Z.W. Optimal cost design of water distribution networks using harmony search. *Eng. Optim.* 2006, *38*, 259–280.
7. Perelman, L.; Ostfeld, A. An adaptive heuristic cross-entropy algorithm for optimal design of water distribution systems. *Eng. Optim.* 2007, *39*, 413–428.
8. Lin, M.-D.; Liub, Y.-H.; Liua, G.-F.; Chua, C.-W. Scatter search heuristic for least-cost design of water distribution networks. *Eng. Optim.* 2007, *39*, 857–876.
9. Geem, Z.W. Particle-swarm harmony search for water network design. *Eng. Optim.* 2009, *41*, 297–311.
10. Mohan, S.; Jinesh Babu, K.S. Optimal water distribution network design with honey-bee mating optimization. *J. Comput. Civil Eng.* 2010, *24*, 117–126.
11. Zheng, F.; Simpson, A.R.; Zecchin, A.C. A decomposition and multi-stage optimization approach applied to optimization of water distribution systems with multiple sources. *Water Resour. Res.* 2013, *49*, 380–399.
12. Abebe, A.; Solomatine, D. Application of global optimization to the design of pipe networks. In Proceedings of International Conference on Hydroinformatics, Copenhagen, Denmark, 24–26 August 1998; pp. 989–995.
13. Alfonso, L.; Jonoski, A.; Solomatine, D.P. Multiobjective optimization of operational responses for contaminant flushing in water distribution networks. *J. Water Res. Plan. Manag.* 2010, *136*, 48–58.
14. Geem, Z.W.; Kim, J.-H.; Jeong, S.-H. Cost efficient and practical design of water supply network using harmony search. *Afr. J. Agric. Res.* 2011, *6*, 3110–3116.
15. Rossman, L.A. *EPANET 2: Users Manual*; US Environmental Protection Agency: Cincinnati, OH, USA, 2000.
16. Samani, H.M.; Mottaghi, A. Optimization of water distribution networks using integer linear programming. *J. Hydraul. Eng.* 2006, *132*, 501–509.
17. Walski, T.; Chase, D.V.; Savic, D.; Grayman, W.M.; Beckwith, S.; Koelle, E. *Advanced Water Distribution Modeling and Management*; Haestead Press: Waterbury, CT, USA, 2003.
18. Todini, E. Looped water distribution networks design using a resilience index based heuristic approach. *Urban Water* 2000, *2*, 115–122.
19. Prasad, T.D.; Park, N.-S. Multiobjective genetic algorithms for design of water distribution networks. *J. Water Resour. Plan. Manag.* 2004, *130*, 73–82.
20. Prasad, T.D.; Tanyimboh, T.T. Entropy based design of "Anytown" water distribution network. In Proceedings of Water Distribution Systems Analysis, Kruger National Park, South Africa, 17–20 August 2008; pp. 1–12.
21. Ghajarnia, N.; Haddad, O.B.; Mariño, M.A. Reliability based design of water distribution network (WDN)

considering the reliability of nodal pressures. In Proceedings of World Environmental and Water Resources Congress, Kansas City, MS, USA, 17–21 May 2009; pp. 1–9.

22. Liu, H.X.; Savic, D.; Kapelan, Z.; Zhao, M.; Yuan, Y.X.; Zhao, H.B. A diameter-sensitive flow entropy method for reliability consideration in water distribution system design. *Water Resour. Res.* 2014, *50*, 5597–5610.

23. Atkinson, S.; Farmani, R.; Memon, F.A.; Butler, D. Reliability indicators for water distribution system design: Comparison. *J. Water Res. Plan. Manag.* 2014, *140*, 160–168.

24. Awumah, K.; Goulter, I.; Bhatt, S. Assessment of reliability in water distribution networks using entropy based measures. *Stoch. Hydrol. Hydraul.* 1990, *4*, 309–320.

25. Tanyimboh, T.T.; Templeman, A.B. Calculating maximum entropy flows in networks. *J. Oper. Res. Soc.* 1993, *44*, 383–396.

26. Farmani, R.; Walters, G.A.; Savic, D.A. Trade-off between total cost and reliability for anytown water distribution network. *J. Water Res. Plan. Manag.* 2005, *131*, 161–171.

27. Di Nardo, A.; Greco, R.; Santonastaso, G.F.; Di Natale, M. Resilience and entropy indices for water supply network sectorization in district meter areas. In Proceedings of the 9th International Conference on Hydroinformatics, Tianjing, China, 7–11 September 2010.

28. Raad, D.N.; Sinske, A.N.; Van Vuuren, J.H. Comparison of four reliability surrogate measures for water distribution systems design. *Water Resour. Res.* 2010, *46*, W05524.

29. Geem, Z.W.; Kim, J.H.; Loganathan, G.V. A new heuristic optimization algorithm: Harmony search. *Simulation*2001, *76*, 60–68.

30. Saka, M.P. Optimum geometry design of geodesic domes using harmony search algorithm. *Adv. Struct. Eng.*2007, *10*, 595–606.

31. Cheng, Y.M.; Li, L.; Lansivaara, T.; Chi, S.C.; Sun, Y. J. An improved harmony search minimization algorithm using different slip surface generation methods for slope stability analysis. *Eng. Optim.* 2008, *40*, 95–115.

32. Mun, S.; Geem, Z.W. Determination of viscoelastic and damage properties of hot mix asphalt concrete using a harmony search algorithm. *Mech. Mater.* 2009, *41*, 339–353.

33. Ayvaz, M.T. Application of harmony search algorithm to the solution of groundwater management models. *Advanc. Water Res.* 2009, *32*, 916–924.

34. Geem, Z.W. Multiobjective optimization of time-cost trade-off using harmony search. *J. Constr. Eng. Manag.*2010, *136*, 711–716.

35. Geem, Z.W. Novel derivative of harmony search algorithm for discrete design variables. *Appl. Math. Comput.*2008, *199*, 223–230.

36. Geem, Z.W.; Sim, K.-B. Parameter-setting-free harmony search algorithm. *Appl. Math. Comput.* 2010, *217*, 3881–3889.

37. Hasancebi, O.; Erdal, F.; Saka, M.P. Adaptive harmony search method for structural optimization. *J. Struct. Eng.*2010, *136*, 419–431.

38. Geem, Z.W.; Cho, Y.-H. Optimal design of water distribution networks using parameter-setting-free harmony search for two major parameters. *J. Water Res. Plan. Manag.* 2011, *137*, 377–380.

39. Alperovits, E.; Shamir, U. Design of optimal water distribution systems. *Water Resour. Res.* 1977, *13*, 885–900.

40. Lay-Ekuakille, A.; Vendramin, G.; Trotta, A. Spirometric measurement post-processing: Expiration data recovery. *Sens. J. IEEE* 2010, *10*, 25–33.

41. Lay-Ekuakille, A.; Trotta, A. Processing stabilometric data for electronic knee: Training and calibration. In Proceedings of IEEE Instrumentation and Measurement Technology Conference, Austin, TX, USA, 3–6 May 2010; pp. 882–885

CHAPTER 15 OPTIMIZATION OF UPSTREAM DETENTION RESERVOIR FACILITIES FOR DOWNSTREAM FLOOD MITIGATION IN URBAN AREAS

Thi Thuy Ngo, Do Guen Yoo, Yong Sik Lee and Joong Hoon Kim

Keywords: Urban upstream detention reservoir, flood mitigation, meta-heuristics optimization

INTRODUCTION

In the context of urbanization and climate change, flooding has received much attention in sustainable development activities. Because of an increase in impervious areas, such as rooftops, squares and roads, surface runoff flow in urban watersheds is higher than in natural watersheds. During the rainy season, when surface runoff in sewer systems exceeds the designed capacity, storm water overflows onto the land surface through manholes or pumping reservoirs and flooding occurs [1]. To address these storm water floods, the traditional and continuing solution has been to improve the capacity of conduits, pumping facilities or outlet structures to accommodate the higher drainage flow [2]. The construction and maintenance costs required for these system modifications are extremely high, especially for metropolitan cities. In such a situation, detention storage is considered as an effective means for flood peak reduction in order to minimize flood damage [3].

Generally, detention reservoirs can be constructed by damming a channel or by excavating into the existing ground to temporarily capture storm water for slow release through controlled outlets. The detention basins can be engineered tanks and natural ponds located at the downstream outlet of drainage systems, but can also be on-site or off-site locations in upstream sub-catchments on both local and regional scales [4]. In an urban area, detention storage is mostly dry and will be temporarily filled during severe rainfall events lasting several hours. However, the delayed drainage in order to mitigate peak flow and flood volume downstream may coincide with storm water from upstream, aggravating potential flooding situations. Since the storage in detention basins is designed according to critical frequencies, their capacities may be exceeded in severe flood events. In this case, inundation occurs in the detention storage areas. Therefore, effective operation of these detention reservoirs is imperative.

The optimization of the efficiency of detention basins has been discussed for several decades. In most literature, the size and location of detention basins have been determined to minimize flood damage using different algorithms, such as nonlinear programming [5], dynamic programing [6,7,8,9], and meta-heuristics algorithms such as the genetic algorithm [10], simulated annealing [11], the nondominated sorting genetic algorithm [12], and tabu search [13]. More recently, the optimal operation of urban detention basins has received much attention. These studies optimized the flood reduction effect of urban detention storage by finding suitable designs for outlet structures consisting of orifices and/or weir combinations [10,11], optimal pumping schedules [14,15,16,17], or optimal gate operation [18]. However, the majority of these studies have evaluated downstream detention ponds located at the outlet of the drainage system and have applied reservoir routing to simulate flow through the detention ponds.

The present study proposes the optimal detention facilities of an upstream pond in drainage networks to minimize flood damage in downstream areas by coupling optimization techniques with a flood routing model. Because land for

Citation: Thi Thuy Ngo, Do Guen Yoo, Yong Sik Lee and Joong Hoon Kim, Optimization of Upstream Detention Reservoir Facilities for Downstream Flood Mitigation in Urban Areas, doi:10.3390/w8070290.

detention basins might otherwise be used for more economical purposes and construction costs are expensive, the size of the detention storage is set as the same as the existing area. The factors considered in this study are, therefore, the inlet and outlet structures.

METHODOLOGY

A detention reservoir is a flow retardation structure in an urban drainage system. It provides space to temporarily store incoming flow to reduce peak flows downstream. The flood flow impounded in a detention reservoir is then released back to the sewer system by various types of outlets, such as weirs or orifices, or by pumping [19]. In general, the detention reservoir facilities are designed to store and release flood water according to a certain frequency. Despite reducing the flood damage downstream by releasing flood water slowly, flooding risk may occur at the area of the detention basin because of flood events that exceed the basin design's capacity. When severe flooding occurs, the detention reservoir and its facilities cannot convey flood water, and local flooding may happen. Furthermore, insufficient draining to the main conduit system may coincide with the peak flow downstream. For these scenarios, instead of flood reduction, the detention reservoir could aggravate potential flooding. The traditional solutions for the aforementioned problems are enlarging the sizes of the storage units and outlet structures, which requires monetary investment and labor. In this study, the existing storage units and outlets are considered to determine the optimal conditions for the inlet and outlet facilities.

In this study, a side-weir inlet and a pump station outlet of the detention reservoir are considered (Figure 1). These detention facilities aim to capture flood water from the main conduits and then drain it back when the peak flood period has passed to reduce the water level at control node F. When the water depth in conduit A is higher than the critical level, the flood water will flow over weir B to storage unit C and is temporarily impounded there until it is drained to the main conduit E by pumping station D.

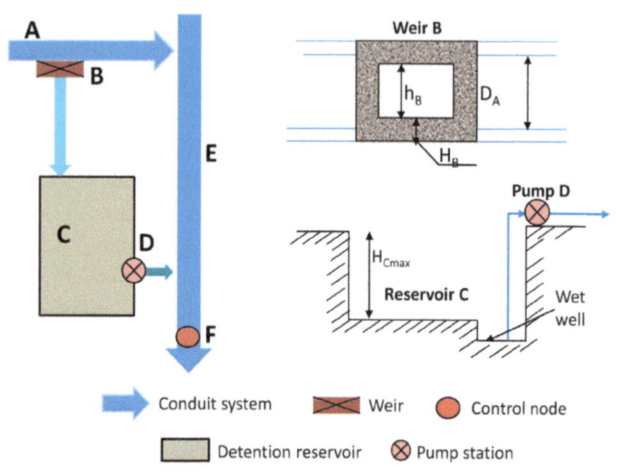

Figure 1: Schematic representation of detention reservoir facilities.

The pumps in the detention reservoir work until the storage unit is empty and without switches, which is related to the maintenance cost [20]. Since detention reservoirs are normally dry and only used in flooding situations, the pump stations in urban drainage systems often work only a few hours during storm events [14]. Therefore, the energy cost is negligible compared to damages from inundation. Considering the flooding risk in the detention area, the aim of the pump station in this case is to minimize flood damage downstream while avoiding flooding at the detention area.

In the pump station, there are several pump units and each of them has its own working level. The operation of a pump station is a combination of pump units in the ON/OFF status at set water levels [16]. For side-weir detention basins, the design parameters include crest height, weir length, and opening height and are determined based on a defined frequency. In fact, similar to storage units, the existing weirs may not satisfy emergency flood scenarios and may cause flooding in the main conduits while the detention storage is not filled. This study, therefore, not only proposes an optimal pumping operation but also determines the optimal side-weir parameters to minimize flood damage in sewer systems and the detention area without changing the size of the storage unit.

We suppose that the pump station D includes n pump units and each of them works at a certain water level associated with possible ranges of the front-pool depth of the detention reservoir. The depth of the detention reservoir C is discretized into m increments ranging from zero to a maximum allowable depth (H_{Cmax}). The search space for pumping includes m^n combinations. For a weir facility, we assume that the weir B exists with a defined length and height. To improve the draining ability of the weir, the optimal crest elevation H_B needs to be obtained. The possible height H_B ranges in k discrete increments from zero to DA−hB, in which D_A is the diameter of conduit A, and h_B is the weir height (Figure 1). The decision variables in this study are the depths at which the pump units turn on and the crest depth of the side-weir with the total size of the search space equaling $k \times m^n$.

Problem Development

As discussed above, the purpose of the proposed optimization model is to search for the optimal pumping schedule and weir crest height to minimize flood damage downstream considering inundation risk in the detention area. In this case, the terminology "flood damage" can be represented by "maximum water depth" at downstream control node F on the main conduit E, at which deviation impacts of the pumping are negligible. The "inundation risk in detention" can be considered as the total flooded volume in this detention area. The objective function can be expressed as follows:

$$Min\left\{Max\left[H(t)\right] + p \times \sum H_{Cflood}(t)\right\} \quad (1)$$

This is subject to:

$$\sum H_{Cflood}(t) = \frac{\sum V_{Cflood}(t)}{Area_C} \quad (2)$$

$$0 \leqslant H_{p,i} \leqslant H_{Cmax}; i = (1,2,\ldots,n) \quad (3)$$

$$0 \leqslant H_B \leqslant (D_A - h_B) \quad (4)$$

where $H(t)$ is the water level at the downstream control

point at time instant t; p is the penalty coefficient; $H_{Cflood}(t)$ is the flooding depth at storage unit C at time t; $V_{Cflood}(t)$ is the flooded volume at the storage node at time t; $Area_C$ is the surface area of the storage unit C; H_{Cmax} is the maximum allowable water level of the storage unit C; $H_{p,i}$ is the water depth at which pump unit ith operates; H_B is the crest height of the weir B; D_A is the diameter of the conduit A upstream of the detention reservoir; and h_B is the height of the weir in the previous design. The parameters $H(t)$ and $V_{Cflood}(t)$ are calculated using the flood routing software EPA-SWMM 5.1 [21].

The total inundated volume at the detention basin is represented by the second term in Equation (1), which is the total flooding depth at the detention reservoir during the flooding event. This term is calculated by Equation (2) and equals zero in the case of no flooding in the detention area. The operating depths of the pump units $H_{p,i}$ can be taken as discrete values between zero and the maximum depth in the detention storage area. The range of crest height H_B is bounded by the size of upstream conduit D_A and the opening height of the weir h_B as represented in Figure 1.

Optimization Technique

The optimization problem is solved by an improved version of the particle swarm optimization algorithm (PSO) called the extraordinary particle swarm optimization algorithm (EPSO). PSO is a well-known population-based meta-heuristic algorithm that is widely applied in different engineering fields. The algorithm mimics the behavior of animal flocking, such as with fish or birds [22]. The individuals in the PSO are particles that fly through the search space following the current best individuals. The EPSO is a newly improved version of PSO that shows promise for engineering optimization problems [23]. The particles in the EPSO algorithm move toward their determined targets instead following the best particles, i.e., global best and personal best. The targets can be a personal best, a global best, or even the worst. The location of particles is updated by using two user-defined parameters, including a combined coefficient C_0 and a target potential range coefficient α. The upper bound of the target range, $T_{up} = round\ (\alpha \times N_{pop})$, is introduced to control the range of the target solution. If the target of particle ith belongs to the target range, the particle will be guided to its corresponding target; otherwise, a random search method is applied. The main steps of the EPSO are as follows:

Step 1: Setting initial parameters. In the EPSO, there are two user-defined parameters: the combined coefficient C_0 ranging from zero to two, and α, a value indicating the upper bound of a potential target in an entire population, ranging from zero to one.

Step 2: Initializing the swarm population. Particles are randomly generated between the lower and upper bounds of a given problem. The particle jth at iteration t is represented by position $\vec{X}_j(t) = (x_j^1, x_j^2, \ldots, x_j^N)$, where N is the dimension of the search space.

Step 3: Sorting particles based on their cost/fitness.

Step 4: Selecting the target. A target is selected for each particle at each iteration among all particles using the determined index (T) of particles.

$$T = round\left(rand \times N_{pop}\right) \tag{5}$$

where N_{pop} is the population size and rand is the uniform distributed random number [0,1]. The index of target (T) ranges from zero to N_{pop}.

Step 5: Updating particles to their new locations using the following equation:

$$X_j(t+1) = \begin{cases} \vec{X}_j(t) + C_0\left(\vec{X}_{Tj}(t) - \vec{X}_j(t)\right) & if\ T \in (0, T_{up}) \\ \vec{LB}_j + rand \times \left(\vec{UB}_j - \vec{LB}_j\right) & Otherwise \end{cases} \tag{6}$$

where LB and UB are the lower and upper bounds of the search space, respectively.

Step 6: Checking the stopping criterion. If the stopping condition is met, stop. Otherwise, return to *Step 3*.

In this study, the EPSO is coupled with a flood routing simulation model called the Storm Water Management Model (EPA-SWMM 5.1) to obtain the optimal pumping schedule and the crest depth in order to minimize the maximum water level at the control location downstream. The decision variables obtained in the optimization model include the working depths of n pump units $(H_{pi})^n$ and the weir crest height H_B. These will be used to generate pump curves and the parameters of the weir that will be input into the flood routing model, SWMM. The detailed procedure is described in the following flowchart (Figure 2).

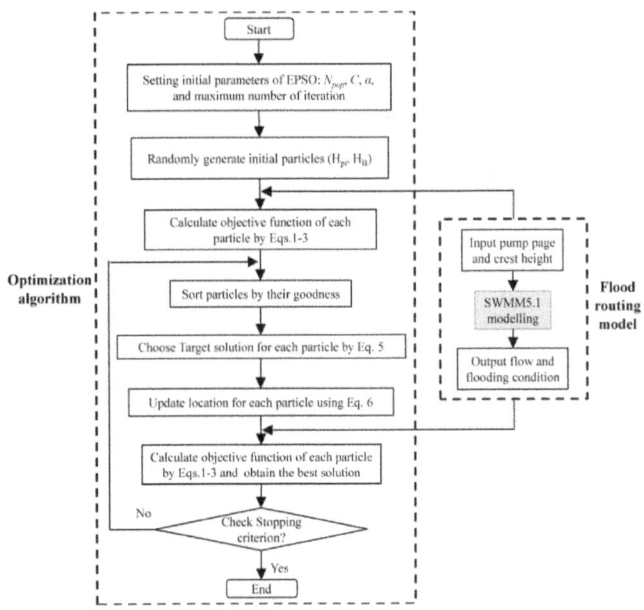

Figure 2: Flowchart of coupled model of EPSO and SWMM5.1.

STUDY AREA

The study area is an urban drainage system located in the Dorim stream watershed, Daerim3. The Dorim stream is one of the branches of the Han River that flows through Seoul, the capital of South Korea (Figure 3). The drainage network occupies an area of 248.5 ha with 1977 pipes and 1576 subcatchments. The area used to suffer inundation in the rainy season due to insufficient conveyance by the conduit system.

In 2009, a detention reservoir was built upstream of the network to temporarily impound flood water and then slowly drain the water into the main sewer network. The detention reservoir consists of a storage unit with a total capacity of 2447 m^3, a side-weir inlet and a pipe transferring flood water from the main conduit to the storage unit, and a pump station that releases water to the main conduit at a rate of 9 m^3/min per pump unit (Figure 4 and Table 1). In flood events, when the water level in the upstream conduit is higher than the crest height of weir B, flood water will flow into storage unit C. Currently, the pump station only operates when either the water level at storage unit C reaches 2.33 m or a rainfall event ends. It stops when the storage is empty. The weir and storage unit are designed according to five-year and 20-year return periods, respectively.

Figure 3: Layout of Daerim3 catchment in Seoul.

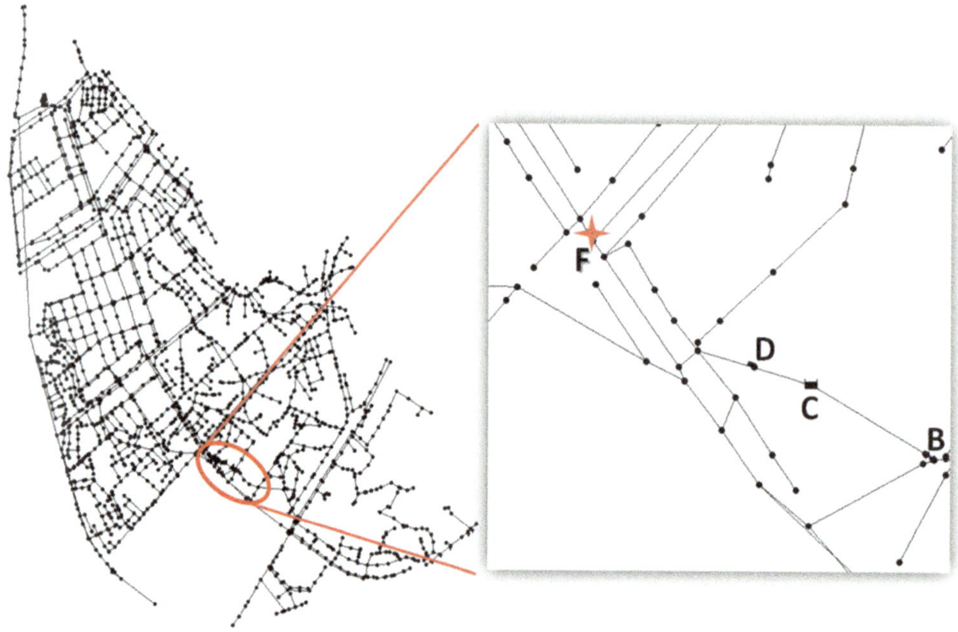

Figure 4: Daerim3 conduit system and layout of Daerim3 detention reservoir.

Table 1: Parameters of the detention reservoir facilities.

Storage Unit		Side-Weir		Pump Station	
Max Depth (m)	Total Volume (m^3)	Upstream Diameter (m)	Weir Height (m)	Discharge Capacity per Pump Unit (m^3/min)	Number of Pumps
3.2	2447	1.5	0.4	9	2

In 2010, a rainfall event on 21 September with a total rainfall of 259.5 mm caused flooding in the Daerim3 area, in particular in the area near the detention reservoir C (Figure 5). The frequency estimations of this rainfall event corresponding to the Huff method [24], the practical method applied in Seoul, are different for various rainfall durations. The return period estimated for a 60 min duration is around 20 years, but those of 90 min and 120 min durations are 50 and 120 years, respectively. In Korea, short rainfall durations, less than or equal to an hour, are normally applied for the design of urban drainage system facilities [25]. Even though the detention storage was designed for 20 year storms, during this storm event, the capacity of the detention reservoir was exceeded, causing flooding in the area of the detention reservoir. The optimal operating conditions and design of detention reservoir facilities are necessary to mitigate flood damage in the neighboring area. Because of the limitation of the area in the urban catchment, the crest height of the weir and the pumping schedule are considered as decision variables in this study. As discussed in Section 2, the aim of the optimization model is minimizing the maximum flood water depth at a controlled downstream node known as point F in Figure 4 while considering the flood risk in the detention area. This controlled node F is located on the main conduit at which the peak water level is stable during the pumping period from the detention reservoir. Flooding at this node may cause significant damage to roads and commercial buildings.

Figure 5: Flooding area (blue hatch pattern) in Daerim3 catchment in September 2010 [26].

The storage maximum depth (H_{Cmax} = 3.2 m) is discretized into 0.1 m levels to assign the operation of two pump units. For the side-weir, the crest height is determined based on 0.1 m increments ranging from zero to 1.1 m (i.e., DA−hB). The size of the total search space in this case study is 12×32^2 = 12,288 combinations. This total number of combinations coupled with the flood routing model will cause cumbersome computational efforts to seek the best design and schedule of detention reservoir facilities. Therefore, EPSO is applied to find the optimal pumping schedule and weir crest height, significantly reducing the time required. The SWMM model, after calibration, is coupled with EPSO to obtain the optimal solution. The 10 min time series of the rainfall event in September 2010 is used as a typical storm with various return periods for different rainfall durations for this optimization model following the scheme shown in Figure 2.

RESULTS AND DISCUSSION

Optimization of Pump Operation

The typical flood event of September 2010 is modeled by SWMM version 5.1 to simulate the flooding situation in the study area. For the current operation, the simulation time is around 260 s, and it consumes approximately 887 h to process a total of 12,288 combinations of crest height and pumping schedules. Using the proposed optimization model, 12 crest height cases were considered to find the optimal pumping schedule. The runtime of the optimization model coded in MATLAB with 500 function evaluations on an Intel Core i5 3.4 GHz system with 8 GB RAM was around 24 h for all 12 cases of side-weir crest height. The recommended parameters used to solve the constraint problem shown in Equations (1)–(3), including a penalty coefficient, a combined coefficient, and the potential range, are set as $p = 10^3$, $C_0 = 0.3$, and $\alpha = 0.8$, respectively, according to information in the literature [23].

The optimal solution obtained from the optimization model

is shown in Table 2 and Figure 6, in which the optimal crest height H_B is 0.4 m and the working depths of the pumps start at 0.2 m and 1.7 m for the first and second pump units, respectively. Figure 7 presents a comparison between two operating scenarios in term of water levels at the detention storage and the control node F in the flood event on 21–22 September 2010. In the current design and pumping strategy, the water depth at the detention storage unit exceeded the maximum level of the pond and caused extreme flooding for 150 min in this area. With the optimal operation rule and design obtained from the optimization model (Figure 7a), the water depth in the detention pond is maintained lower than the allowable level, avoiding inundation in the neighboring area. In this case, the peak water depth in the detention pond is dramatically smaller than that in present case (a decrease of 0.19 m). For the control node in Figure 7b, the water depth in the optimal strategy shows a decrease of 0.11 m compared to that of the current operation. The total flooded volume and the number of flooded nodes with optimal facilities were also remarkably reduced from 6.729×10^3 to 1.426×10^3 m^3(79%) and from 18 to 16 nodes (11%), respectively (Table 3).

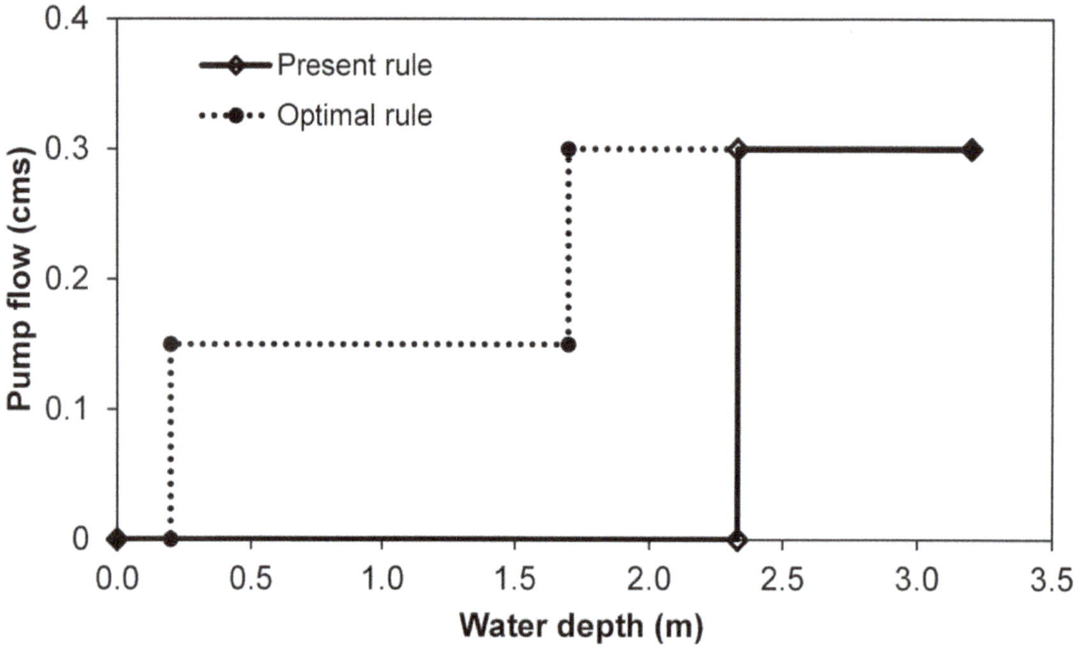

Figure 6: Present and optimal rule curves of pump station D in the detention basin.

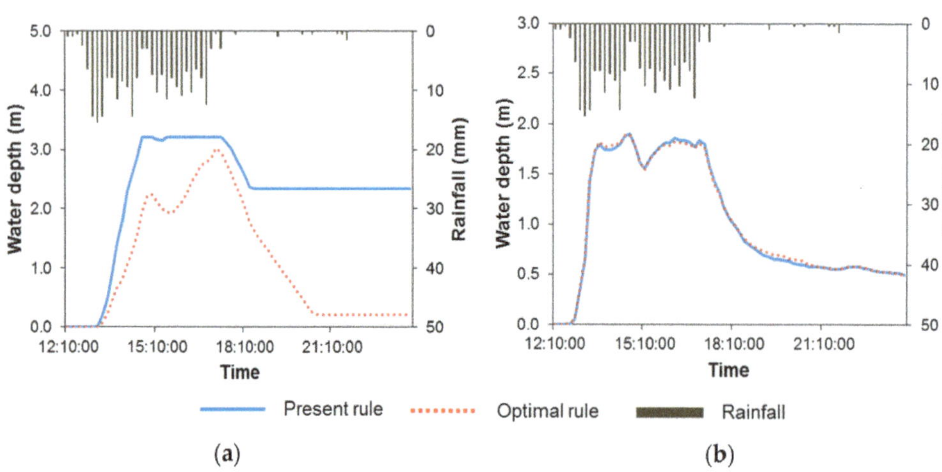

Figure 7: Comparison of current and optimal detention reservoir facilities in terms of water depth at: (**a**) detention storage unit and (**b**) control node in September 2010.

Table 2: Optimal decision variables of the detention facilities.

Side-Weir	Pump Operation	
Crest Height (m)	Working Depth (m)	
	Pump 1	Pump 2
0.4	0.2	1.7

Table 3: Comparison of present and optimal detention reservoir facilities according to flood damages in 2010.

Criteria	2010	
	Present Facilities	Optimal Facilities
Max depth at C (m)	3.20	3.01
Flooded duration at C (10 min)	15	0
Max depth at F (m)	1.90	1.89
Total flooded volume (10^3 m^3)	6.729	1.426
No. flooded nodes	18	17

Notes: C = Detention reservoir; F = Control node.

Comparing the present and optimal scenarios, it can be seen that the present pump operation and weir crest height in the detention basin are inefficient at controlling flooding. The current low crest height of weir B drains flood water to detention reservoir C, while the high working levels of the pump station compounds this amount of water in the detention reservoir instead of releasing it to the main conduit E. This causes flooding in the detention area despite utilizing the whole detention volume while the main pipe is still able to convey more flood water. The functional efficiency of the detention facilities is improved under the optimal operation and design conditions. The higher crest height and lower starting level of the pump units help the pump station drain more water to the main conduits without increasing the water depth at the downstream control node. The storage volume of the detention reservoir C is also successfully utilized to reduce flooding in the neighboring area. Although only 95% of the storage volume of the detention reservoir (according to a water depth of 3.01 m) is used, flood damage is reduced in terms of the inundation in the detention area and the flood peak at the downstream control point.

Verification of Optimization Techniques

As mentioned in Section 3, there are 12 × 32^2 = 12,288 combinations of potential operating conditions for the detention facilities in this study, which consumes 887 h (around 37 days) for the simulation. Due to this time, the simulation of all possible alternatives is regarded as an impractical approach. The optimization technique, namely EPSO, is used instead to determine the optimal detention basin facility parameters without the time challenge. It should be noted that the optimizer cannot guarantee finding the global optimum, especially in engineering problems. This section, therefore, verifies the efficiency of the proposed optimization method by comparing its result with that obtained from a straightforward approach.

All possible cases are evaluated using SWMM to find the global optimum. The simulated results, including the peak water depth at control node F and the total flooded depth at detention storage C, are used to calculate the objective function in Equation (1). The optimal solution is determined by finding the minimum objective function. Table 4 illustrates the results obtained from the simulation of all alternatives.

Table 4: Global optimum solution for the detention facilities.

Rule	Weir Crest Height (m)	Working Depth (m)		Maximum Water Depth at Node F (m)	Total Flooded Volume at C (10^3 m^3)	Objective Function
		Pump 1	Pump 2			
Present rule	0.3	2.33	2.33	1.90	3.304	4217.8
Optimal rule	0.4	0.2	1.7	1.89	0	1.89

Notes: C = Detention reservoir; F = Control node.

The optimal conditions derived from the simulation of all possible combinations are weir crest height at 0.4 m and working depths for the first and second pumps at 0.2 and 1.7 m, respectively. This global optimum is equivalent to that obtained by the optimization technique. Therefore, the proposed method of integrating the optimizer with the simulation modeling is efficient in seeking the optimal detention facility conditions to minimize flood damage.

Performance of Optimal Pump Operation

To evaluate the efficiency of the optimal facility conditions, a recorded rainfall event and several designed rainfall scenarios are used to simulate the drainage system Daerim3 while the detention facilities are operated with the optimal conditions. It is noted that the optimal conditions increase the threshold for transferring flood water to the detention storage of weir B and reduce the starting level of pump station D. These conditions may raise the water depth at control node F if more flood water is drained to the main conduit E. The criteria consisting of the maximum water depth in the storage unit and control node, total flooded volume, and number of flooded nodes are considered to assess the performance of the new optimal operation conditions.

A rainfall event in 2011 lasted two days (26–28 July) and resulted in extreme inundation in the southern part of Seoul. The estimated return periods of this event vary with different rainfall durations; e.g., there is a 10-year return period for a 1 h rainfall, but more than 50 years for a 6 h or greater rainfall [25,27]. The implementation of the optimal conditions and design in this flood event in 2011 is represented in Figure 8 and Table 5.

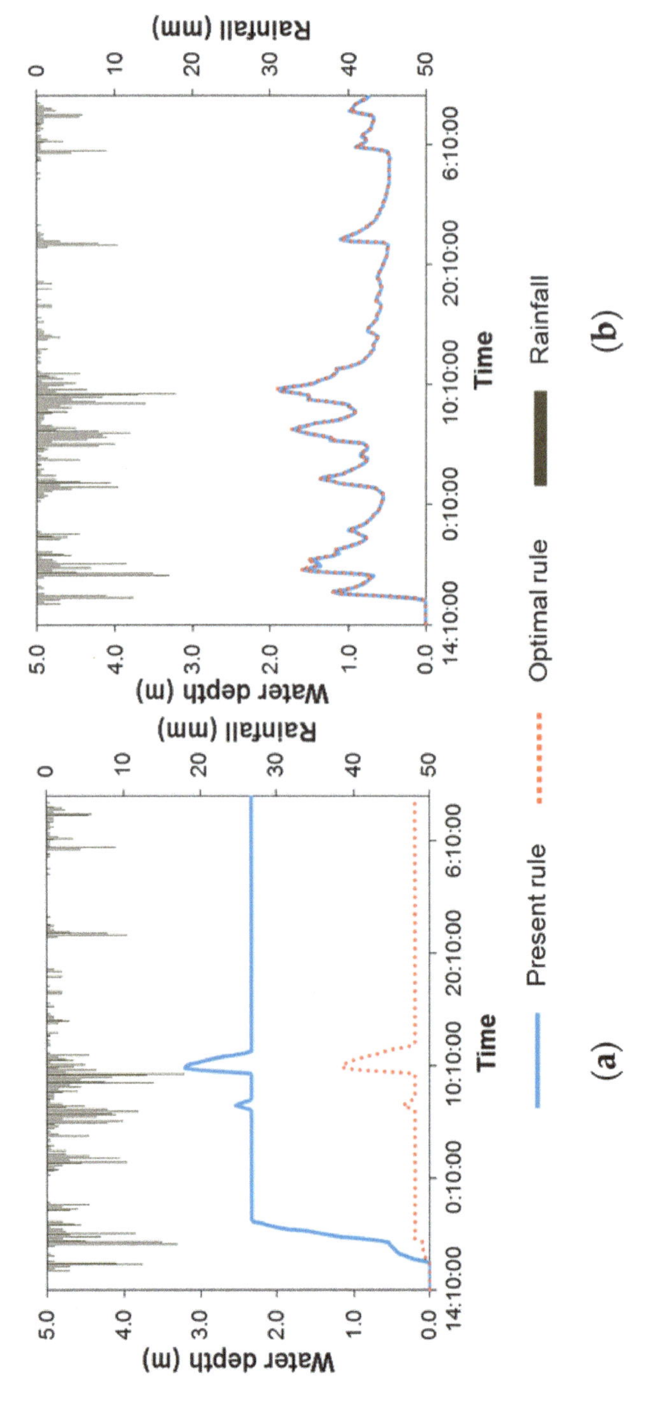

Figure 8: Comparison of current and optimal detention reservoir facilities in terms of water depth at (**a**) detention storage unit and (**b**) control node in July 2011.

Table 5: Comparison of the present and optimal detention reservoir facilities regarding flood damage in 2011.

Criteria	2011	
	Present Facilities	Optimal Facilities
Max depth at C (m)	3.20	1.17
Flooded duration at C (10 min)	1	0
Max depth at F (m)	1.89	1.91
Total flooded volume (10^3 m^3)	0.607	0.491
No. flooded nodes	14	14

Notes: C = Detention reservoir C; F = Control node F.

In general, the optimal facility conditions outperform the present conditions in terms of flood peak at the detention storage and flooded volume in the whole catchment. The water depth in the storage unit dramatically decreases by 2.03 m (63% of maximum water depth in the present operation), preventing inundation in the neighboring area and reducing energy consumption by using only one pump unit during the rainfall event. Under optimal operation, flooding occurs at 14 nodes with total flood water of 0.491×10^3 m^3, a decrease of 19% of the flooded volume compared to the present operation. These results for severe flood events in 2010 and 2011 with different return periods for different rainfall durations show that the optimal conditions surpass the present operation in terms of flood damage reduction for both the detention area and the entire watershed in extreme flood events.

To comprehensively verify the efficiency of the optimal facilities in both normal and severe flooding situations, various designed rainfall scenarios are built for different durations (60, 90, 120, and 180 min) and return periods (20, 30, and 50 years). The Huff method [24] is used to generate rainfall hyetographs corresponding to pairs of rainfall depth and duration. The hydrological model, SWMM 5.1, is applied to simulate flood events caused by the synthetic rainfall events. A full comparison between the present and optimal facilities under the different designed rainfall scenarios is made and the results in terms of the maximum water level at storage unit C and control node F, total flooded volume, and number of flooded nodes in the entire watershed are shown in Table 6.

Table 6: Comparison of present and optimal detention reservoir facilities regarding flood damages in designed scenarios.

Scenarios		Max Depth at C (m)		Max Depth at F (m)		Total Flooded Volume (10^6 m^3)		No. Flooded Nodes	
		Pre.	Opt.	Pre.	Opt.	Pre.	Opt.	Pre.	Opt.
20 year	60 min	2.33	1.74	1.95	2.10	1.404	2.263	27	33
	90 min	2.66	1.88	1.95	2.00	2.271	2.275	33	34
	120 min	3.07	2.28	1.96	1.95	2.502	2.504	32	30
	180 min	3.20	2.47	1.95	1.95	2.907	2.421	24	22
30 year	60 min	2.33	1.52	1.94	1.99	2.398	2.400	32	30
	90 min	2.87	1.99	2.02	2.06	3.701	3.697	40	40
	120 min	3.20	2.33	2.00	2.05	4.175	4.105	39	41
	180 min	3.20	2.75	1.95	1.97	4.491	3.745	34	33
50 year	60 min	2.53	1.77	2.08	2.11	3.996	4.003	40	44
	90 min	3.20	2.43	2.21	2.16	6.373	6.335	54	52
	120 min	3.20	2.85	2.20	2.20	7.516	7.066	54	59
	180 min	3.20	3.19	2.07	2.06	7.342	6.323	50	50

Notes: C = Detention storage; F = Control node; Pre. = Present facilities; Opt. = Optimal facilities.

Under the optimal conditions, the maximum water level at the detention storage is dramatically reduced in most of the rainfall scenarios except for only minor reductions in the 50-year 180 min scenario. In severe scenarios with 30- and 50-year return periods, the water depth reduction is about 0.5–1.0 m, which significantly mitigates flooding in the detention area that occurs when flood water exceeds conveying ability. At control node F, water depths in the rainfall scenarios show a small fluctuation, slightly increasing in low return periods and short durations and gradually decreasing in severe scenarios (with a magnitude of around 5 cm). The total flooded volumes in the whole watershed are reduced using the optimal operating conditions compared to those in the present operation except in normal scenarios with low return periods and short durations. The flooding decrease is significant for the designed flood events lasting 180 min (around 15% of the total flooded volume for 20-year, 30-year, and 50-year return periods). Moreover, the number of flooded nodes using the optimal operating conditions is reduced in most of the rainfall scenarios except for the short-duration rainfall events.

CONCLUSIONS

The study presents a promising method for optimizing detention facilities for flood control purposes. The new pumping operation and weir design are derived from an approach that couples optimization techniques with a flood routing model and outperforms the present operation in terms of flooding in both the detention area and the entire watershed. The straightforward approach of finding the global optimum is conducted to verify the efficiency of the optimization method. This comparison shows that the optimizer is successful in searching for the optimal detention facility conditions in the study area. The optimal solution is then assessed by applying it to a severe historical flood event in 2011 and various designed rainfall scenarios. The optimal conditions outperformed the present in both historical flood events and designed rainfall scenarios. With regard to flooding mitigation in the detention area, the application of the new design and operation would be significant.

Even though this method has advantages of flood reduction and reduces computation time, it still contains uncertainties in simulation modeling due to calibration methods and data availability. The spatial and temporal variability of recorded rainfall data could result in uncertainty of designed rainfall scenarios. Moreover, the quantity and quality of rainfall data are also considered as uncertain sources. Further research will be aimed at handling these uncertainties. The constrained objective function evaluated in this study also needs to be improved in future work.

ACKNOWLEDGMENTS

This research was supported by a grant (13AWMP-B066744-01) from the Advanced Water Management Research Program funded by the Ministry of Land, Infrastructure, and Transport of the Korean government.

AUTHOR CONTRIBUTIONS

Do Guen Yoo and Thi Thuy Ngo conceived and designed the study; Thi Thuy Ngo and Yong Sik Lee performed the detailed work; Thi Thuy Ngo and Do Guen Yoo analyzed the results; Joong Hoon Kim guided the entire study and provided suggestions for improvements to the manuscript; Thi Thuy Ngo wrote the paper; and Do Guen Yoo reviewed the manuscript critically.

CONFLICTS OF INTEREST

The authors declare no conflict of interest.

REFERENCES

1. Hsu, M.H.; Chen, S.H.; Chang, T.J. Inundation simulation for urban drainage basin with storm sewer system. *J. Hydrol.* 2000, *234*, 21–37.
2. Bennett, M.S.; Mays, L.W. Optimal design of detention and drainage channel systems. *J. Water Resour. Plan. Manag. ASCE* 1985, *111*, 99–112.
3. Green, C.H.; Parker, D.J.; Tunstall, S.M. *Assessment of Flood Control and Management Options*; WCD Thematic Review, Options Assessment: IV.4; World Commission on Dams: Cape Town, South Africa, 2000.
4. McCuen, R.H. A regional approach to urban storm water detention. *Geophys. Res. Lett.* 1974, *1*, 321–332.
5. Stafford, N.; Che, D.; Mays, L.W. Optimization Model for the Design of Infiltration Basins. *Water Resour. Manag.*2015, *29*, 2789–2804.

6. Mays, L.W.; Bedient, P.B. Model for optimal size and location of detention. *J. Water Resour. Plan. Manag. ASCE*1982, *108*, 270–284.

7. Taur, C.K.; Toth, G.; Oswald, G.E.; Mays, L.W. Austin detention basin optimization model. *J. Hydraul. Eng. ASCE* 1987, *113*, 860–878.

8. Behera, P.K.; Papa, F.; Adams, B.J. Optimization of regional storm-water management systems. *J. Water Resour. Plan. Manag. ASCE* 1999, *125*, 107–114.

9. Travis, Q.B.; Mays, L.W. Optimizing Retention Basin Networks. *J. Water Resour. Plan. Manag. ASCE* 2008, *134*, 432–439.

10. Park, M.; Chung, G.; Yoo, C.; Kim, J. Optimal design of storm water detention basin using the genetic algorithm. *KSCE J. Civ. Eng.* 2012, *16*, 660–666.

11. Oxley, R.L.; Mays, L.W. Optimization-simulation model for detention basin system design. *J. Water Resour. Manag.* 2014, *28*, 1157–1171.

12. Yu, P.C.; Yang, T.C.; Kuo, C.M.; Tai, C.W. Integration of physiographic drainage-inundation model and nondominated sorting genetic algorithm for detention-pond optimization. *J. Water Resour. Plan. Manag. ASCE*2015, *141*, 04015028.

13. Huang, C.L.; Hsu, N.S.; Wei, C.C.; Luo, W.J. Optimal spatial design of capacity and quantity of rainwater harvesting systems for urban flood mitigation. *Water* 2015, *7*, 5173–5202.

14. Hsu, N.S.; Huang, C.L.; Wei, C.C. Intelligent real-time operation of a pumping station for an urban drainage system. *J. Hydrol.* 2013, *489*, 85–97.

15. Wei, C.C.; Hsu, N.S.; Huang, C.L. Two-stage pumping control model for flood mitigation in inundated urban drainage basins. *Water Resour. Manag.* 2014, *28*, 425–444.

16. Yazdi, J.; Kim, J.H. Intelligent Pump Operation and River Diversion Systems for Urban Storm Management. *J. Hydrol. Eng. ASCE* 2015, *20*, 04015031.

17. Yazdi, J.; Choi, H.S.; Kim, J.H. A methodology for optimal operation of pumping stations in urban drainage systems. *J. Hydro-Environ. Res.* 2015, *11*, 101–112.

18. Le, V.C.; Hiramatsu, K.; Harada, M.; Mori, M. Optimal gate operation of a main drainage canal in a flat, low-lying, agricultural area using a tank model. *Paddy Water Environ.* 2008, *6*, 243–256.

19. Iowa Department of Natural Resources. *Iowa Storm Water Management Manual*; Iowa Department of Natural Resources: Des Moines, IA, USA, 2010.

20. Lansey, K.E.; Awumah, K. Optimal pump operations considering pump switches. *J. Water Resour. Plan. Manag. ASCE* 1994, *120*, 17–35.

21. U.S. Environmental Protection Agency (EPA). *Storm Water Management Model User's Manual Version 5.1*; EPA: Cincinnati, OH, USA, 2015.

22. Eberhart, R.C.; Kennedy, J. A new optimizer using particle swarm theory. In Proceedings of the Sixth International Symposium on Micromachine and Human Science, Nagoya, Japan, 4–6 October 1995; pp. 39–43.

23. Ngo, T.T.; Sadollah, A.; Kim, J.H. A cooperative particle swarm optimizer with stochastic movements for computationally expensive numerical optimization problems. *J. Comput. Sci.* 2016, *13*, 68–82.

24. Huff, F.A. *Time Distribution of Heavy Rainstorms in Illinois*; ISWS.CIR-173/90, Circular 173; Department of Energy and Natural Resources, State of Illinois: Champaign, IL, USA, 1990.

25. Yoo, C.; Park, C.; Jun, C. Evaluation of the Concept of Critical Rainfall Duration by Bivariate Frequency Analysis of Annual Maximum Independent Rainfall Event Series in Seoul, Korea. *J. Hydrol. Eng. ASCE* 2015, *21*, 05015016.

26. Safe Seoul. Available online: http://safecity.seoul.go.kr:8070/scmyn_cf/map/ (accessed on 20 November 2015).

27. Lee, D.I.; Ryu, C.S. A Comparative Study of Rain Intensities Retrieved from Radar and Satellite Observations: Two Cases of Heavy Rainfall Events by Changma and Bolaven (TY15). *J. Korean Earth Sci. Soc.* 2012, *33*, 569–582.

CHAPTER 16

DAILY RESERVOIR RUNOFF FORECASTING METHOD USING ARTIFICIAL NEURAL NETWORK BASED ON QUANTUM-BEHAVED PARTICLE SWARM OPTIMIZATION

Chun-tian Cheng, Wen-jing Niu, Zhong-kai Feng, Jian-jian Shen and Kwok-wing Chau

Keywords: Quantum-behaved particle swarm optimization (QPSO), daily runoff, reservoir forecasting, artificial neural network, hybrid forecast

INTRODUCTION

Accurate daily runoff forecasting is extremely important for hydropower operation control and power grid operation scheduling [1,2,3,4,5]. Over the past decades, there have been abundant traditional classical research works for the daily runoff forecasting, which can be broadly separated into two categories: process-based model like Xin'anjiang model [6,7], and data-based model such as autoregressive model and moving average model [8,9]. These approaches entail exogenous input and rely on a number of assumptions for natural environment, leading to some deviation from the exact results in most cases. In recent years, with the booming development of heuristic methods, many researchers pay attention to applying them in daily runoff forecasting or the parameter selection of the hydrologic model, including artificial neural network [10,11], SCE-UA algorithm [12,13], support vector machine [14,15] and other hybrid methods [16,17].

As a typical artificial neural network, back propagation neural network (BP) can nearly simulate any complex linear or non-linear functional relationship without knowing the correlation between the input data and the output data [18,19]. After learning from the training data set, BP can be used to predict a new output data effectively with the corresponding input data. Compared with other methods, BP has high fault tolerance, strong robustness and easy adaptability to online learning. BP has been widely used in many practical areas, including load prediction, wind speed prediction and daily runoff forecasting. However, BP has some drawbacks, such as long computing time, slow convergence and easy to encounter local minimum. Hence, a variety of hybrid optimization methods using such global optimization algorithm like particle swarm optimization (PSO) are developed to improve the generalization ability of the artificial neural network [20,21]. These hybrid optimization methods can improve the BP forecasting performance in varying degrees. However, when PSO is applied for the ANN parameters selection, it may be trapped into the local optima of the objective function because PSO is restricted by search capability. Therefore, the promotion space is still large for ANN parameter selection using an evolutionary algorithm [22,23,24].

In recent years, a novel particle swarm optimization variant called quantum-behaved particle swarm optimization algorithm (QPSO) was proposed by Sun et al. [25] in 2004. In QPSO, the global optimal solution in the whole searching space can be guaranteed theoretically. Moreover, simulation results of numerous complex benchmark functions showed that QPSO has better global searching ability than the basic PSO [26,27]. Hence, QPSO are widely used to solve the complex optimization problems which includes hydrothermal scheduling and economic dispatch problem [28,29]. However, up to now, there are a few reports about using QPSO for parameter calibration of artificial neural network. Therefore, to improve the generalization ability and calculation efficiency of artificial neural network, a hybrid method, coupling artificial neural network and QPSO, is developed for daily reservoir

Citation: Chun-tian Cheng, Wen-jing Niu, Zhong-kai Feng, Jian-jian Shen and Kwok-wing Chau, Daily Reservoir Runoff Forecasting Method Using Artificial Neural Network Based on Quantum-behaved Particle Swarm Optimization, doi:10.3390/w7084232.

runoff forecasting in this research, where QPSO algorithm is selected as training algorithm for artificial neural network to enhance hydrologic forecast accuracy.

The remaining is organized as follows. Artificial neural network is introduced briefly in Section 2. Section 3explains the PSO and QPSO methods in details, then the proposed QPSO-ANN model is presented in Section 4.Section 5 provides the results of the proposed method and other methods and gives discussions on their performances. Finally, conclusions are summarized in Section 6.

ARTIFICIAL NEURAL NETWORK

In general, the relation between input vector and output vector of a nonlinear system can be expressed as $\boldsymbol{Y} = \boldsymbol{H}(\boldsymbol{X})$, where $\boldsymbol{X} = [X_1,\ldots,X_i,\ldots,X_n]^T$ is input vector; X_i is the ith input data; n is the number of input data; $\boldsymbol{Y} = [Y_1,\ldots,Y_j,\ldots,Y_u]^T$is output vector; Y_j is the jth input data; u is the number of output data; $\boldsymbol{H}(\cdot)$ denotes the complex nonlinear relation which can be estimated by some meta-heuristic methods like artificial neural network.

In our study, BP neural network is chosen as the basic approach to estimate input-output relations $\boldsymbol{H}(\cdot)$ of the nonlinear hydrological system. The main advantage of BP is that it can reflect the complex nature of underlying process with less information than other traditional methods. A typical BP neural network consists of three layers, input layer, hidden layer and output layer, whose structure is shown in Figure 1. Each layer is composed of a series of interconnected processing nodes. As a specific neuron, each node in any layer uses a nonlinear transfer function to calculate the inner product of input vector and weight vector to get a scalar result. Two neighboring layers are connected via the weights of the nodes between these layers. The input layer receives and transmits input data to hidden layer. The hidden layer may contain a single layer or multi-layer that receives values from the previous layer. Each hidden layer is responsible for the input information conversion and then delivers them to the next hidden layer or output layer. The output layer presents the simulated results and has only one single layer with one or several nodes. In a single calculation, the BP neural network can obtain overall error between the estimated output values and the target output values, then loss function gradient is calculated. The gradient-descent algorithm is fed to update weights and thresholds to minimize loss function. The connection weights and thresholds between any two feed forward-connected neurons will be unceasingly adjusted until the error meets the termination conditions. Then, the optimized BP neural network can be used to forecast the target value with the corresponding input vector.

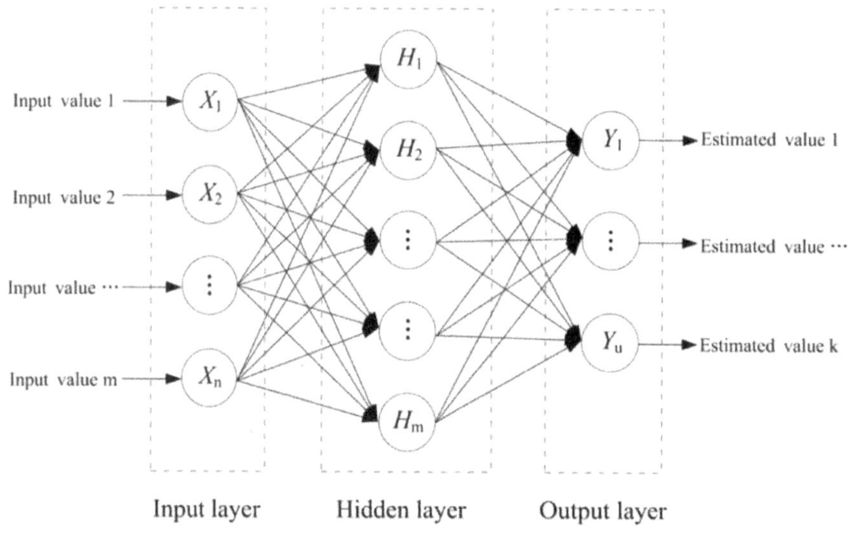

Figure 1: Schematic diagram of three-layer back propagation (BP) neural network.

Due to the defects of gradient descent method, the result may find a locally optimal solution instead of global optimum in most cases due to the existence of many local optima on the error surface. The convergence in back propagation learning cannot be guaranteed. Moreover, the computation speed is rather slow because the gradient descent algorithm requires small learning rates for stable learning.

QUANTUM-BEHAVED PARTICLE SWARM OPTIMIZATION

Particle Swarm Optimization

Particle Swarm Optimization (PSO), proposed by Kennedy and Eberhart in 1995 [30], is a well-known evolutionary population-based algorithm for global optimization problems [31,32]. The main concept of PSO is from analogy of biological and sociological feeding behavior of bird swarm, and its basic variant is a population (called a swarm) of candidate solutions (called particles). PSO is initialized with random solutions which search for optima by flying through problem space from generation to generation. The flight of each particle is guided continuously by its own best known position, ***pbest***, and the best known position of the whole population, ***gbest***. Each particle has position vector and velocity vector, and explores in the searching space by a few simple formulas.

There is extensive and profound homology or resemblance between PSO and other evolutionary computation techniques like genetic algorithm. Compared with genetic algorithm, PSO has faster convergence speed since it has no evolution operators like crossover and selection. Moreover, PSO has few parameters to adjust and depends directly upon function values rather than derivative information. In the past few years, PSO has been successfully applied to many research and application areas. However, the main defect of PSO is that global convergence cannot be guaranteed, especially when the number of decision variables or dimensions to be optimized are large. In other words, PSO is easily trapped into local optimum although it may have fast convergent rate [32,33].

Quantum-Behaved Particle Swarm Optimization

In order to deal with disadvantages of PSO, quantum-behaved particle swarm optimization (QPSO) was developed by Sun et al. [25] in 2004. From the quantum mechanics perspective, QPSO considers the particle possess quantum behavior and cannot determine the exact values of position vector and velocity vector simultaneously according to uncertainty principle [25,26]. Hence, there is no velocity vector in the particle of QPSO, and particle state is associated with an appropriate time-dependent Schrödinger equation and can be characterized by wave function ψ instead of position and velocity [27,28], where $|\psi|^2$ is the probability density function of its position. Let M particles in ddimensional space with $\bar{k}$ maximum generations, the position vector of ith particle at kth generation can be expressed as $\boldsymbol{x}_i(k) = [x_{i,1}(k), x_{i,2}(k), \ldots, x_{i,d}(k)]^T$. There will be $\boldsymbol{gbest}(k) = [gbest_1(k), gbest_2(k), \ldots, gbest_d(k)]^T$ and $\boldsymbol{pbest}_i(k) = [pbest_{i,1}(k), pbest_{i,2}(k), \ldots, pbest_{i,d}(k)]^T$. Employing Monte-Carlo method, particle moves according to the following iterative equation:

$$x_{i,j}(k+1) = \begin{cases} p_{i,j}(k+1) + a(k) \times |mbest_j(k+1) - x_{i,j}(k)| \times \ln(1/r_1), if\ r_2 \geq 0.5 \\ p_{i,j}(k+1) - a(k) \times |mbest_j(k+1) - x_{i,j}(k)| \times \ln(1/r_1), if\ r_2 < 0.5 \end{cases} \quad (1)$$

$$p_{i,j}(k+1) = r_3 \times pbest_{i,j}(k) + (1-r_3) \times gbest_j(k) \quad (2)$$

$$mbest_j(k+1) = \frac{1}{M}\sum_{i=1}^{M} pbest_{i,j}(k) \quad (3)$$

$$a(k) = \frac{(a_1 - a_2) \times (\bar{k} - k)}{\bar{k}} + a_2 \quad (4)$$

for i = 1,2,...,M; j = 1,2,...,d; k = 1,2,..., $\bar{k}$. Where $x_{i,j}(k)$ is position for jth dimension of ith particle in kth generation; r_1, r_2, r_3 are random variables distributed uniformly in [0,1]; a(k) is contraction-expansion coefficient in kth generation which controls the convergence speed of the particle; a_1, a_2 are maximum and minimum value of a(k), respectively; and there are usually $a_1 = 1.0$, $a_2 = 0.5$ [28]; $p_{i,j}(k)$ is the jth dimension of local attractor i in kth generation; ***mbest***represents the mean best position defined as mean of all ***pbest*** position of the whole population.

PARAMETERS SELECTION FOR ARTIFICIAL NEURAL NETWORK BASED ON QPSO ALGORITHM

In order to obtain better forecast accuracy, the novel QPSO algorithm is employed for parameters selection of BP neural network. In this paper, there is only one node in output layer, which is the daily runoff forecast value. If the node number of input layer and hidden layer are n and m, respectively, the architecture of the ANN neural network isn-m-1. The flow chart of the proposed method is shown in Figure 2, and the fundamental idea of the proposed method can be described as follows:

Step 0: Set basic parameters for the proposed method.

Step 0.1: Set maximize iterations $\bar{k}$ and population size M in QPSO.

Step 0.2: Divide data into training and testing sets.

Step 0.3: Define transfer function of neurons, which is a sigmoid function in this paper, i.e.:

$$f[x] = \frac{1}{1+e^{-x}} \quad (5)$$

Step 1: The input and output data in both training and testing sets are normalized to ensure the quality of forecast results.

$$X = \{X_i'\} = a \times \frac{X_i - X_i^{\min}}{X_i^{\max} - X_i^{\min}} + b \quad (6)$$

where X_i' and X_i is the normalized value and real value of each vector, respectively; $X_i^{\min}$ and $X_i^{\max}$ are the minimum and maximum value of input or output arrays; a and b are the positive normalized parameters, respectively. Based on large numbers of numerical experiments, we found that when the variable a = 0.2 and b = 0.6 are adopted to normalize the raw data, the forecasting models performs better. Hence, we use the variable a = 0.2 and b = 0.6 for data normalization in this paper.

Step 2: The QPSO algorithm is employed to select the parameters of BP neural network. The ith particle in the kth generation is denoted by $\boldsymbol{x}_i(k) = \{\boldsymbol{w}_i, \boldsymbol{b}_i\}$. Here, $\boldsymbol{w}_i$ and $\boldsymbol{b}_i$ represent the connection weights and bias matrix between any two layers of the BP neural network, respectively.

Step 3: Set k = 1, and initialize the parameters w and b of every particle randomly in the searching space, which are the connection weight and bias on each node, respectively.

Step 4: Use the parameters to calculate the fitness of each particle. Here, the fitness is the forecasting error between the output values and the target ones.

Step 4.1: Calculate the outputs of all hidden layer nodes for each training sample.

$$y_j = f\left[\sum_{i=1}^{n} x_i w_{ji} + b_j\right], j = 1,2,\cdots,m \quad (7)$$

where w_{ji} represents the connection weight from the input node i to the hidden node j, b_j stands for bias of neuron j,y_j is the output value of the hidden layer node j.

Step 4.2: Calculate the output data of the BP neural network for each training sample.

$$o_1 = f\left[\sum_{j=1}^{m} y_i w_{1j} + b_1\right] \quad (8)$$

where w_{1j} represent the connection weight from hidden node j to the output node 1, b_1 stands for the bias of the neuron; o_1 stands for the output data of network.

Step 4.3: Step 4.1 and 4.2 are repeated until all the training set samples are calculated. Then the forecasting error Fis regarded as the fitness of the particle $x_i(k)$.

$$F[x_i(k)] = \frac{1}{S}\sum_{s=1}^{S}(o_s - t_s)^2 \quad (9)$$

where o_s and t_s is the sth normalized output value and target value in the training data, respectively. S is the number of training set samples.

Step 5: Update the best known position of each particle and the best known position of the whole population according to the following two formulas:

$$p_i(k) = \begin{cases} x_i(k) & if(k=1) \\ \arg\min\{F[p_i(k-1)], F[x_i(k)]\} & otherwise \end{cases} \quad (10)$$

$$gbest(k) = \begin{cases} \arg\min\{\min_{1 \le i \le M}\{F[p_i(k)]\}, F[gbest(k-1)]\} & if(k>1) \\ \arg\min\{\min_{1 \le i \le M}\{F[p_i(k)]\}\} & otherwise \end{cases} \quad (11)$$

Step 6: Calculate the mean best position and the contraction-expansion coefficient with Equations (3) and (4), respectively.

Step 7: Update current position of each particle by the Equation (1).

Step 8: Set k = k+1, if the maximum iterations $\bar{k}$ reached, the flow will go to Step 9, else go back to Step 4.

Step 9: Output the optimal parameter of the BP neural network, which will be used for new data forecasting process.

Step 10: Before starting the forecasting process, the input vector are needed to be normalized by Equation (6), then transmit the processed data into the calibrated artificial neural network model to obtain predictive value. The predictive data need to be renormalized to the original range of output data by Equation (12).

$$\hat{Y} = \{\hat{Y}_i\} = \frac{(Y_i' - b)\times(Y_i^{max} - Y_i^{min})}{a} + Y_i^{min} \quad (12)$$

where Y_i' and $\hat{Y}_i$ are normalized forecasting value and real forecasting value of the output vector, respectively; Y_i^{min} and Y_i^{max} are minimum and maximum value of the output arrays, respectively.

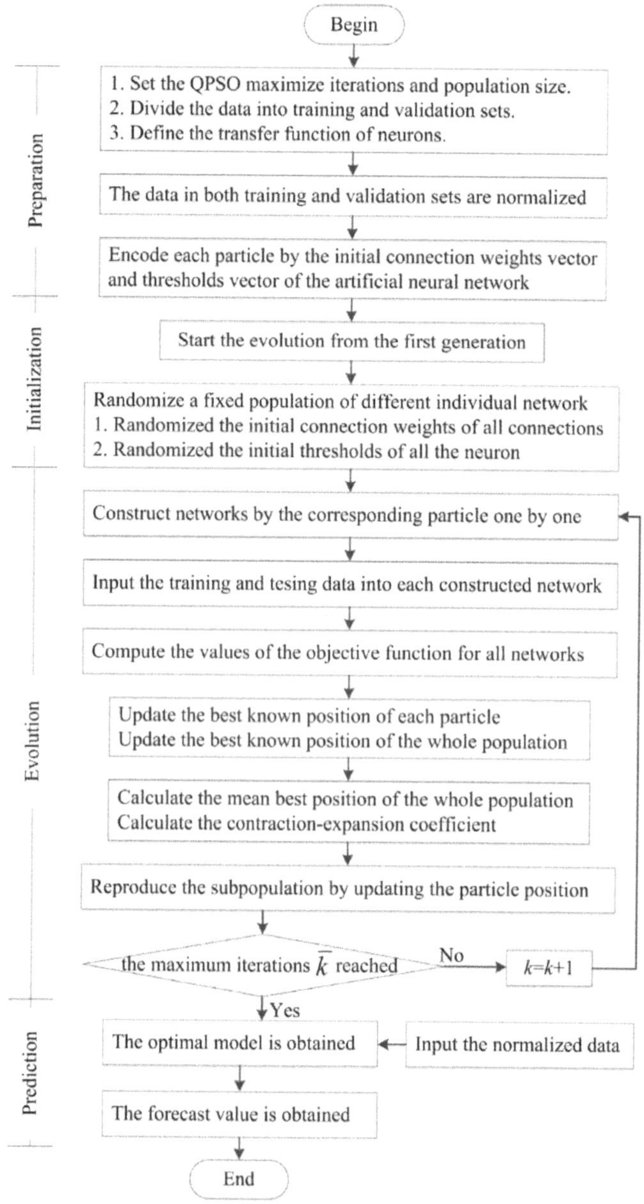

Figure 2: Flow chart of the proposed method.

SIMULATIONS

Study Area and Data Used

The study site is Hongjiadu reservoir in Wu River of Guizhou Province in southwest China. The Wu River is the biggest branch at southern bank of Yangtze River. Hongjiadu reservoir is the leading one with multi-year regulation ability of eleven cascade hydropower stations on Wu River. Rainfall produces most of the runoff. Its drainage area is 9900 km^2 and the mean annual runoff is 155 $m^3 \cdot s^{-1}$ at the dam site. The total reservoir storage is 4.95 billion cubic meters and the regulated storage is 3.36 billion cubic meters. Locations of Wu River and Hongjiadu reservoir are shown in Figure 3.

Figure 3: Location of study area.

All the daily runoff data collected from Hongjiadu reservoir operators is from 1 January 2006 to 31 December 2013. The data set between 1 January 2006 and 31 December 2012 is used for parameters calibration, while that from 1 January 2013 to 31 December 2013 is for testing. Moreover, before applying the proposed method for daily runoff predication, the data should be normalized to avoid numerical difficulties during calculation and guarantee consistency of greater numeric ranges as well as smaller ones. In the modeling process, input and output data sets are linearly scaled to [0.2, 0.8] as shown in Equation (6).

Performance Assessment Measures

Many performance measure methods have been developed to assess the forecast accuracy. However, so far, there is no unified standard since each measure can reflect one or more characteristics of the forecasting method. Hence, four commonly used metrics are selected to evaluate the forecast results: coefficient of correlation (R), Nash-Sutcliffe efficiency coefficient (NSE), root mean squared error (RMSE) and mean absolute percentage error (MAPE). Revaluates the linear relation between two data sequences, while NSE shows the capability of the model in predicting values away from the mean. The larger the value of R and NSE, the better the performances of the forecasting model.RMSE and MAPE measure the residual error and the mean absolute percentage error between the observed and forecasted data, respectively. The smaller the value of RMSE and MAPE, the better the performances of the forecasting model.

$$R=\sum_{i=1}^{n}\left[(Y_i-\bar{Y})(\hat{Y}_i-\bar{\hat{Y}})\right]\Big/\sqrt{\sum_{i=1}^{n}(Y_i-\bar{Y})^2(\hat{Y}_i-\bar{\hat{Y}})^2} \quad (13)$$

$$NSE=1-\sum_{i=1}^{n}(Y_i-\hat{Y}_i)^2\Big/\sum_{i=1}^{n}(Y_i-\bar{Y})^2 \quad (14)$$

$$RMSE=\sqrt{\frac{1}{n}\sum_{i=1}^{n}(Y_i-\hat{Y}_i)^2} \quad (15)$$

$$MAPE=\frac{1}{n}\sum_{i=1}^{n}\left|\frac{Y_i-\hat{Y}_i}{Y_i}\right|\times 100 \quad (16)$$

where Y_i and $\hat{Y}_i$ are the observed value and predictive value of ith data, respectively. $\bar{Y}$ and $\bar{\hat{Y}}$ represent the mean value of the observed value and predictive value, respectively. n is the total number of data set used for performance evaluation and comparison.

ANN Model Development

For ANN, proper selection of input variables will be helpful to find the best-fitted model. Based on the analysis of cross correlation coefficient and autocorrelation coefficient, four input combinations with different antecedent rainfalls and runoffs were developed for a comparative purpose. Table 1 summarizes these ANN models used for this study, where Runoff(t), Rainfall(t) are the runoff and rainfall value at the t-th period, respectively. To ensure the generalization capability of ANN, we use the trial and error method to determine the optimal network architecture. According to Chau et al. in 2005, the training process needs to be stopped when the error of the testing set starts to increase and that of the training set is still decreasing. Figure 4 shows the performance for the testing set against various numbers of neurons for model 1. The optimal ANN architecture adopted for model 1 is 2-4-1. The other three models have the same procedures as that for model 1. Table 2 shows the architecture and indices of various ANN forecasting models for Hongjiadu reservoir. We can find that antecedent two-day rainfalls and antecedent two-day runoff should be chosen as predictors, and ANN with the architecture of 4-7-1 performs best at this situation.

Table 1: Inputs and relation for various artificial neural network (ANN) forecasting models.

Model	Inputs	Relation between Output Variable and Input Variables
1	*Runoff(t-1),Rainfall(t-1)*	*Runoff(t)=**H**[Runoff(t-1),Rainfall(t-1)]*
2	*Runoff(t-1),Rainfall(t-1),Rainfall(t-2)*	*Runoff(t)=**H**[Runoff(t-1),Rainfall(t-1),Rainfall(t-2)]*
3	*Runoff(t-1),Runoff(t-2),Rainfall(t-1)*	*Runoff(t)=**H**[Runoff(t-1),Runoff(t-2),Rainfall(t-1)]*
4	*Runoff(t-1),Runoff(t-2),Rainfall (t-1),Rainfall(t-2)*	*Runoff(t)=**H**[Runoff(t-1),Runoff(t-2), Rainfall(t-1),Rainfall(t-2)]*

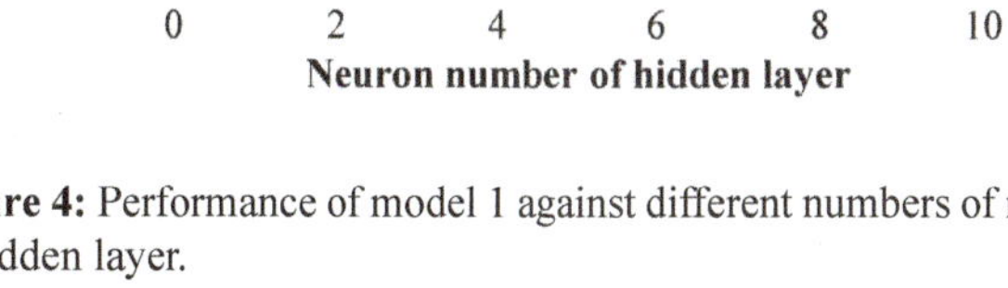

Figure 4: Performance of model 1 against different numbers of nodes in hidden layer.

Table 2: Architectures and indices of various ANN forecasting models for Hongjiadu reservoir.

Model	Model Architecture	Training				Testing			
		R	*NSE*	*RMSE* ($m^3 \cdot s^{-1}$)	*MAPE* (%)	*R*	*NSE*	*RMSE* ($m^3 \cdot s^{-1}$)	*MAPE* (%)
1	2-4-1	0.892	0.740	82.490	38.150	0.883	0.747	63.562	35.912
2	3-6-1	0.891	0.737	82.989	37.903	0.903	0.757	62.321	38.417
3	3-5-1	0.893	0.742	82.090	36.493	0.903	0.761	61.792	37.190
4	4-7-1	0.907	0.783	75.286	34.866	0.904	0.773	60.252	35.680

Comparison of Different Methods

In order to verify the effectiveness of the proposed method, the same training and verification samples are used for these two models, ANN and ANN-QPSO, and the above four quantitative indexes are employed to evaluate their performances. According to the above-mentioned analysis, the neural network architectures of both ANN and ANN-QPSO are 4-7-1 for Hongjiadu reservoir. Moreover, for ANN-QPSO, the number of population is set to be 300 whilst the maximize iterations is 500. The two algorithms are implemented by adopting JAVA language.

Table 3 presents the statistics results using various models developed for Hongjiadu study area. It can be seen that these two methods have different performances during both training and testing periods. Compared to the basic ANN, the proposed method is able to produce better forecast results for the daily runoff forecasting in Hongjiadu reservoir. In the training phase, the ANN-QPSO model improved the ANN forecasting ability with about 28.18% and 48.08% reduction in RMSE and MAPE values, respectively. The improvements of the forecasting results regarding theR and NSE were approximately 3.97% and 13.41%, respectively. In the testing phase, when compared with that of ANN, the statistical values of R and NSE of the proposed method increases by 5.42% and 17.46% respectively, while the value of RMSE and MAPE decreases by 36.34% and 28.81%. Figure 5 shows the convergence characteristic for objective functions of two methods in Hongjiadu reservoir. The objective function of the proposed method uses about 4 s to converge to a small neighborhood of the final result, whilst the ANN nearly stops the searching process since 1 second. In addition, in term of total computing time from Table 3, the ANN-QPSO decreases by 66.56% when compared to that of ANN. Thus, it can be concluded that the proposed method needs less computation time and has higher forecasting accuracy degree and global search capability than conventional ANN.

Table 3. Performance indices of two methods for Hongjiadu reservoir.

Method	Time(s)	Training				Testing			
		R	*NSE*	*RMSE* ($m^3 \cdot s^{-1}$)	*MAPE* (%)	*R*	*NSE*	*RMSE* ($m^3 \cdot s^{-1}$)	*MAPE* (%)
ANN-QPSO	10.1	0.943	0.888	54.074	18.102	0.953	0.908	38.354	25.401
ANN	30.2	0.907	0.783	75.286	34.866	0.904	0.773	60.252	35.680

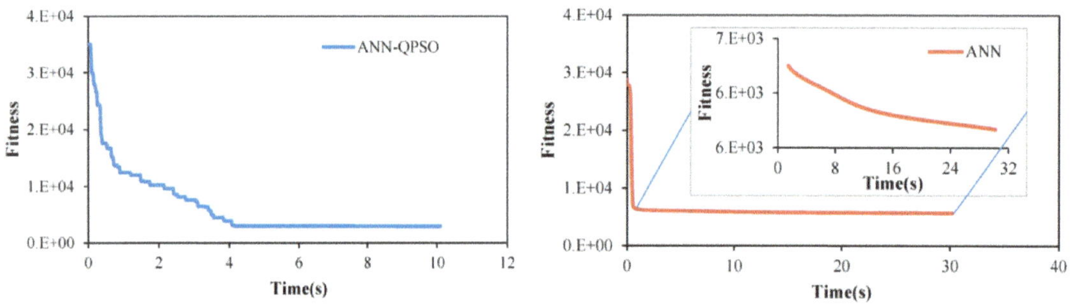

(a) (b)

Figure 5: Convergence characteristic of the objective function of two methods for Hongjiadu reservoir (**a**) ANN-QPSO; (**b**) ANN.

The observed peak flow and forecasted peak flow of two models for Hongjiadu reservoir are shown in Table 4. Both ANN-QPSO and ANN models forecast the maximum peak discharge as 1641.8 $m^3 \cdot s^{-1}$ and 1258.3 $m^3 \cdot s^{-1}$ instead of the observed 1696.4 $m^3 \cdot s^{-1}$, corresponding to about 3.2% and 25.8% underestimation, respectively. Furthermore, the absolute averages of the relative error of the ANN-QPSO and ANN models for forecasting the 8 peak flow are 11.6% and 30.9%, respectively. In summary, the ANN-QPSO method performs better than ANN in term of peak flow estimation.

Table 4: Observed peak flow and forecasted peak flow of two models for Hongjiadu reservoir.

Period	Date	Observed Peak ($m^3 \cdot s^{-1}$)	Forecasted Peak ($m^3 \cdot s^{-1}$)		Relative Error (%)	
			ANNP-QPSO	QPSO	ANNP-QPSO	QPSO
Training	2006-06-30	854.3	792.7	747.6	−7.2	−12.5
Training	2007-07-30	1435.8	1234.2	1108.2	−14.0	−22.8
Training	2008-06-22	1663.8	1514.5	861.5	−9.0	−48.2
Training	2009-08-04	628.5	456.4	407.4	−27.4	−35.2
Training	2010-07-11	1076.0	1053.8	806.7	−2.1	−25.0
Training	2011-06-23	561.9	471.6	426.6	−16.1	−24.1
Training	2012-07-26	1696.4	1641.8	1258.3	−3.2	−25.8
Testing	2013-06-09	1343.0	1151.9	622.5	−14.2	−53.6
Average (absolute)					11.6	30.9

The above analysis indicates that, for daily runoff forecasting, the proposed model can obtain better results than basic ANN model with significant improvements in terms of four different statistical indicators. Figure 6 and Figure 7demonstrate the scatter plots of observed data versus forecasted data using ANN and ANN-QPSO models during the training and testing period. The performances of both prediction models in the training period and testing period are respectively shown in Figure 8 and Figure 9. From Figure 6, Figure 7, Figure 8 and Figure 9, it can be clearly seen that the forecasting results of ANN model are inferior to that by the proposed ANN-QPSO model. For example, the model proposed in this paper can obtain the approximate maximum flows which is about 1700 $m^3{\cdot}s^{-1}$ during the training period while an obvious deviation exists between the original observed data and the forecasted data in ANN model. Hence, the ANN-QPSO model can mimic daily runoff better than that by ANN model.

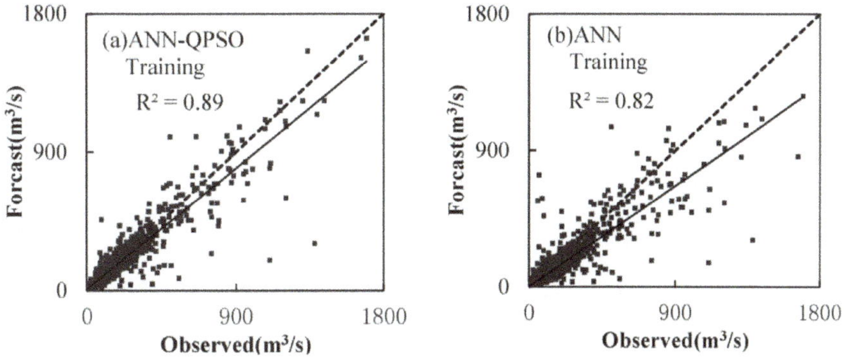

(a) (b)

Figure 6: Scatter plots of observed data vs. forecasted data during the training period. (**a**) ANN-QPSO; (**b**) ANN.

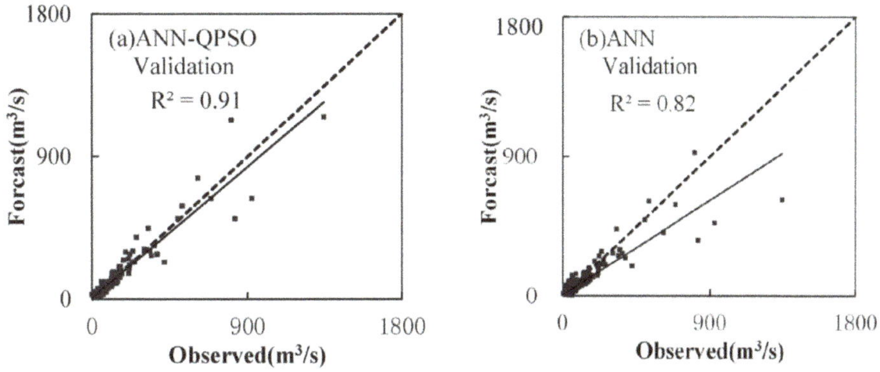

(a) (b)

Figure 7: Scatter plots of observed data vs. forecasted data during the testing period. (**a**) ANN-QPSO; (**b**) ANN.

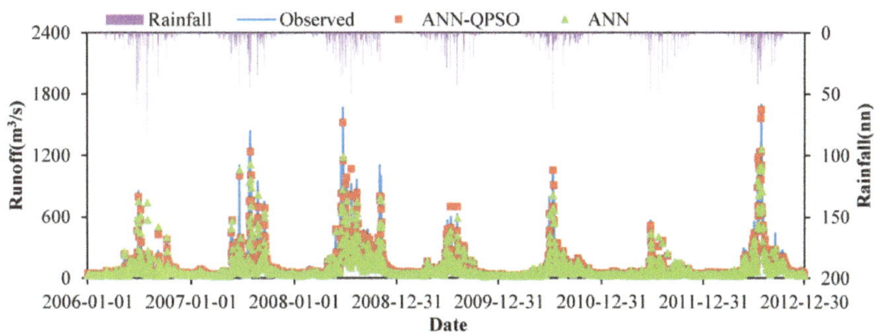

Figure 8: ANN, ANN-QPSO forecasted data and observed runoff data during the training period.

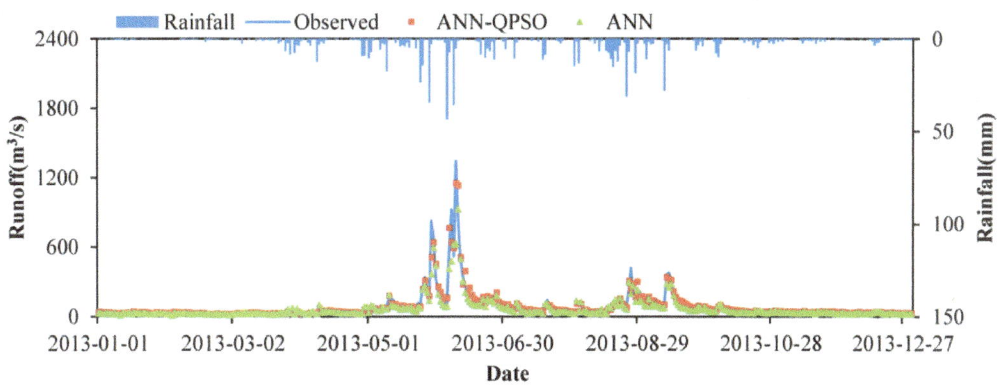

Figure 9: ANN, ANN-quantum-behaved particle swarm optimization (QPSO) forecasted data and observed runoff data during testing period.

CONCLUSIONS

Globally, hydropower energy has become one of the promising growing clean and renewable energy sources. It is of great significance to accurately predict daily reservoir runoff for integration of hydropower energy in power system. In this research, a novel method called ANN-QSPO, which is based on artificial neural network (ANN) and quantum-behaved particle swarm optimization (QPSO), was developed for daily reservoir runoff forecasting to help reservoirs plan and manage in a more sustainable manner. In the proposed ANN-QSPO method, QPSO was employed to select the ANN optimal parameters and the ANN was used for the prediction after the training process. The proposed approach was compared with ANN model for daily runoff forecasting of Hongjiadu reservoir in southeast China. From the experiment, the results show that the proposed method achieves much better forecast accuracy than basic ANN model. Compared with the statistical values of R and NSE of ANN, the improvements of the proposed method were approximately 3.97% and 13.41% in the training phase, while respectively increases by 5.42% and 17.46% in the testing phase. In term of total computing time, the ANN-QPSO decreases by 66.56% when compared to that of ANN. Thus, QPSO algorithm can act as an alternative training algorithm for the ANN parameters selection.

ACKNOWLEDGMENTS

This study is supported by the Major International Joint Research Project from the National Nature Science Foundation of China (51210014) and the National Basic Research Program of China (973 Program) (No. 2013CB035906).

AUTHOR CONTRIBUTIONS

All authors contributed extensively to the work presented in this paper. Chun-tian Cheng contributed to the subject of the research and literature review. Wen-jing Niu and Zhong-kai Feng contributed to modeling, data statistical analysis and finalized the manuscripts. Jian-jian Shen contributed to the manuscripts review and commented. Kwok-wing Chau contributed to the manuscripts revision and supervision.

CONFLICTS OF INTEREST

The authors declare no conflict of interest.

REFERENCES

1. Zhang, J.; Cheng, C.T.; Liao, S.L.; Wu, X.Y.; Shen, J.J. Daily reservoir inflow forecasting combining QPF into ANNs model. Hydro. Earth Syst. Sci. 2009, 6, 121–150.
2. Wang, W.C.; Chau, K.W.; Xu, D.M.; Chen, X.Y. Improving forecasting accuracy of annual runoff time series using ARIMA based on EEMD decomposition. Water Resour. Manag. 2015, 29, 1–21.
3. Duan, Q.Y.; Sorooshian, S.; Gupta, V. Effective and efficient global optimization for conceptual rainfall-runoff models. Water Resour. Res. 1992, 28, 1015–1031.
4. Valipour, M.; Banihabib, M.E.; Behbahani, S.M.R. Comparison of the ARMA, ARIMA, and the autoregressive artificial neural network models in forecasting the monthly inflow of Dez dam reservoir. J. Hydrol. 2013, 476, 433–441.
5. Kneis, D.; Bürger, G.; Bronstert, A. Evaluation of medium-range runoff forecasts for a 500 km^2 watershed. J. Hydrol. 2012, 414, 341–353.
6. Zhao, R.J. The Xinanjiang model applied in China. J. Hydrol. 1992, 135, 371–381.
7. Cheng, C.T.; Ou, C.P.; Chau, K.W. Combining a fuzzy optimal model with a genetic algorithm to solve multi-objective rainfall-runoff model calibration. J. Hydrol. 2002, 268, 72–86.
8. Wang, W.C.; Chau, K.W.; Cheng, C.T.; Qiu, L. A comparison of performance of several artificial intelligence methods for forecasting monthly discharge time series. J. Hydrol. 2009, 374, 294–306.
9. Lin, G.F.; Chen, G.R.; Huang, P.Y. Effective typhoon characteristics and their effects on hourly reservoir inflow forecasting. Adv. Water Resour. 2010, 33, 887–898.
10. Hsu, K.; Gupta, H.V.; Sorroshian, S. Artificial neural network modeling of the rainfall-runoff process. Water Resour. Res. 1995, 31, 2517–2530.
11. Machado, F.; Mine, M.; Kaviski, E.; Fill, H. Monthly rainfall-runoff modelling using artificial neural networks. Hydrol. Sci. J. 2011, 56, 349–361.
12. Duan, Q.Y.; Sorooshian, S.; Gupta, V.K. Optimal use of the SCE-UA global optimization method for calibrating watershed models. J. Hydrol. 1994, 158, 265–284.
13. Duan, Q.Y.; Gupta, V.K.; Sorooshian, S. Shuffled complex evolution approach for effective and efficient global minimization. J. Optim. Theory Appl. 1993, 76, 501–521.
14. Lin, J.Y.; Cheng, C.T.; Chau, K.W. Using support vector machines for long-term discharge prediction. Hydrol. Sci. J. 2006, 51, 599–612.
15. Lima, A.R.; Cannon, A.J.; Hsieh, W.W. Nonlinear regression in environmental sciences by support vector machines combined with evolutionary strategy. Comput. Geosci. 2013, 50, 136–144.
16. Guo, W.J.; Wang, C.H.; Zeng, X.M.; Ma, T.F.; Yang, H. Subgrid parameterization of the soil moisture storage capacity for a distributed rainfall-runoff model. Water 2015, 7, 2691–2706.
17. Kamruzzaman, M.; Shahriar, M.S.; Beecham, S. Assessment of short term rainfall and stream flows in South Australia. Water 2014, 6, 3528–3544.
18. Chau, K.W.; Wu, C.L.; Li, Y.S. Comparison of several flood forecasting models in Yangtz River. J. Hydrol. Eng.2005, 10, 485–491.
19. Piotrowski, A.P.; Napiorkowski, J.J. Optimizing neural networks for river flow forecasting evolutionary computation methods versus the Levenberg-Marquardt approach. J. Hydrol. 2011, 407, 12–27.
20. Chau, K.W. Particle swarm optimization training algorithm for ANNs in stage prediction of Shing Mun River. J. Hydrol. 2006, 329, 363–367.
21. Asadnia, M.; Chua, L.H.C.; Qin, X.S.; Talei, A. Improved particle swarm optimization based artificial

neural network for rainfall-runoff modeling. J. Hydrol. Eng. 2014, 19, 1320–1329.

22. Liu, W.C.; Chung, C.E. Enhancing the predicting accuracy of the water stage using a physical-based model and an artificial neural network-genetic algorithm in a river system. Water 2014, 6, 1642–1661.

23. Wu, C.L.; Chau, K.W. Rainfall-runoff modeling using artificial neural network coupled with singular spectrum analysis. J. Hydrol. 2011, 399, 394–409.

24. Jeong, D.I.; Kim, Y.O. Combining single-value streamflow forecasts—A review and guidelines for selecting techniques. J. Hydrol. 2009, 377, 284–299.

25. Sun, J.; Feng, B.; Xu, W.B. Particle swarm optimization with particles having quantum behavior. In Congress on Evolutionary Computation, 2004 (CEC2004), Proceedings of the 2004 Congress on Evolutionary Computation, Portland, OR, USA, 19–23 June 2004; pp. 325–331.

26. Fang, W.; Sun, J.; Ding, Y.; Wu, X.; Xu, W. A review of quantum-behaved particle swarm optimization. IETE Tech. Rev. 2010, 27, 336–348.

27. Xi, M.; Sun, J.; Xu, W. An improved quantum-behaved particle swarm optimization algorithm with weighted mean best position. Appl. Math. Comput. 2008, 205, 751–759.

28. Feng, Z.K.; Liao, S.L.; Niu, W.J.; Shen, J.J.; Cheng, C.T.; Li, Z.H. Improved quantum-behaved particle swarm optimization and its application in optimal operation of hydropower stations. Adv. Water Sci. 2015, 26, 413–422. (in Chinese).

29. Sun, C.; Lu, S. Short-term combined economic emission hydrothermal scheduling using improved quantum-behaved particle swarm optimization. Expert Syst. Appl. 2010, 37, 4232–4241.

30. Eberhart, R.; Kennedy, J. A new optimizer using particle swarm theory. In Proceedings of the 6th International Symposium on Micro Machine and Human Science, New York, NY, USA, 4–6 October 1995; pp. 39–43.

31. Gaing, Z.L. Particle swarm optimization to solving the economic dispatch considering the generator constraints. IEEE Trans. Pow. Syst. 2003, 18, 1187–1195.

32. Clerc, M.; Kennedy, J. The particle swarm-explosion, stability, and convergence in a multidimensional complex space. IEEE Trans. Evol. Comput. 2002, 6, 58–73.

33. Ren, C.; An, N.; Wang, J.; Li, L.; Hu, B.; Shang, D. Optimal parameters selection for BP neural network based on particle swarm optimization: A case study of wind speed forecasting. Knowl. Based Syst. 2014, 56, 226–239.

CHAPTER 17

OPTIMIZATION OF OPEN CANAL CROSS SECTIONS BY DIFFERENTIAL EVOLUTION ALGORITHM

Mustafa Erkan Turan, Mehmet Ali Yurdusev

Keywords: Open canal, differential evolution algorithm, optimization

INTRODUCTION

Open canals are used in water resources systems to transfer large quantity of water from a river or another source to where it is used. They are essential elements of irrigation and waterpower systems. They are free surface structures, which carry water by gravity. An open canal may require substantial amount of investment depending on its length and cross section, making the optimal sizing essential. Optimal sizing is to find the optimal cross section dimensions at minimum construction cost.

Nonlinear optimization techniques such as Lagrange Multipliers have traditionally been used to undertake optimal sizing of open canal cross sections. Several researches under different conditions have been performed into optimal cross section design. Majority of this research deals with assessing optimal canal sections of different geometries for uniform flow conditions [1], [2], [3], [4]. There are relatively fewer number of studies for this issue considering non-uniform flow conditions [5][6]. Swamee et al. (2000) have obtained the parameters of an optimal canal cross section based on the minimum cost of earthwork, which increases with an increase in excavation depth [6]. Differential evolution algorithm (DEA) is one of the evolutionary algorithms that can be used for an optimization process. There are several DEA studies for different optimization problems including the identification of structural system parameters [7], multiple objective reservoir operation problem [8] and mass minimization truss problem [9].

In this study, optimum cross sections of different canal geometries are obtained using differential evolution algorithm. The different geometries considered include triangular, circular, rectangular and trapezoidal canal sections. The objective function is in the form of cost minimization. The cost figures taken into account are the costs of canal covering and excavation. The findings of optimization exercises carried out by differential evolution algorithm are compared with those of given in related literature. Based on the findings obtained, it can be stated DEA is capable of finding the optimal dimensions for canal cross-sections.

DEFINITION OF THE PROBLEM

The solution to an optimization problem aims to find global optima. In the particular problem of canal cross section problem, the optimization problem comprises an objective function in the form of minimum cost subject to the flow requirements to convey a specific discharge in the canal considered. The decision variables, the dimensions of canal cross sections, are side slope, bottom width, flow depth, and radius. Geometric properties of a generic canal cross section are given in Figure 1.

Citation: Mustafa Erkan Turan and Mehmet Ali Yurdusev, Optimization of Open Canal Cross Sections by Differential Evolution Algorithm, doi:10.3390/mca16010077.

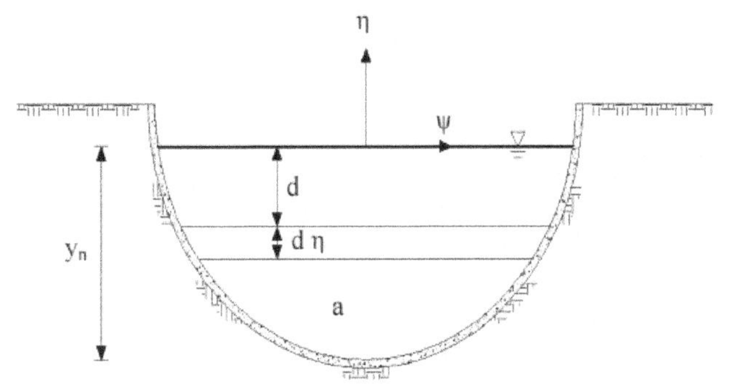

Figure 1: Generic Type of Canal Section

Given the type of canal lining and slope, the cost function for a lined canal can be written as follows:

$$C = \beta_L P + \beta_E A + \beta_A \int_0^A a\,d\eta \tag{1}$$

where b_L : unit lining cost in TL/L^2 , b_E : unit excavation cost in TL/L^3 , b_A : additional cost of excavation per unit depth in TL/L^4 , a: flow area at height η, P: wetted perimeter, A: flow area and L: length. Such a total cost expression for a canal assumes that the canal is constructed for the same soil conditions and cost of excavation increases with the increase in excavation depth.

Assuming the uniform flow conditions apply in the canal, Manning's uniform flow equation is used to define uniform flow as follows:

$$Q = \frac{1}{n} A R^{2/3} S^{1/2} \tag{2}$$

where n: Manning roughness coefficient, A: flow area, R: hydraulic radius and S: bottom slope of canal

The optimization problem for any canal cross section is as follows: minimize

$$C = \beta_L P + \beta_E A + \beta_A \int_0^A a\,d\eta \tag{3}$$

subject to

$$Q - \frac{1}{n} A R^{2/3} S^{1/2} = 0 \tag{4}$$

Several dimensionless parameters can be defined to examine the effects of the variables and to compare with the results given in related literature using the length scale

$$\lambda = \left(\frac{Qn}{S^{1/2}}\right)^{3/8} \tag{5}$$

The dimensionless parameters defined are as follows:

$$C_* = C / \beta_E \lambda^2 \tag{6}$$

$$\beta_{L*} = \beta_L / \beta_E \lambda^2 \tag{7}$$

$$\beta_{A*} = \beta_A \lambda / \beta_E \tag{8}$$

$$A_* = A / \lambda^2 \tag{9}$$

$$P_* = P / \lambda \tag{10}$$

$$b_* = b / \lambda \tag{11}$$

$$y_{n*} = y_n / \lambda \tag{12}$$

$$r_* = r / \lambda \tag{13}$$

Using these dimensionless parameters, the optimization problems stated in (3) and (4) can be rewritten as follows:

minimize

$$C_* = \beta_{L*} P_* + A_* + \frac{\beta_{A*} \int_0^{y_n} a\,d\eta}{\lambda^3} \tag{14}$$

subject to

$$1 - A_*^{5/3} P_*^{-2/3} = 0 \tag{15}$$

The optimization problem given in (14) and (15) is arranged for triangular, rectangular, trapezoidal and circular canal cross sections, of which geometric dimensions are given in Figure 2, 3, 4 and 5 as follows:

Optimization problem for triangular cross section:

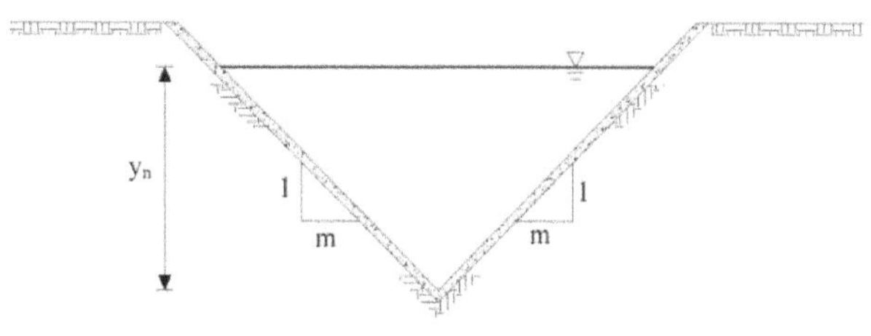

Figure 2: Geometric Dimensions for Triangular Cross Section

minimize

$$C_* = 2\beta_{L*} y_{n*} \sqrt{1+m^2} + m y_{n*}^2 + \frac{\beta_{A*} y_{n*}^3}{3} \tag{16}$$

subject to

$$1 - \frac{\left(m y_{n*}^2\right)^{5/3}}{\left(2 y_{n*} \sqrt{1+m^2}\right)^{2/3}} = 0 \tag{17}$$

Optimization problem for rectangular cross section:

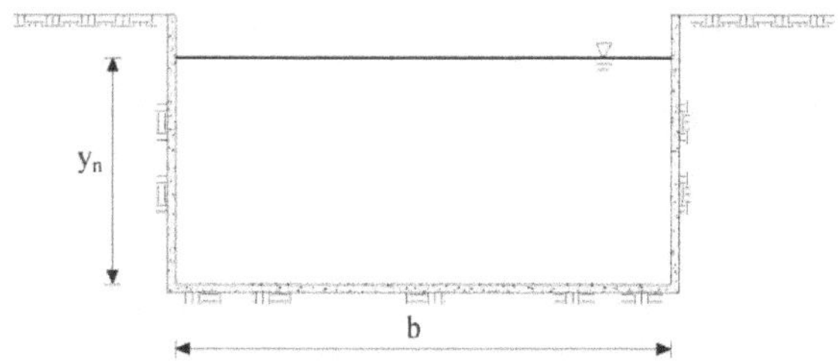

Figure 3: Geometric Dimensions for Rectangular Cross Section

minimize

$$C_* = \beta_{L*}(2y_{n*} + b_*) + b_*\, y_{n*} + \frac{\beta_{A*}\, b_*\, y_{n*}^2}{2} \tag{18}$$

subject to

$$1 - \frac{(b_*\, y_{n*})^{5/3}}{(2y_{n*} + b_*)^{2/3}} = 0 \tag{19}$$

Optimization problem for trapezoidal cross section:

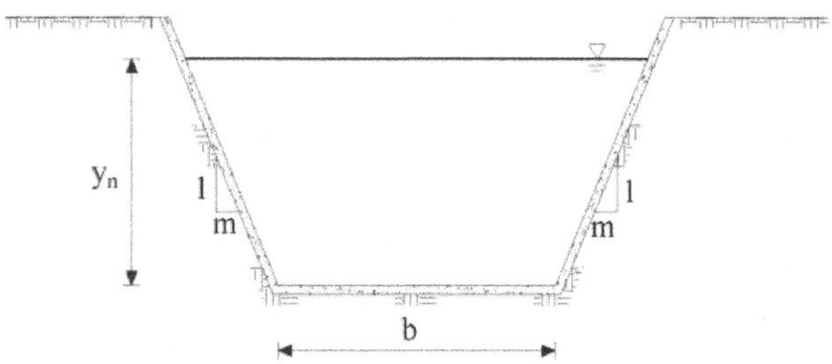

Figure 4: Geometric Dimensions for Trapezoidal Cross Section

minimize

$$C_* = \beta_{L*}\left(2y_{n*}\sqrt{1+m^2} + b_*\right) + \left(b_*\, y_{n*} + m\, y_{n*}^2\right) + \beta_{A*}\left(\frac{b_* y_{n*}^2}{2} + \frac{m\, y_{n*}^3}{3}\right) \tag{20}$$

subject to

$$1 - \frac{\left(b_*\, y_{n*} + m\, y_{n*}^2\right)^{5/3}}{\left(2y_{n*}\sqrt{1+m^2} + b_*\right)^{2/3}} = 0 \tag{21}$$

Optimization problem for circular cross section:

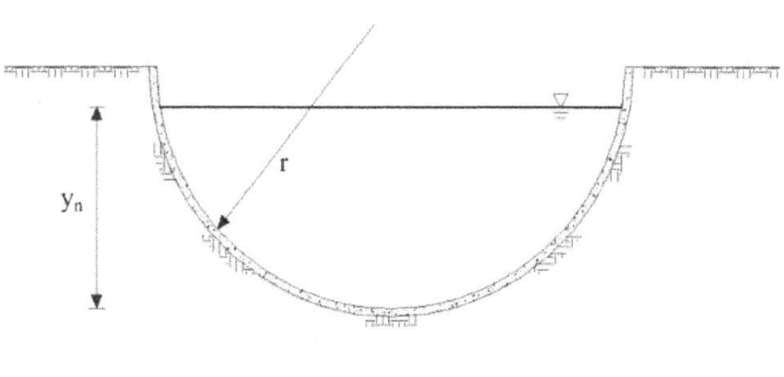

Figure 5: Geometric Dimensions for Circular Cross Section

minimize

$$\begin{aligned} C_* = {} & \beta_{L*}\left\{r_*\left[\pi - 2\arcsin\left(\frac{r_* - y_{n*}}{r_*}\right)\right]\right\} + \frac{r_*^2}{2}\left[\pi - 2\arcsin\left(\frac{r_* - y_{n*}}{r_*}\right) - 2\left(\frac{r_* - y_{n*}}{r_*}\right)\frac{\sqrt{r_*^2 - (r_* - y_{n*})^2}}{r_*}\right] \\ & + \beta_{A*}\left(\frac{r_*}{2}\right)\left\{-(r_* - y_{n*})\left[\pi - 2\arcsin\left(\frac{r_* - y_{n*}}{r_*}\right)\right] + 2\sqrt{r_*^2 - (r_* - y_{n*})^2}\left[\frac{2r_*^2 + (r_* - y_{n*})^2}{3r_*^2}\right]\right\} \end{aligned} \tag{22}$$

subject to

$$1 - \frac{\left\{\frac{r_*^2}{2}\left[\pi - 2\arcsin\left(\frac{r_* - y_{n*}}{r_*}\right) - 2\left(\frac{r_* - y_{n*}}{r_*}\right)\frac{\sqrt{r_*^2 - (r_* - y_{n*})^2}}{r_*}\right]\right\}^{5/3}}{\left\{r_*\left[\pi - 2\arcsin\left(\frac{r_* - y_{n*}}{r_*}\right)\right]\right\}^{2/3}} = 0 \tag{23}$$

THE DIFFERENTIAL EVOLUTION ALGORITHM (DEA)

The differential evolution algorithm (DEA), which was proposed by Storn and Price [10], is an evolutionary optimization algorithm like genetic algorithms.

DEA executes population basis. The individuals constituting population generate new populations through mutation, crossover and selection operators. Each individual in the population is a vector of D dimension. The dimension of the vector, D, is the same number as the number of variables in optimization problem [11].

DEA reaches to optima by the following steps;

1. Initialization

Lower (b_{jL}) and upper (b_{jU}) bounds of each variables are determined. The initial value of the jth parameter of the ith vector is calculated by uniformly distributed via

$$x_{j,i,o} = \text{rand}_j(0,1)(b_{j,U} - b_{jL}) + b_{j_L} \tag{24}$$

2. Mutation

The solution point that the vector represents moves in solution space by this operator. This process can be realized by 3 randomly selected vectors. The mutation vector can be obtained by;

$$v_{j,g} = x_{r3,g} + F(x_{r1,g} - x_{r2,g}) \tag{25}$$

3. Crossover

A trial vector is obtained through the crossover of two vectors, mutant vector and target vector, as follows:

$$u_{j,g} = u_{j,i,g} = \begin{cases} v_{j,i,g} & \text{if } \text{rand}_j(0,1) \leq \\ x_{j,i,g} & \text{otherwise} \end{cases} \tag{27}$$

C_r is called crossover probability and takes the values between 0 and 1.

Selection

The individuals to form the new generation are selected at this stage. Each trial vector obtained through crossover is compared with the target vector. The vector with the lowest objective function value survives into the next generation g+1. This process is expressed as follows:

$$x_{i,g+1} = \begin{cases} v_{j,g} & \text{if } f(u_{i,g}) \leq f(u_{i,g}) \\ x_{j,g} & \text{otherwise} \end{cases} \tag{27}$$

Termination Criteria

Since the process is iterative, the termination criteria are normally required to stop the process. For this purpose, the sufficiently small difference, namely error, between the values obtained or a specific number of iteration are used [12].

APPLICATIONS

The optimization problems set forth in Section 2 for four different shapes of canal cross sections have been solved by differential evolution algorithm and the section variables obtained. During the operation of DEA, some user-defined parameters are required. These parameters include the number of individuals in population, NP, scale rate, F, and crossover probability Cr. These parameters should be selected in such a way that the solution process is speeded up. In this study, following values were used: Np=10, F=0.85 and Cr=0.5.

The results obtained in this study by DEA have been compared with those given in Aksoy and Sakarya for the same section types.

First of all, classical optimum cross section problem where excavation cost does not change with excavation depth, e.g. $\hat{a}_{a*}$ =0, was solved. This problem was solved in Literature by Langrange Multipliers (LM) and the results of such solution are taken into account for comparing the results of DEA as shown in Table (1) and (2).

Table 1: The results of LM and DEA for optimum nondimensional section variables for $b_{A*} = 0$

Section Variables	Triangular section		Rectangular section	
	LM	DEA	LM	DEA
Side slope, m*	1.000	1.000	-	-
Bottom width, b*	-	-	1.834	1.83358
Flow depth, y_{n*}	1.297	1,297	0.917	0.91721

Section Variables	Trapezoidal section		Circular section	
	LM	DEA	LM	DEA
Side slope, m*	0.577	0.577	-	-
Bottom width, b*	1.118	1.117	-	-
Flow depth, y_{n*}	0.968	0.968	1.004	1.004
Radius, r*	-	-	1.004	1.00387

As a second exercise, the problem is solved by DEA for the different values ofb $_{A*}$ andb $_{L*}$ section variables obtained. Aksoy and Sakarya solved the same problem by a numerical optimization (NO) technique. For triangular cross section, the results of both DEA and NO have been presented in Figure 6 and 7 as the results for other types are not traceable in Aksoy and Sakarya.

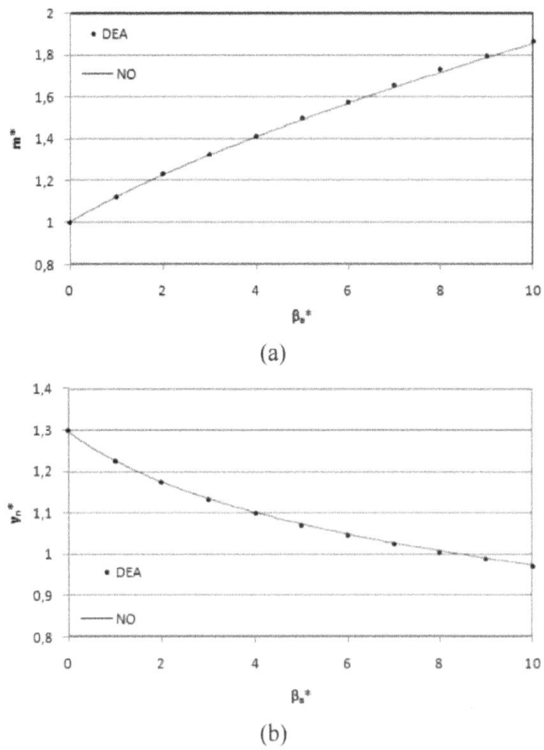

Figure 6: Variation of (a) optimum nondimensional side slope and (b) normal depth of a triangular section with $b_{L*} = 1.0$

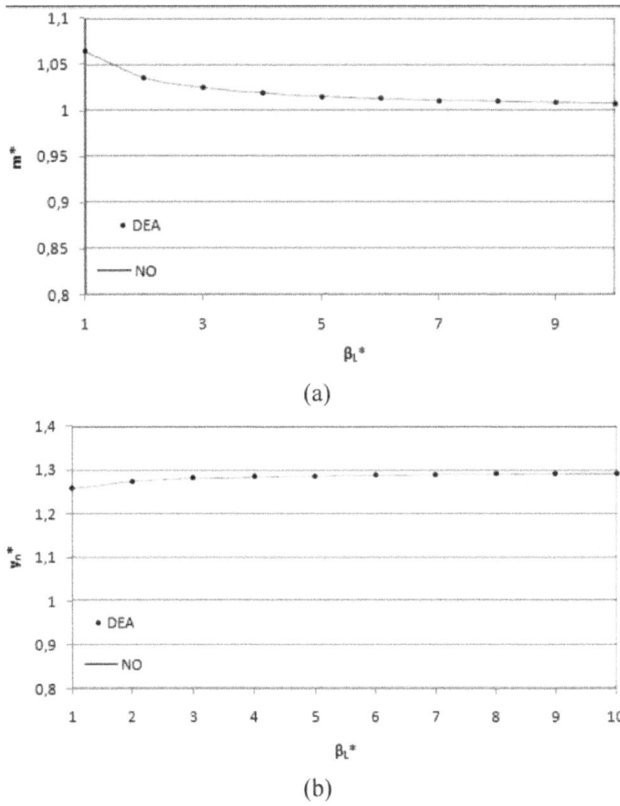

Figure 7: Variation of (a) optimum nondimensional side slope and (b) normal depth of a triangular section with $b_{A*} = 0.5$

CONCLUSION

The classical problem of optimal canal cross section design has been revisited. The problem is solved by Differential

Evolutionary Algorithm for the different types of section types including circular triangular, rectangular and trapezoidal shapes. The problem is considered the same as that formulated in related literature so as to make one-to-one comparisons to see the capability of DEA results of DEA have confirmed the results obtained in the literature. Therefore, it has been concluded that DEA can successfully be applied to the problem and also is capable of satisfactory results.

REFERENCES

1. Chow, V.T. 1959. Open-channel hydraulics. McGraw-Hill Book Company, New York.
2. Babaeyan-Koopaei, K., Valentine, E. M., and Swailes, D. C. 2000. Optimal design of parabolic-bottomed triangle canals." Journal of Irrigation and Drainage Engineering. 126(6), 408–411.
3. Babaeyan-Koopaei, K. 2001. Dimensionless curves for normal depth calculations in canal sections. Journal of Irrigation and Drainage Engineering. 127(6), 386–389.
4. Chahar, B.R. 2007. Optimal Design of a special Class of Curvilinear Bottomed Channel Section. Journal Hydraulic Engineering, ASCE. 133(5), 571-576.
5. Swamee, P.K. 1995. Optimal irrigation canal sections. Journal of Irrigation and Drainage Engineering. 121, 467–469.
6. Swamee, P.K., Mishra, G.C., and Chahar, B.R. 2000. Minimum cost design of lined canal sections. Journal of Water Resources Management. 14, 1–12.
7. Tang H, Xue S and Fan C. 2008. Differential evolution strategy for structural system identification. Computers and Structures. 86, 21-22
8. Reddy, M.J. and D.N. Kumar, 2007. Multiobjective differential evolution with application to reservoir system optimization. J. Comput. Civil Eng. 21, 136-146.
9. Hull P. V., Tinker M. L., Dozier G. 2006. Evolutionary optimization of a geometrically refined truss. Structural and Multidisciplinary Optimization. 31, 311– 319
10. Storn R, Price K. (1995). Differential evolution – a simple and efficient adaptive scheme for global optimization over continuous spaces. Technical Rep. No. TR-95-012, International Computer Science Institute, Berkley (CA).
11. Storn R., Price K., Lampinen J. A. 2005. Differential Evolution: A Practical Approach to Global Optimization. Germany: Springer-Verlag Berlin Heidelberg.
12. Ali, M. M., Törn, A., 2004. Population Set-Based Global Optimization Algorithms: Some Modifications and Numerical Studies. Computer & Operations Research. 31, 1703-1725.
13. Aksoy, B., and Altan-Sakarya, A. B. 2006. Optimal lined channel design. Canadian. J. Civ. Eng., 33(5), 535–545.

CHAPTER 18 SURFACE WATER QUALITY MONITORING SITE OPTIMIZATION

Hua Wang, Mengan Wu , Yanqing Deng, Chunyan Tang, and Rui Yang

Keywords: Poyang Lake, inter-annual stability, correlation coefficient, optimization scheme, numerical model

INTRODUCTION

Water quality monitoring site arrangement is the foundation to obtain the first-hand data for water environment management. A reasonable water quality monitoring network should not only meet the needs for long-term data accumulation, water quality assessment and trend analysis, but also reflect in a timely way the dynamic status of the water environment and water pollution to provide scientific guidance for water resources management and water environment protection [1,2,3]. In the 1970s, some scholars started related researches on water environment monitoring site arrangement. In 1976, the U.S. Geological Survey Bureau and Water Resources Committee cooperated to divide the whole basin in the United States into 21 parts and arranged 352 units for regular environment monitoring [4]. In the early 1980s, Sanders published a monograph giving a more detailed introduction on environment monitoring site arrangement [5]. Relevant research started in China in the late 1980s. Originally, the government set up some water environment monitoring sites just to enhance environment protection for some major rivers and lakes [6,7]. Most of these sites were laid in the absence of background information and without scientific argumentation [8]. In addition, with the accelerated industrialization and increased population, the pollutant loads surrounding the water bodies have varied in both spatial and temporal scale, and thorough research is needed to indicate that whether the primary monitoring site network could adapt well to the changed conditions. Currently, few studies have been conducted on water quality monitoring site optimization. In this paper, we propose a coupled method to optimize the monitoring sites for a huge freshwater lake based on field investigations, mathematical analysis, and numerical simulation tests. Poyang Lake, the largest freshwater lake in China, was selected as the research area.

Citation: Hua Wang, Mengan Wu, Yanqing Deng, Chunyan Tang and Rui Yang, Surface Water Quality Monitoring Site Optimization, doi:10.3390/ijerph111111833.

MATERIALS AND METHODS

Study Area

Poyang Lake (28°25′–29°45′ N, 115°50′–116°44′ E) is located on the south bank of the middle-lower Yangtze River in Jiangxi Province, China (Figure 1). It is the largest freshwater in China and is quite a valuable ecological resource in the world. The lake can be divided into the south area and the north area. The north area is a river-style lake connecting the external Yangtze River, of which the average size is 40 km in length, 3–5 km in width, and 2.8 km at its narrowest. The south area is the main lake, of which the size is 133 km in length and 74 km at its widest. Due to the waterfront location, the lake regulates the water volume from the upper main five rivers, Ganjiang River, Fuhe River, Xinjiang River, Raohe River and Xiuhe River, to the downstream Yangtze River, so both the water level and water area of the lake evidently fluctuate with the seasons. The multi-year average water level of Poyang Lake is about 13.30 m, and the corresponding water area and volume are 2291.9 km^2 and 2.1×10^9 m^3 respectively. The average annual rainfall at Poyang Lake is about 1632 mm, of which 74.4% is concentrated from March to August. Due to its huge water surface, the lake covers many districts, including Nanchang, Xinjian, Jinxian, Yugan, Boyang, Duchang, Hukou, Jiujiang, Xingzi, De'an, and Yongxiu. In 1979, to reflect the water quality of the lake, 19 water quality monitoring sites were arranged without any scientific rationale (Figure 1). However, in recent years, the increased population and accelerated industrialization have resulted in a changed pollutant distribution surrounding the lakes. The population of the basin in 2012 has increased to about 4.7×10^7 people. According to the statistical data, the sub-basin of the Ganjiang River is characterized by the highest pollutant load transported into the lake, and the loads of COD, NH_3-N and TP, contributed to the lake in a year are respectively 97,625 t, 18,255 t and 7494 t, on average. The loads of the Xinjiang sub-basin rank in second place, and the average annual amounts of transported pollutants are separately 19,525 t, 3651 t and 1499 t, on average [9]. In view of the enhanced pollutant load, it is necessary to study whether the primary monitoring sites, which maintained the status quo without any optimization till now, could adapt to the changed conditions.

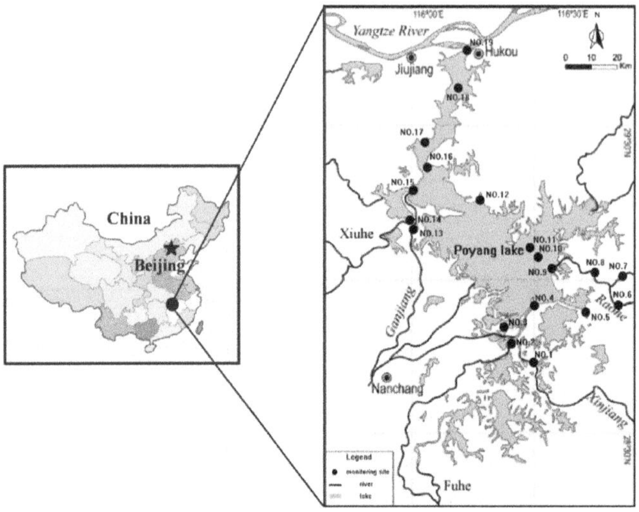

Figure 1: General view of the research area.

Field Investigation

Water samples from the surface layer at the 19 sites were collected artificially in the middle of each month from 2008 to 2012. The analyzed parameters included pH, DO, COD, TN, TP, NH_3-N, Cr VI, volatile phenol, total arsenic, heavy metal, chlorophyll, *etc.* The water samples were collected at 0.5 m under the water surface with a 2 L organic glass sampler. During the sampling process, the sites were positioned by GPS and the water samples were collected and measured three times to minimize the errors associated with the measurements. The average values were then considered as the water quality data of different parameters. Water samples of COD were stored in 500 mL glass bottles, and the samples of TN, TP were stored in 500 mL polyethylene bottles. All the bottles were disinfected by 70% alcohol in the lab before sampling. To eliminate the interference of substances adhering to the bottle walls such as adsorbed organic matter and oil, the bottles were washed sequentially with acid and deionized water. DO was directly measured by a portable water analysis instrument in the field. The samples of COD, NH_3-N and TP were added with sulfate until pH ≤ 2 and transferred to the Water Environment Monitoring Center of Poyang Lake for analyzing within 24 hours. The three parameters were measured with the typical dichromate method, Nessler's reagent colorimetry, and ammonium molybdate spectrophotometric method, respectively. The methods for water sample collection, storage and test were referred to the Water and Exhausted Water Monitoring Analysis Method (fourth edition), proposed by the China State Environmental Protection Administration in 2002 [10].

Mathematic Analysis

The optimization scheme was proposed on the basis of field investigations and mathematical analysis. The water quality data of 19 present monitoring sites in recent consecutive five year (2008–2012) was collected for quantitative research, and we selected the inter-annual variation coefficient and water quality correlation coefficient as the optimization basis. The inter-annual variation coefficient reflected the water quality stability of a certain site. The higher the value is, the more the causal factors influence the water quality. If the inter-annual variation coefficient exceeded a limit value, it was considered that the water quality of the site fluctuated too widely to reflect the actual water quality. According to some related results, we selected 70% as the critical value of variation coefficient to determine the inter-annual water quality stability of each site [11,12].

The water quality correlation coefficient reflected the water quality relationship between two monitoring sites. The higher the value is, the closer the two sites are. In contrast, the lower the value is, the farther the two sites are. Here, when the correlation coefficient exceeded the upper limit, it was considered that the studied two sites were duplicated and when the correlation coefficient was less than the lower limit, it was considered that the two sites were arranged with a spatial gap and a new site should be added between the two sites for a more reasonable monitoring network. The greater absolute value of correlation coefficient indicates the stronger correlation. According to the empirical test value [13], the correlation intensity can be induced to the following four sections: $0.8 \leq |r| < 1$, indicates the high-strong relevance; $0.6 \leq |r| < 0.8$, indicates the strong relevance; $0.4 \leq |r| < 0.6$, indicates the mid relevance, and $|r| < 0.4$, indicates the weak relevance [14,15]. If the water quality correlation coefficient between two sites is very high, the point with higher inter-annual variation coefficient will be suggested to be removed. In contrast, if the water quality correlation coefficient between two sites is at a significantly low level, a new monitoring site will be proposed to be added and the detailed location should be considered with the factors including the site distribution density, correlation with the nearby sites and cost for regular monitoring.

It is a fact that water stratification influences the vertical concentrations in a lake. Nevertheless, in this paper the field investigated data were all the concentrations of the water samples taken 0.5 m under the water surface, and the vertical fluctuations were not adequately considered for the following two reasons: firstly, unlike an isolated lake, Poyang Lake is a typical river-connected lake, with a water exchange period of only 21 days. The frequent water exchange between the lake and external rivers results in good hydrodynamic conditions in the lake, especially in the south and north parts. The average velocities of the lake could reach 1.28 $m \cdot s^{-1}$ and 1.85 $m \cdot s^{-1}$ in the flood seasons and dry seasons, respectively. Due to the dynamic behavior in the lake, the vertical concentrations don't differ with water depth as evidently as at some other lakes. In addition, for the deposition of the sediment transported into the lake from the external rivers, the bottom land of the lake is gradually upraised. Although the water level of the lake fluctuates markedly in a year, the average water depth of the whole lake is only 6.5–7.0 m, even in the flood season with the highest water level [16,17], so we considered the concentration of the water sample under the water surface as approximately the depth-averaged value in this study.

2-D Unsteady Numerical Model

The numerical model was established to obtain the corresponding water quality data in the areas without field investigation, which was of importance to subsequent rationally analyze the optimized monitoring network. Here, we developed the model in the framework of finite volume method (FVM) and the flux vector splitting (FVS) scheme was employed to calculate the numerical normal flux of variables across the interface between grids to guarantee the simulation accuracy.

Basic Controlling Equation

The conservation forms of 2-D shallow water equations and the convection-diffusion equations are written as follows [18,19,20]:

$$\begin{cases} \frac{\partial(hu)}{\partial t} + \frac{\partial\left(hu^2+gh^2/2\right)}{\partial x} + \frac{\partial(huv)}{\partial y} = gh\left(s_{0x} - s_{fx}\right) \\ \frac{\partial(hv)}{\partial t} + \frac{\partial(huv)}{\partial x} + \frac{\partial\left(hv^2+gh^2/2\right)}{\partial y} = gh\left(s_{0y} - s_{fy}\right) \\ \frac{\partial(hC)}{\partial t} + \frac{\partial(huC)}{\partial x} + \frac{\partial(hvC)}{\partial y} = \frac{\partial}{\partial x}\left(D_x h \frac{\partial C}{\partial x}\right) + \frac{\partial}{\partial y}\left(D_y h \frac{\partial C}{\partial y}\right) - KhC + S \end{cases} \quad (1)$$

where h is the water depth, m; t is time, s; u and v are respectively the depth-averaged velocity components in x and y directions, $m \cdot s^{-1}$; g is the acceleration of gravity, $m \cdot s^{-2}$; S_{ox} and S_{fx} are the bed slope and friction slope in the x direction; S_{oy} and S_{fy} are the bed slope and friction slope in the y direction; F_x and F_y are the friction force components in the x and y directions, which reflect the wind stress; f is the Coriolis parameter; D_x and D_y are the dispersion coefficients of pollutants in the x and y directions under dynamic conditions, $m^2 \cdot s^{-1}$; C is the water pollutant concentration, $mg \cdot L^{-1}$; K is the degradation coefficient, d^{-1}; S is the source-sink vector, in which the internal sediment release, biological impacts and atmospheric deposition are involved.

Equations Solution

The 2-D water flow and quality equations can be combined to be calculated, and the equations (1) can be written as the following unified form [21,22,23]:

$$\frac{\partial q}{\partial t} + \frac{\partial f(q)}{\partial x} + \frac{\partial g(q)}{\partial y} = b(q) \quad (2)$$

where q is the vector of the conserved physical quantities; $f(q)$ and $g(q)$ are the flux vectors in the x and y directions, respectively; $b(q)$ is the source-sink vector. Equation (2) was integral-discrete under any shape unit by the finite volume method, and the FVS format was applied to solve the numerical flux in the normal direction. Detailed steps are documented in [24,25].

Calibration and Verification

The numerical model was calibrated and validated against the field measured data in April and August 2010. Calculated region was the whole lake. According to topographic character, the lake area was divided into 6239 quadrilateral grids (average mesh size is 700 m × 700 m) and 7533 nodes by Gambit. The measured water flow and water quality in the upper five rivers and water level in the inlet of Yangtze River were selected as the calculation boundaries. To guarantee the stability and precision of the model the calculation time step was taken as 0.1 s. It was found that the calculated results were close to the measured data and the average relative error was approximately 10%. The established model could therefore reasonably reflect the water current and water quality processes in Poyang Lake.

RESULTS AND DISCUSSION

Inter-Annual Stability Analysis for Each Site

According to the field investigation results, the factors of DO, COD, NH_3-N and TP were characterized by the highest pollutant share rates. Because DO and TP are important for trophic state characterization, they were directly selected for the subsequent optimization study. In terms of the indexes of COD and NH_3-N, although they are the indicators of pollutants from urban or livestock sources, and not conservative, we also considered them in the subsequent studies for the following two reasons: firstly, though many physical, chemical and biological factors are involved in the water quality driving mechanism, the boundary input always plays the most important role, and due to the accelerated industrialization and increased population, massive amounts of pollutants from urban or livestock sources have been transported to the lake directly or indirectly by the upstream rivers, so the COD and NH_3-N loads in the lake were evidently enhanced. In addition, in this paper we placed more emphasis on the water quality data analysis than on the driving mechanisms of different factors on it and the studied water quality data are all the final results under the integrated impacts of series of influencing factors, including physical, chemical and biological ones.

Based on the historical monitoring data from 2008 to 2012, the inter-annual variation coefficients of the indexes at each site were calculated by SPSS. The results indicated that the indexes of DO and COD fluctuated less evidently than NH_3-N and TP. The variation coefficients of DO and COD remained at a relatively lower level, and the average values in the whole lake were 8.38% and 8.73%, respectively. DO at Fuhe Outlet (No. 2) and COD at Changjiang (No. 7) varied with the widest range and the variation coefficients were 12% and 19.48%, respectively. The most stable DO and COD sites were Hukou (No. 19) and Zhuxikou (No. 16), and the variation coefficients were separately only 2.91% and 2.31%. NH_3-N and TP at the sites fluctuated more evidently with years as a result of rainfall intensities and non-point sources. The average variation level of NH_3-N and TP in the whole lake were approximately 42.7% and 37.5%. The most significant variations of NH_3-N and TP in the lake were both detected at Piaoshan (No. 10), and the variation coefficients reached 99% and 73.9%, respectively. The variable amplitudes of NH_3-N at Xinjiang West Branch (No. 1) and TP at Ganjiang Main Branch (No. 13) were at the lowest level, and the variation coefficients were 21.8% and 14.0%, respectively. In terms of spatial distribution, water quality in the outlet area and north area of the lake was more stable than that in the middle and the south areas. The average inter-annual variation coefficients of DO, COD, NH_3-N and TP of the five monitoring sites were 5.13%, 8.75%, 38.35% and 27.58%, respectively. The water quality in the south area of the lake varied more widely than that in the north area and outlet area, especially for the indexes of DO and TP. The average variation coefficients of DO and TP at sites No. 1 to No. 6 could reach 10.35% and 42.52%, respectively. Water quality at the sites in the middle area of the lake fluctuated with the highest level. The inter-annual variation coefficients of DO, COD, NH_3-N, and TP could averagely reach 8.93%, 9.36%, 48.08%, and 39.82 % (Figure 2).

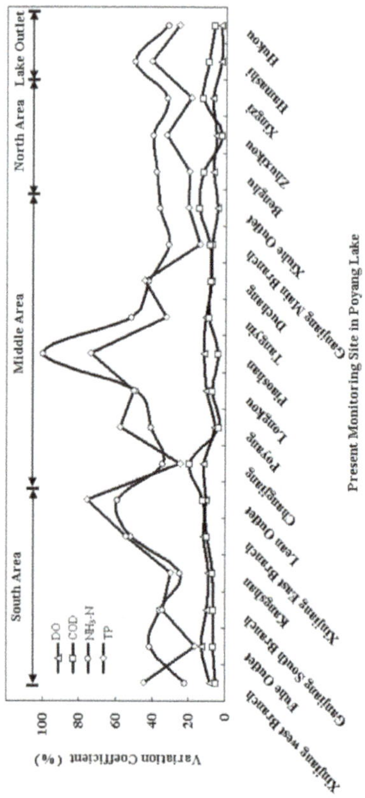

Figure 2: Inter-annual variation coefficients at each present monitoring site in Poyang Lake.

Correlation Analysis for Adjacent Sites

Considering the location of the present 19 monitoring sites, we figured out 16 groups of adjacent sites for study. Based on the field investigated data from 2008 to 2012, the correlation coefficient for each group between adjacent sites was calculated by the Bivariate Process in the correlation analysis module of SPSS software. The results are shown in Figure 3. Generally, the water quality correlation intensities varied with groups and indexes. In the south area of the lake, except TP, the indexes of DO, COD, and NH_3-N at No. 1–No. 2 were in high-strong relevance, and the correlation coefficients reached 0.91, 0.90, and 0.83, respectively. The correlation intensities of NH_3-N, COD, TP, and DO at No. 2–No. 3 transformed from weak relevance to high-strong relevance, and the correlation coefficients were 0.026, 0.48, 0.799 and 0.87, respectively. For No. 4–No. 9, DO and NH_3-N were more closely related than COD and TP. The correlation coefficients of DO and COD at No. 5–No. 6 were 0.81 and 0.79 in strong relevance, while TP and NH_3-N were in weak relevance. In the middle area of the lake, the water quality correlation intensities at the groups of No. 10–No. 11, No. 12–No. 16, and No. 13–No. 14 were higher than those at other groups, and the average coefficients were 0.80, 0.71, and 0.82, respectively. The strongest correlation level was detected at No. 10–No. 11 where the correlation coefficients of DO, COD, NH_3-N, and TP could separately reach 0.66, 0.90, 0.94, and 0.71. The water quality at the groups of No. 7–No. 8 and No. 9–No. 10 were related at the lowest level, and the correlation coefficients were 0.41 and 0.50. In the north area of the lake, among the three groups of adjacent sites, No. 15–No. 16, No. 16–No. 17, and No. 17–No. 18, COD and DO was in weak and mid relevance respectively, while TP and NH_3-N were basically in strong relevance and high-strong relevance, separately. At No. 18–No. 19, the water quality was closely related, and all the indexes were in strong relevance or high-strong relevance. The correlation coefficients of DO, COD, NH_3-N, and TP could reach 0.75, 0.96, 0.90, and 0.76, respectively.

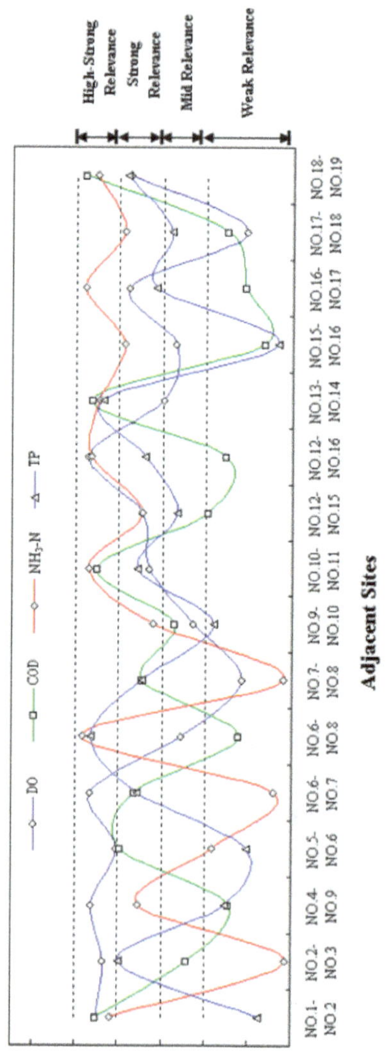

Figure 3: Correlation analysis for present adjacent monitoring sites in Poyang Lake.

Optimization for Water Quality Monitoring Sites

According to the analysis on the inter-annual variation coefficients of 19 sites and correlation coefficients of 16 groups of adjacent sites, the optimization scheme for the present water quality monitoring network was proposed with consideration of the lake morphology. In the north lake, the distance between the monitoring site No. 17 and No. 18 was about 20.8 km. In view of the spatial distance, the annual water quality stability, and the weak water quality relevance of the two sites, a new monitoring site was suggested to be added in the midpoint of No. 17 and No. 18, which could be named Changling, A1 (29°53′77″ N, 116°12′46″ E) in Figure 4. In the south lake, the three present monitoring sites No. 4, No. 9 and No. 11 were distributed basically in the east, and the water quality correlations at No. 4–No. 9 and No. 9–No. 11 were relatively weak. We propose to add a new monitoring site to the west of the area covered by the present three sites in the south-north direction, which could be named Nanjiang, A3 (29°00′95″ N, 116°42′93″ E) in Figure 4.

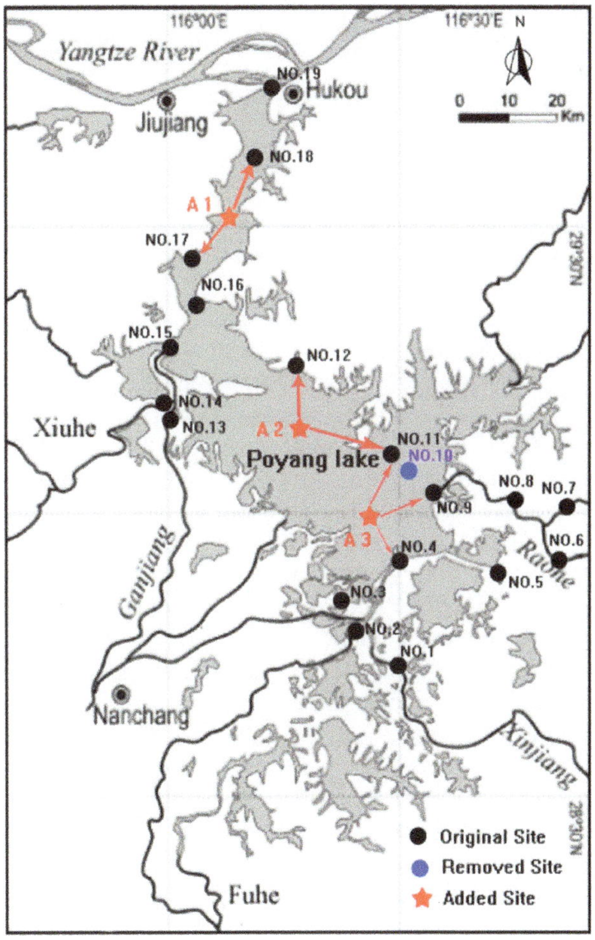

Figure 4: Optimized water quality monitoring network in Poyang Lake.

Due to the close spatial distance, the water quality at No. 10 was extremely closely related to that at No. 11. Considering the poor water quality stability at No. 10, we suggest removing the present No.10 site. The high correlation coefficients at the groups No. 18–No. 19, No. 1–No. 2, and No. 13–No. 14 indicated that the water qualities at these monitoring sites were closely related and to some extent the sites were arranged with repeatability. However, these sites were set at the inflowing or outflowing sections of the lake, making them of importance to obtain the boundary water quality, so it is suggested that these boundary monitoring sites be retained. The present No. 11 site was 20.8 km from No. 12 in the north and 37.5 km from No. 13 in the west, and no site was in the broad water area to the west of No. 11. To improve the representativeness of the monitoring network, it is suggested that a new monitoring site added in the middle of No. 11 to No. 13 in the core of the triangle formed by No. 11, No. 12 and No. 13, which could be named Huzhong, A2 (29°15′25″ N, 116°19′75″ E) in Figure 4.

Rationality Test for the Optimized Network

In the optimized water quality monitoring network for Poyang Lake, one primary site was removed and three new sites were added. Further verification was needed to answer whether the proposed adjustment was scientific and reasonable. Because there was no historical monitoring data at the added sites, the established 2-D unsteady model was used to estimate the corresponding water quality. The water current and water quality processes from 2008 to 2012 were simulated year by year to get the corresponding data for the three added sites, Changling, Huzhong, and Nanjiang. The inflow boundaries of the model were the upper Yangtze River, Xiuhe River, Ganjiang River, Fuhe River, Xinjiang River and Raohe River. The outflow boundary was the downstream Yangtze River section. The field water current and water quality at the boundaries from 2008 to 2012 were determined according to the measured data from the Hydrology Bureau of Jiangxi Province and Bureau of Hydrology, Changjiang Water Resource Committee. Consistent with the model calibration, Poyang Lake was divided into 6239 quadrilateral unit grids and 7533 nodes. The time step was taken as 1 s for simulation accuracy. The initial roughness of the area where the aquatic plants existed was taken as 0.04, and that of the remaining lake area was 0.025. Based on the calculated results, the inter-annual variation coefficients of the three added sites and the correlation coefficients between the new adjacent sites were studied. We found that the water quality of the three added sites varied generally in a reasonable range. Variation coefficients of NH_3-N and TP were relatively higher than that of DO and COD, but all of them were lower than the critical standard of 70% (Figure 5). The correlation coefficients among Changling, Huzhong, Nanjiang and the corresponding adjacent sites were in mid relevance (Figure 6). The correlation coefficients of DO and NH_3-N between A2-No. 11 were a bit higher but still in the acceptable range. In general, the 21 optimized water quality monitoring sites could scientifically replace the primary network. The stability and representativeness of the new monitoring network were improved.

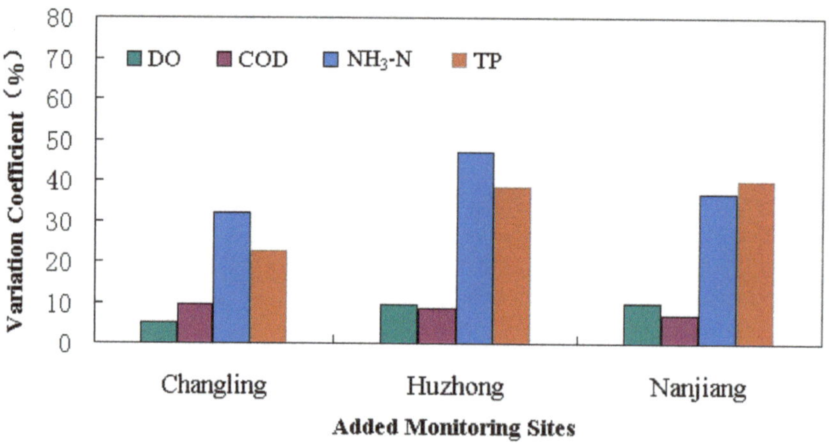

Figure 5: Inter-annual variation analysis for the added sites in Poyang Lake.

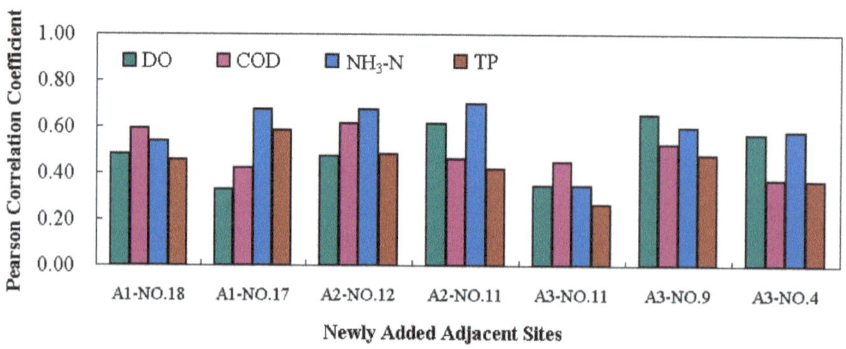

Figure 6: Correlation analysis for the added sites and adjacent sites in Poyang Lake.

Implication and Limitation

At present, although a few experts have conducted some studies on lake water quality monitoring site optimization, additional efforts are needed to establish a mature systematic method to optimize lake water quality monitoring sites. The methodology proposed here, with field investigations, mathematical analysis, and numerical simulation tests being considered systematically, can provide an important reference on monitoring network optimization for some other shallow lakes in the Yangtze River basin, such as Taihu Lake and Dongting Lake. However, there is still uncertainty in the method due to the lack of the water quality data at the added sites. Although some corresponding data can be obtained by numerical simulation, the calculated results may differ from the actual values to some extent. Nevertheless, what we aim at in this paper is to propose a new framework and method for site optimization and put forward a more scientific and reasonable scheme to adjust the primary monitoring network. The results may not be the absolute ultimate plan, but it points out a research direction for site optimization in the subsequent in-depth study.

CONCLUSIONS

This paper proposed a coupled method to optimize the surface water quality monitoring sites for a huge freshwater lake based on field investigations, mathematical analysis, and numerical simulation tests. Poyang Lake, the largest freshwater lake in China, was selected as the research area. Based on the water quality data at the current 19 monitoring sites in consecutive five years (2008–2012), the water quality inter-annual stability at each site and the correlation intensities between adjacent sites were analyzed. In view of the lake morphology, an optimized scheme was proposed to improve the present monitoring network. The primary 19 site network in Poyang Lake was suggested to be adjusted to 21 sites, with the primary Piaoshan site being removed and the new Changling, Huzhong, and Nanjiang ones added. By numerical simulation the corresponding water quality data were obtained for a rationality test on the optimized network. The results demonstrated that the optimized water quality monitoring network in Poyang Lake could replace the primary one well. The study would be beneficial for water environment management in research area and provide guidance for analogous optimization work in other lakes. Finally we would like to add two points that: (1) in the succeeding studies, water stratification should be considered for a more scientific and reasonable optimization; (2) additional attention should be paid to the combination of different means for water quality monitoring, such as satellite observations and the continuous monitoring using buoys, which may evidently reduce the cost.

ACKNOWLEDGMENTS

This work was supported by the National Natural Science Foundation of China (No. 51309082) and Major Science and Technology Program for Water Pollution Control and Treatment of China (No. 2012ZX07506-002 & No. 2012ZX07101-001).

AUTHOR CONTRIBUTIONS

Hua Wang and Mengan Wu made conception and design of the study, and completed the writing of the paper. Yanqing Deng provided data for the paper. Chunyan Tang and Rui Yang made revisions for some intellectual content.

CONFLICTS OF INTEREST

The authors declare no conflict of interest.

REFERENCES

1. Harmancioglu, N.B.; Ozkul, S.D.; Alpaslan, M.N. Water Quality Monitoring and Network Design. In *Environmental Data Management*; Water Science and Technology Library: New York, NY, USA, 1998; Volume 27, pp. 61–106.
2. Timmerman, J.G.; Beinat, E.; Termeer, K.; Cofino, W. Analyzing the data-rich-but-information-poor syndrome in Dutch water management in historical perspective. *Environ. Manag.* 2010, *45*, 1231–1242.
3. Wen, C.G.; Lee, C.S. A neural network approach to multiobjective optimization for water quality management in a river basin. *Water Resour. Res.* 1998, *34*, 427–436.
4. Davies-Colley, R.J.; Smith, D.J.; Ward, R.C.; Bryers, G.G.; McBride, G.B.; Quinn, J.M.; Scarsbrook, M.R. Twenty years of New Zealand's national rivers water quality network: Benefits of careful design and consistent operation. *J. Am. Water Resour. Assoc.* 2011, *47*, 750–771.
5. Wang, J. Huangshui River water environment monitoring section optimization settings. *J. Qinghai Environ.* 2002, *3*, 27–30.
6. Wang, Z.P.; Meng, X.L.; Huo, Q.G. Optimization principles of environmental monitoring sections for surface water and their applications in Henan province. *China Environ. Monit.* 2003, *19*, 15–27.
7. Yang, L. How to optimize the distributing location of surface water monitor. *Guangzhou Environ. Sci.* 1996, *11*, 23–26.
8. Wang, J. The Study on the Optimization of Surface Water Environmental Monitoring Network of Chaohu Lake. Master's Thesis, Hefei University of Technology, Hefei, China, 2012.
9. Mo, M.H.; Yang, J.; Gu, S.; Song, Y.J. Estimation of non-point source pollution load in the area around Poyang Lake.*Yangtze River* 2010, *41*, 51–53.
10. Wei, F.S. *Water and Exhausted Water Monitoring Analysis Method*, 4th ed.; China Environmental Science Press: Beijing, China, 2002.
11. Wang, J.X.; Li, S.B.; Zhang, W.M. Study on atmospheric environmental excellent layout monitoring point of province-scale. *Environ. Monit. China* 2001, *17*, 18–21.
12. Chen, H.Y. Correlation analysis of interannual variabilities in SST in SCS and Nin0/DMI indice. *Adv. Mar. Sci.* 2007, *25*, 20–26.
13. Fang, K.T. *Applied Multivariate Statistical Analysis*; East China Normal University Press: Shanghai, China, 1989.
14. Chen, Z. Methods of factor analysis and cluster analysis using SPSS software. *Mark. Res.* 2006, *6*, 45–48.
15. Zhang, Y.L.; Cao, H.M.; Jiang, Y.J.; Cai, Z.M. Application of SAS in analysising experimental data of fruit trees. *J. Liaoning Agric. Coll.* 2009, *11*, 7–10.
16. Gu, P.; Wan, J.B. hydrology character of Poyang Lake and its influence on water quality. *Environ. Pollut. Control* 2011,*33*, 15–19.
17. Lai, X.J.; Jiang, J.H.; Huang, Q.; Xu, L.G. Two-dimensional numerical simulation of hydrodynamic and pollutant transport for Lake Poyang. *J. Lake Sci.* 2011, *23*, 893–902.
18. Zhao, D.H.; Shen, F.X. A 2D sediment transport model based on the FVM with FDS for tidal rivers. *J. Hydrodyn.* 2004,*19*, 99–102.
19. Garcia, R.; Kahawita, R.A. Numerical solution of the St. Venant Equations with the MacComack finite-difference scheme. *J. Numer. Methods Fluids* 1986, *6*, 259–274.
20. Shi, Y.; Hu, S.Y. A finite volume method for numerical modeling of 2-D flow and sediment movements on unstructured grids. *Adv. Water Sci.* 2002, *13*, 409–413.
21. Zhao, D.H.; Shen, H.W.; Lai, J.S. Approximate Riemann solvers in FVM for 2-D hydraulic shock waves modeling. *Hydraul. Eng. ASCE* 1996, *122*, 693–702.
22. Zhao, D.H.; Shen, H.W.; Tabios, G.Q. Finite-volume two-dimensional unsteady-flow model for river basins. *Hydraul. Eng. ASCE* 1994, *120*, 863–883.
23. Zhao, D.H.; Qi, C.; Yu, W.D. Finite volume method and Riemann solver for depth-averaged two-dimensional flow-pollutants coupled model. *Adv. Water Sci.* 2000, *11*, 368–374.
24. Ding, L.; Pang, Y.; Wu, J.Q.; Li, Y.P. Three second-order schemes for simulation water quality jump. *J. Hydraul. Eng.*2004, *35*, 50–55.
25. Spekeijse, S.P.; Hemker, P.W. Multigrid Solution of Steady Euler Equations. Available online: http://link.springer.com/chapter/10.1007%2F978-3-663-14245-4_4 (accessed on 15 September 2014).

INDEX